Health Informatics

Other related titles:

You may also like

- PBHE017 | Leonard Goldschmidt and Rona Margaret Relova | Patient-Centered Digital Healthcare Technology: Novel applications for next generation healthcare systems | 2020
- PBHE041 | Mohamed Abouhawwash, Sudeep Tanwar, Anand Nayyar, and Mohd Naved | Innovations in Healthcare Informatics: From interoperability to data analysis | 2023
- PBHE058 | Karthik Ramamurthy, Suganthi Kulanthaivelu, Kulandairaj Martin Sagayam, Soumi Dutta, and Paryati | Technologies for Healthcare 4.0: From AI and IoT to blockchain | 2023

We also publish a wide range of books on the following topics:
Computing and Networks
Control, Robotics and Sensors
Electrical Regulations
Electromagnetics and Radar
Energy Engineering
Healthcare Technologies
History and Management of Technology
IET Codes and Guidance
Materials, Circuits and Devices
Model Forms
Nanomaterials and Nanotechnologies
Optics, Photonics and Lasers
Production, Design and Manufacturing
Security
Telecommunications
Transportation

All books are available in print via https://shop.theiet.org or as eBooks via our Digital Library https://digital-library.theiet.org.

IET HEALTHCARE TECHNOLOGIES SERIES 066

Health Informatics

Technologies and applications

Edited by
Gaurav Gupta, Manju Khari and Shakir Khan

The Institution of Engineering and Technology

About the IET

This book is published by the Institution of Engineering and Technology (The IET).

We inspire, inform and influence the global engineering community to engineer a better world. As a diverse home across engineering and technology, we share knowledge that helps make better sense of the world, to accelerate innovation and solve the global challenges that matter.

The IET is a not-for-profit organization. The surplus we make from our books is used to support activities and products for the engineering community and promote the positive role of science, engineering and technology in the world. This includes education resources and outreach, scholarships and awards, events and courses, publications, professional development and mentoring, and advocacy to governments.

To discover more about the IET, please visit https://www.theiet.org/.

About IET books

The IET publishes books across many engineering and technology disciplines. Our authors and editors offer fresh perspectives from universities and industry. Within our subject areas, we have several book series steered by editorial boards made up of leading subject experts.

We peer review each book at the proposal stage to ensure the quality and relevance of our publications.

Get involved

If you are interested in becoming an author, editor, series advisor, or peer reviewer please visit https://www.theiet.org/publishing/publishing-with-iet-books/ or contact author_support@theiet.org.

Discovering our electronic content

All of our books are available online via the IET's Digital Library. Our Digital Library is the home of technical documents, eBooks, conference publications, real-life case studies, and journal articles. To find out more, please visit https://digital-library.theiet.org.

In collaboration with the United Nations and the International Publishers Association, the IET is a Signatory member of the SDG Publishers Compact. The Compact aims to accelerate progress to achieve the Sustainable Development Goals (SDGs) by 2030. Signatories aspire to develop sustainable practices and act as champions of the SDGs during the Decade of Action (2020–2030), publishing books and journals that will help inform, develop, and inspire action in that direction.

In line with our sustainable goals, our UK printing partner has FSC accreditation, which is reducing our environmental impact to the planet. We use a print-on-demand model to further reduce our carbon footprint.

British Library Cataloguing in Publication Data
A catalogue record for this product is available from the British Library

ISBN 978-1-83724-088-3 (hardback)
ISBN 978-1-83724-089-0 (PDF)
ISBN 978-1-83724-990-9 (EPUB3)

Cover image: koto_feja/E+ via Getty Images

Contents

Preface

The integration of healthcare with digital technologies has opened new frontiers for improving diagnostics, treatment, and patient care. *Health Informatics: Technologies and Applications* is designed to serve as a comprehensive guide for undergraduate and postgraduate students, faculty, researchers, and industry professionals from disciplines such as Information Technology, Computer Science, Electronics and Communication Engineering, and allied domains. The book explores the role of machine learning, big data, and biomedical signal processing in shaping the future of healthcare.

The volume begins by laying a strong foundation in health data analytics and big data applications, setting the stage for understanding the challenges and opportunities in managing complex medical data. It then transitions into the application of machine learning techniques for diagnosing critical neurological and physiological conditions such as brain tumors, Parkinson's disease, Amyotrophic Lateral Sclerosis (ALS), and obstructive sleep apnea (OSA). These chapters emphasize the importance of accurate feature extraction, biomarker identification, and model optimization in building reliable diagnostic systems.

Subsequent chapters expand the discussion to AI-driven signal reconstruction, demonstrating how audio signals can be used to detect chronic respiratory conditions and emotional state. Recognizing the growing reliance on cloud-based health systems and IoMT (Internet of Medical Things), the book includes focused discussions on data privacy, federated learning, and secure data sharing. These chapters address crucial concerns related to maintaining the confidentiality and integrity of sensitive health data while leveraging distributed computing environments.

The book also introduces theoretical frameworks for breast cancer detection, explores proteomics techniques for microbial protein analysis, and revisits OSA diagnosis using advanced machine learning models. Each chapter, while distinct in focus, contributes to a holistic understanding of health informatics and its role in modern medicine.

By bringing together theory, application, and real-world challenges, this book provides readers with both conceptual insights and practical tools. It aims to bridge the gap between healthcare practitioners and technologists, encouraging interdisciplinary collaboration for innovative, ethical, and patient-centered healthcare solutions.

About the editors

Gaurav Gupta is an professor at Yogananda School of Artificial Intelligence Computers and Data Science, Faculty of Engineering and Technology, Shoolini University, India. He has 15 years of teaching experience, and his academic contributions includes over 100 conference papers, journal articles, book chapters, and books. His area of interest includes machine learning, IoT, and cloud computing for healthcare. He is an active senior member of the Institute of Electrical and Electronics Engineers (IEEE) and the International Association of Engineers (IAENG).

Manju Khari is a professor in the School of Computer and Systems Sciences at Jawaharlal Nehru University in New Delhi, India. Her research interests include information security, software security testing, computer networks, the Internet of Things, cyber forensics, and deep learning. Before joining JNU, she worked with Netaji Subhas University of Technology, Delhi. Her prolific research portfolio includes over 190+ published papers in premier refereed national and international journals and conferences, including highly regarded publishers such as IEEE, ACM, Springer, Inderscience, and Elsevier. Dr Khari has also co-authored 2 books published by NCERT and co-edited 20 edited books, further solidifying her status as a leading scholar in her field.

Shakir Khan is an professor at the College of Computer and Information Sciences, Imam Mohammad Ibn Saud Islamic University (IMSIU), Saudi Arabia. He has over 15 years of national and international teaching, research and IT experience. He has published more than 130 conference papers, journal articles, book chapters, and books as well as 3 patents. He has contributed to many international conferences in the capacity of committee member and keynote speaker. His research interests include ML/AI, cloud computing, IoT, big data analytics, and bioinformatics. He has been acknowledged by Imam University and received a research excellence award in 2019.

Chapter 1

Health data analytics and big data applications

*Jaya Vardhan Reddy[1], Ajay Sharma[1,2],
Devendra Babu Pesarlanka[1] and Shamneesh Sharma[2]*

Abstract

For the last few years, a large amount of data has been produced in the health sector around the globe. This generated data is of various types, namely structured, unstructured, and semi-structured data. Traditional study of medical data is too complex to process and analyze data, so it is necessary for advanced technology to handle massive amounts of data, unearth valuable insights, and process this large amount of data. Big data is introduced here to manage massive data produced by different sources. Big data can be able to deal with different characteristics such as volume, velocity, veracity, and variability. This produced a large amount of healthcare data need to be stored, processed, and analyzed for progressive results. Across the globe, the use of big data for early diagnosis of any type of disease by processing and analyzing the data is used at a large scale. In this chapter, the author has discussed the integration of big data in healthcare, types of data generated in the health sector, tools, and methodologies used to store, process, and analyze data, the role of big data in healthcare, and applications of big data in healthcare.

Keywords: Big data; Hadoop; Data Analytics; Applications of Big Data; Healthcare

1.1 Introduction

In this digital era, an enormous amount of data is generated every day, transforming industries and generating an immense number of opportunities in several sectors, even in healthcare. As we belong to the digital era we must store or gather information for our future evolution. It is necessary to pile a large amount of healthcare data such as electronic health records (EHRs), clinical trials, imaging data, administrative data, and patient surveys to investigate and develop current and emerging healthcare issues. Studying these data offers vast potential for

[1]Department of Computer Science and Engineering, Lovely Professional University, India
[2]Academic Operations and Quality Control, byteXL TechEd Private Limited, Hyderabad, Telangana, India

intensifying care delivery, upgrading treatment outcomes, and reducing costs. However, this data deluge also presents critical challenges, especially due to its heterogeneity, sheer scale, and unstructured nature (Lv and Qiao, 2020).

Traditional data analysis approaches are unable to deliver the "Five Vs" nee-ded to prevail over big data with regard to volume, velocity, veracity, variety, and value, and that is where advanced analytics turns out to be essential. Deep learning and machine learning (ML) algorithms such as Convocational Neural Networks (CNN) and Recurrent Neural Networks (RNN) have heightened the ability to unearth practical insights in the vast domains of data. Big data analytics is not only a technological advancement but an elemental shift in approach. Big data analytics detects numerous diseases in the initial phase, lightens the risks related to post-poned treatment, and enhances the care of patients by digging out insights and predictive outcomes. With real-time data analysis, this capacity is enlarged to turn raw data into actionable intelligence, particularly during emergencies when appropriate decisions can be lifesaving. The current work focuses on the role of data analytics and big data applications in healthcare from an evolving viewpoint. Applications, tools, and transformative capabilities will be discussed in depth up to advanced analytical techniques for transformed healthcare delivery and bioinformatics.

1.2   Overview of health data

Health data plays a crucial role in enhancing the patient's treatment, enabling data-driven decisions, and improving patient results (Lee *et al.*, 2017). The increasing amount of health data, fueled by technical advancements and the increased use of digital health tools, highlights the need for effective classification and manage-ment. Health data is generated in different formats, which are classified by their method and source of collection (Mikalef *et al.*, 2019). This type of comprehension is important for handling and evaluating the data. Health data is classified into three primary types (Figure 1.1).

1. Structured data
2. Unstructured data
3. Semi-structured data

Every category has its characteristics, challenges, and use cases, which means that researchers, data scientists, and healthcare professionals need a detailed under-standing of these classifications.

1.2.1   Structured data

Structured data refers to well-arranged information that is easy to store, search, and analyze in predetermined formats such as rows and columns within a database. It is organized into relational databases. Tools like Structured Query Language (SQL) databases are used to query and process structured data (Dubey *et al.*, 2019).

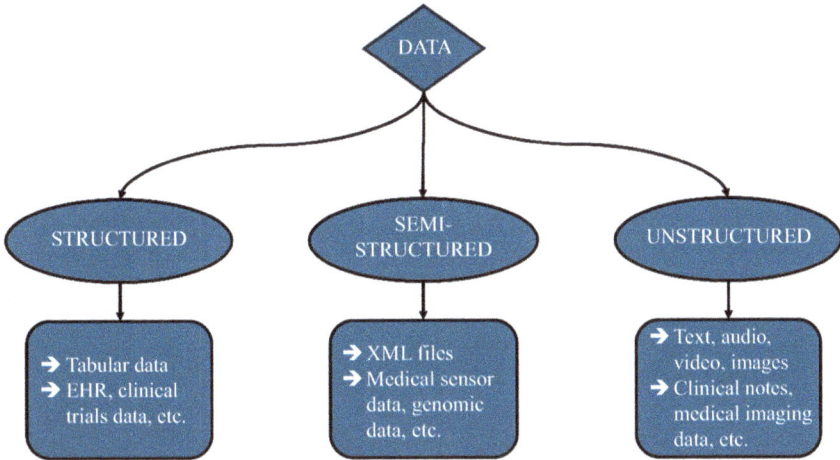

Figure 1.1 The flow chart (diagram) shows different types of data and their examples in the healthcare industry

Examples

- Electronic health records.
- Clinical trials data.
- Insurance claims.

1.2.2 Unstructured data

Unstructured data refers to raw data, unorganized information that cannot conform easily to tabular format or schemas. It requires advanced tools for analyzing and storing data. There is no fixed format or predefined structure. It is challenging to store in traditional relational databases.

Examples

- Clinical notes.
- Medical imaging data.
- Patient feedback data.

1.2.3 Semi-structured data

Semi-structured data is a mixture of both structured data and unstructured data. It does not conform to well-organized data as required for structured data but consists of markers and tags that provide some level of organization. It is organized using tagged elements or key-value pairs. It can be stored in NoSQL databases like MongoDB or Cassandra (Graham *et al.*, 2019).

Examples

- XML or JSON files.
- Medical sensor data.
- Genomic data.

Table 1.1   Table determines the features of different data types used in healthcare

Feature	Structured	Unstructured	Semi-structured
Format	Rigid, predefined	Free-form, raw	Tagged, flexible
Examples	EHR, clinical trials data	Clinical notes, MRIs	HL7 message, JSON
Ease of analysis	High	Low	Moderate
Storage tools	Relational databases	Object storage (e.g., S3)	NoSQL databases

1.3   Big data analytics in health informatics

The variation between big data health analytics from traditional health analysis is mainly in the implementation of computer programming. Traditional healthcare analytics systems are more dependent on various industries for big data analysis. Many healthcare beneficiaries or investors have faith in information technology due to its valuable outcomes, their operating systems are capable of transforming the data into standardized form and work more efficiently. In the healthcare sector, the field of big data analytics is ensuring to provide valuable insights for the healthcare system with tremendous growth. The bulk amount of data generated by the system is stored in hard copies, which must be digitized for further analysis. Big data is responsible for providing the following advantages in the healthcare sector, it enhances healthcare delivery, helps to minimize medication costs, and helps in enhancing healthcare outcomes (Sharma *et al.*, 2021, 2022).

1.4   Four Vs of big data in healthcare

Four important characteristics align with big data: Volume, Velocity, Variety, and Veracity. These four main characteristics are responsible for understanding and resolving the challenges as well as maintaining and analyzing very large datasets. Healthcare data is very huge to deal with a large amount of data, we need these terms which play a crucial role in big data analytics. These core characteristics of big data analytics will help us understand how to leverage data efficiency. By properly managing these aspects, we can gain valuable insights, drive innovation (Wang *et al.*, 2018), and increase operational efficiency. Addressing these four will make organizations exist competitive in the data-driven world. Let us know about these things in detail. The 4Vs of big data in healthcare are:

1. Volume
2. Velocity
3. Variety
4. Veracity

1.4.1 Volume

Big data, as the name specifies, refers to a large amount of data. Every day huge amount of data is generated by the health sector. It is necessary to analyze the huge amount of collected data to improve the level of medication and improve the standards of the healthcare sector. Big data can be capable of dealing with these huge amounts of collected data. There is no fixed limit for the amount of data it is dealing with. This term is usually correlated with enormous amounts of data that should be managed, stored, and analyzed, often more than what can be handled by traditional databases and data processing systems. The amplifying volume of data produced by the latest IT and healthcare systems has been increasing and is driven by minimizing the cost of data storage and processing systems. There is a growing need to unearth valuable information from this data to enhance the healthcare system, improve efficiency in medication, and offer optimized services to health-care professionals. Healthcare organizations need systems that can store and process large amounts of data like petabytes, terabytes, or exabytes of data.

1.4.2 Velocity

Velocity refers to the speed by which the data is generated, processed, and ana-lyzed. Velocity, which symbolizes the main reason for the tremendous growth of data, indicates the speed at which data is collected. Healthcare systems are pro-ducing a large amount of data at a faster rate. It is very important to generate and process the data and analyze it timely. It is crucial in sectors or domains like financial services, stock trading, and fraud detection. Real-time processing of data enables faster response times, enhancing the experience of healthcare professionals and healthcare efficiency (Fisher *et al.*, 2003).

1.4.3 Variety

Variety deals with which form of data it is. There are different formats and different types of data, i.e., structured, semi-structured, and unstructured. Healthcare data is in terms of different formats such as texts, medical imagery, video, audio, and sensor data. Structured data consists of patient clinical data, which is easy to col-lect, store, process, and analyze in a single device. The healthcare sector experi-ences a very low percentage of structured data namely 5–10% of the data. A very huge portion is shared by both unstructured and semi-structured data. These unstructured and semi-structured data consist of e-mails, photos, audio, videos, and other health-related data. These health data also consist of medical records, phy-sician's notes, paper prescriptions, and radiograph films. Healthcare organizations can gather data from various sources like emails, social media, IoT devices, and logs. Dealing with and managing these different types of data will guarantee a comprehensive approach to analytics.

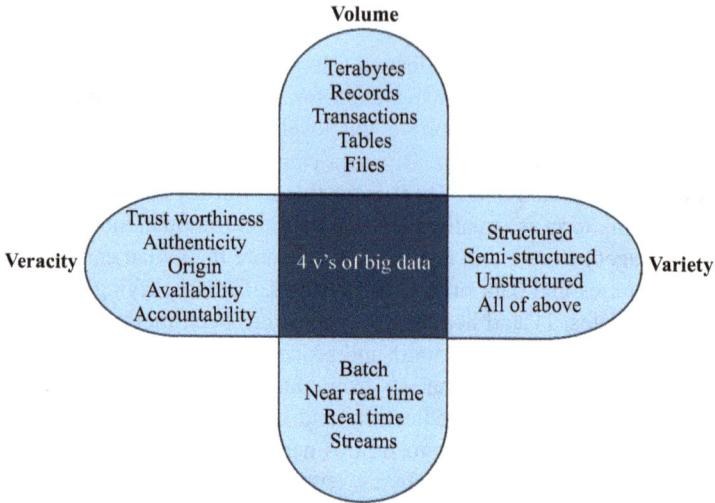

Figure 1.2   Four Vs of big data in healthcare

1.4.4   Veracity

Veracity ensures the quality of the data. It is the degree of assurance that the data is consistent. Consistent data gives us accurate insights and trustable information that helps for making decisions. Data consistency is important for maintaining the data integrity and reliability of the information being analyzed. The result of the big data analytics should be accurate and credible. In healthcare it is crucial to get accurate results, some ML algorithms make decisions that are used by automated machines with the help of data that is inconsistent and misleading. Healthcare analytics needs accurate results and should unearth helpful insights for treating patients and making the most accurate possible decisions. Inconsistent data can lead to useless insights and wrong decision-making. We should ensure the data is clean, consistent, and reliable for analyzing the healthcare data (Fisher *et al.*, 2003). It is crucial in sectors like healthcare, where we need accurate information for treating patients and diagnosis.

1.5   Technologies driving big data analytics in healthcare

Big data analytics in healthcare depends on many advanced technologies. Integration of advanced technologies with big data gives us more efficient results, improves decision making, and provides us with real-time intelligence.

Some of the key technologies that can be integrated with big data:

1.5.1   Artificial intelligence and Machine learning

Artificial intelligence (AI) and machine learning (ML) play a major role in deriving valuable insights from patient healthcare datasets.

AI healthcare analytics

AI algorithms help in deriving the trends and patterns of the patient healthcare data that help us to get accurate results and suggestions for good recommendations. By using AI algorithms, we can upscale the healthcare treatment and provide proper medication to the patient. It is most useful for doctors to get real-time intelligence by using AI algorithms. AI is used in many real-world applications (Rumsfeld *et al.*, 2016), IBM Watson Health uses AI for recommendations of cancer treatment.

Applications:

- Predictive analytics: Predicting the occurrence of a disease or outbreak, and predicting the admission of the patient again in the hospital.
- Diagnostic assistance: AI-driven tools enable advanced image recognition algorithms to find abnormalities in diagnostic scans such as CT scans, X-rays, and MRI scans.

Machine learning techniques

There are various ML algorithms, supervised ML algorithms are helpful to predict diseases and analyze drug responses. Unsupervised learnings are used to cluster the patients based on their symptoms for personalized treatment plans. Reinforcement learning helps in optimizing the healthcare delivery process which includes robotic surgeries. ML models are employed for the prediction of sepsis in ICU patients (Rumsfeld *et al.*, 2016).

1) **Supervised machine learning Algorithms:**
   Supervised machine learning is used to train the model based on the labeled datasets, where the relationship between input–output is predefined. This approach proves to be beneficial for predictive tasks in the healthcare industry.

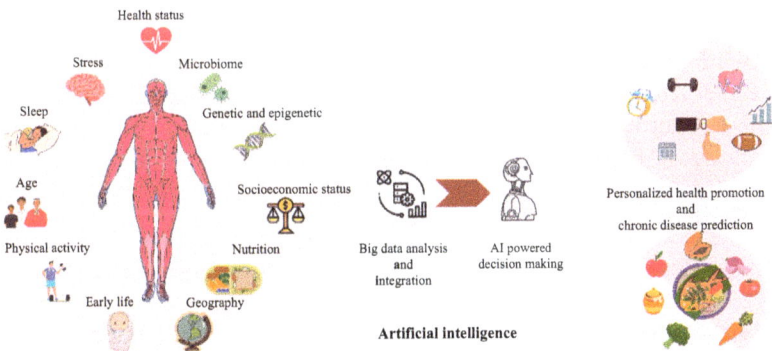

Figure 1.3   This figure shows the integration of big data analytics with artificial intelligence for AI-powered decision making.

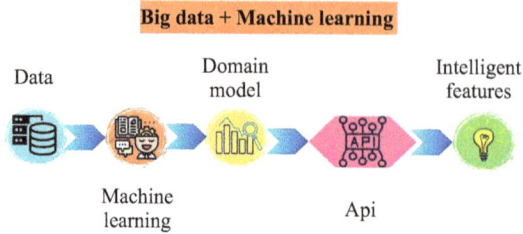

Figure 1.4   Integration of big data with machine learning.

Applications:

- **Disease prediction**: Popular algorithms like decision trees, support vector machines, and neural networks are utilized to predict diseases like diabetes, cancer, and heart conditions based on the patient's information.
- **Drug response analysis:** Models are used to analyze the response of patient's specific medications, allowing tailored treatment plans.
- **Risk stratification**: Models identify high-risk patients for advanced interventions.

2) Unsupervised machine learning algorithms:
Unsupervised ML algorithms are used to train the model without any predefined labels, which mainly focus on finding patterns, clusters, or associations.

Applications:

- **Patient clustering:** Based on symptoms, genetic data, or medical history, patients are grouped to enhance personalized treatment plans.
- **Anomaly detection:** This is used to identify outliers in medical records like rare disease markers and unusual lab results.
- **Healthcare resource allocation:** It optimizes resource distribution by analyzing the usage patterns.

3) Reinforcement learning:
Reinforcement learning (RL) is a trial-and-error technique in which agents learn about optical actions by incorporating them with their environment to maximize rewards (Andreu-Perez *et al.*, 2015).

Applications:

- **Health process optimization:** Resources management, treatment workflows, and improving patient scheduling.
- **Robotic surgeries:** RL algorithms are used to guide robotic systems to execute accurate surgical procedures, improving accuracy, and minimizing tasks.
- **Adaptive therapies:** Adjusting the treatment protocols based on patient responses dynamically.

Table 1.2   Different types of machine learning algorithms used for different techniques

Prediction	Linear regression, logistic regression. Decision tree, support vector machine (SVM), Naive Bayes, K-means, random forest, KNN.
Classification	Linear classifiers, logistic regression. Decision tree, Naive Bayes classifier, SVMs, boosted trees, Random forest, neural networks, nearest neighbor.
Clustering	K-means, mean-shift clustering, DBSCAN (density-based spatial clustering of applications with noise), EM (expectation-maximization), GMM (Gaussian mixture models), agglomerative hierarchical clustering.

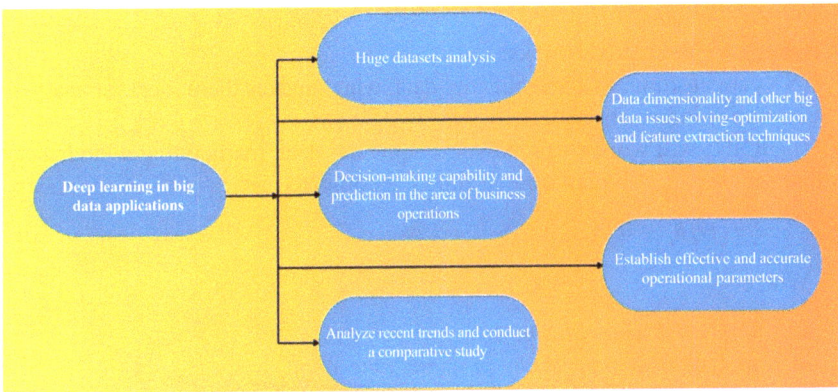

Figure 1.5   Deep learning in big data applications review.

Deep learning in healthcare

Deep learning consists of two types of networks: Convocational Neural Networks (CNN) and Recurrent Neural Networks (RNN). Deep learning makes the disease diagnosis very quick and effortless. CNNs are used for image analysis, which is the most efficient way of analyzing diseases using images. RNNs are used for time-series patient data such as monitoring heart rate and glucose levels. Using these networks or models is very easy (Dash *et al.*, 2019) and can be accessed by anyone, but building this kind of advanced models and networks needs a huge range of learning (Sharma *et al.*, 2024).

1) Convolutional Neural Networks:

CNNs are customized for processing image data, making them precious for medical image analysis.

The formula for a convolution operation network is:

$$s(t) = (x * w)(t) = \Sigma_r x(\tau) \cdot w(t - \tau)$$

Where:

- $x(t)$: *Input image or feature map.*
- $w(t)$: *Kernal (filter).*

- $s(t)$: *Output feature map* (*convolved result*).
- t: *Position in input data*.

Applications:

- **Medical imaging diagnostics**: This analyses X-rays, CT scans, MRIs, and mammograms for conditions such as fractures, organ abnormalities, and tumors.
- **Pathology image analysis:** This identifies cancerous tissues from histo-pathological slides.
- **Retinal disease detection:** This is used for screening for diabetic retino-pathy and other eye diseases.

2) Recurrent Neural Networks:
RNNs are customized for sequential data, which makes them ideal for analyzing time-series information.

RNNs will work by processing sequential data, maintaining a hidden state that evolves. The hidden state at time t is computed as:

$$h_t = f(W_h h_{t-1} + W_x x_t + b_h)$$

Where:

- h_t: *Hidden state at time t.*
- h_{t-1}: *Hidden state at the previous time step.*
- x_t: *Input at time t.*
- W_h, W_x: *Weight matrices for the hidden state and input, respectively.*
- b_h: *Bias term.*
- f: *Activation function* (e.g., tanh *or ReLU*).

Applications:

- **Patient monitoring:** This tracks heart rate, glucose levels, and blood pressure over time.
- **Disease progression prediction:** Which models the progression of chronic illnesses like Parkinson's or Alzheimer's.
- **Electronic health record analysis:** Extracting insights from temporal patient data.

1.5.2   Cloud computing

Cloud computing enables scalability, and cost-effective data storage and processing. The features of cloud computing make the Big Data application more advanced and efficient. Some of the key features are scalability, remote access, and cost-efficiency.

Scalability helps assist a large volume of healthcare data. Remote access helps healthcare providers or patients to access data anywhere, which improves collaboration. Cost efficiency reduces the use of on-premises data centers.

Cloud-based solutions in healthcare:
Data storage and management are used to centralize EHR systems hosted on the cloud. Cloud also provides data analytics platforms. Platforms like Amazon Web

Services (AWS), Microsoft Azure, and Google Cloud provide tools like machine learning for analyzing healthcare data. The cloud is also used for telemedicine integration, where the cloud enables secure remote monitoring and video consultations. Cloud integrates data security by implementing encryption and multi-factor authentication (Kumar and Singh, 2019; Guleria *et al.*, n.d.; Gupta *et al.*, 2022), which enhances the data security for our applications that keeps healthcare data safe from misuse. Cloud computing also addresses the Latency Issues by using a technique called edge computing, which processes data closer to the source.

1.5.3 Natural Language Processing

Natural Language Processing (NLP) is focused on unearthing meaningful information from unstructured text data in healthcare. The main role of NLP is to process the large amount of unstructured data of healthcare such as clinical notes, radiology reports, and discharge summaries. NLP is used to convert unstructured information into structured or well-arranged information. NLP transforms free-text patient information into structured information like analyzable formats.

Applications of NLP in Healthcare:

1) NLP is used in clinical documentation by automating the generation of patient records from dictated or written notes.
2) NLP enables sentiment analysis which monitors patient feedback and satisfaction. This helps healthcare professionals to enhance the services and can treat patients based on their feedback.
3) NLP is used in disease surveillance which analyses public health data for early outbreak detection (Rehman *et al.*, 2021).
4) Speech recognition in NLP is used by health providers for virtual assistants.

1.5.4 Blockchain for secure data sharing

Blockchain enables security, transparency, and tamper-proof in the management of healthcare data. Features of blockchain make big data applications more efficient. Some key features are decentralization, immutability, and smart contracts. Decentralization in blockchain will store data across a distributed network, which reduces risks of single-point failure. Immutability maintains the records that cannot be changed or edited without permission from the network participants. Blockchain has a feature of smart contracts which securely automates data sharing agreements. Blockchain is used in many real-time applications. Medi Ledger is a blockchain network that is used for pharmaceutical supply chain tracking. Guardtime provides blockchain solutions for securing healthcare records (Galetsi *et al.*, 2020; Sharma *et al.*, 2019, 2024).

Applications of blockchain in healthcare:

1) Blockchain allows patient data management, which allows healthcare professionals to control access and own their healthcare records.
2) Blockchain facilitates the sharing of data among stakeholders, which maintains a secure transfer of data among hospitals, labs, and insurance companies (Mehta and Pandit, 2018).

3) Drug supply chain management enables the tracking of the origin and authentication of medications to prevent counterfeit drugs.
4) Clinical trials suggests the transparency and accountability of research data.

1.6   Hadoop-based system

A Hadoop-based system is a distributed computing framework that is used to store and process enormous amounts of data across clusters of standard hardware. The two main components of a Hadoop-based system are the Hadoop Distributed file system (HDFS) and MapReduce. The use of the HDFS is to store the data in a distributed manner, and for parallel data processing it uses MapReduce (Karatas *et al.*, 2022). These types of systems are scalable, cost-effective, and can be able to handle all types of data such as structured, semi-structured, and unstructured data, making them ideal for big data applications.

Hadoop-based system vs traditional system:
Hadoop-based systems are suitable for huge-scale, diverse datasets and contain scalable, cost-effective, and fault-tolerant solutions for the latest big data applications.

Traditional systems are suitable for smaller, structured datasets, but here there is no scalability and flexibility needed for big data. Here are some of the differences between the traditional system and the Hadoop-based system.

Table 1.3   Differences between Hadoop-based systems and traditional systems

Traditional system	Hadoop-based system
A centralized system that depends on relational databases for data processing.	A distributed framework for storing and processing large data across multiple nodes.
Vertically scalable by upgrading hardware (e.g., CPU, RAM).	Horizontally scalable by adding more nodes to the cluster.
It primarily handles structured data in a relational database.	Capable of handling structured, semi-structured, and unstructured data.
Fault tolerance is managed by backup systems or redundancy measures.	Built-in fault tolerance with data replication across nodes.
It is more expensive due to reliance on high-performance hardware.	It is more cost-effective, running on commodity hardware and scaling out.
It shows better performance for small to medium datasets with transactional processing.	It shows lower performance for small datasets but excels with large datasets.
Sequential transactional processing, optimized for small datasets.	Batch processing (MapReduce), real-time processing (Apache Spark), and parallel data processing.
Strict ACID compliance, ensuring data consistency and reliability in transactions.	Eventual consistency, prioritizing high availability, and fault tolerance.
Use cases are transactional applications like banking and inventory systems.	Use cases are big data analytics, machine learning, data warehousing, and log analysis.

1.7   Need for Hadoop in healthcare data solutions

Charles Boicey, an information Solutions Architect at the University of California (UCI), states the ground making an impact of Big Data and Hadoop on healthcare information management. According to Boicey,

> Hadoop is the only technology that allows healthcare to store data in its native form. If Hadoop didn't exist. We would still have to make decisions about what can come into our data warehouse or the electronic medical record (and hear cannot). Now we can bring everything into Hadoop, regardless of the data format or speed of ingest. If I find a new data source I can start storing it the day that I learn about it. We leave no data behind (Vassakis *et al.*, 2018; Sharma and Kumar, 2022).

The importance of this statement is emphasized by the significant increase in health records, which by the end of 2016 was expected to reach tens of billions. As a result, computing technology and infrastructure should be capable of providing cost-effective results that meet the following requirements:

- Unconstrained parallel data processing.
- Storage for a large amount of data such as billions and trillions of unstructured data sets.
- Fault tolerance with high system availability.

Hadoop technology can successfully address the challenges in the healthcare sector due to the features of its MapReduce and HDFS, which can be capable of processing thousands of terabytes of data. Additionally, Hadoop uses affordable commodity hardware which makes it a budget-friendly expense for healthcare industries.

Beyond its scalability and cost-effectiveness, the Hadoop ecosystem delivers various tools like Apache Hive, Apache HBase, and Apache Spark, which improve querying the data (Bouhriz and Chaoui, 2015), processing in real-time, and advanced analytics. This resilient framework enables healthcare professionals to exploit the full potential of a huge volume of datasets, which is useful for enhancing patient treatment, streamlining operations, and pioneering research.

1.8   Hadoop-based applications for the healthcare industry

Healthcare data generally exists in printed form, we cannot be able to analyze these types of data. To process and analyze we need dynamic digitization of data which is in printed form. The health sector consists of majorly semi-structured and unstructured types of data, so we must extract useful information from this unstructured type of data. This data consists of information about patient care, clinical operations, and research. The combination of software utilities known as the Hadoop ecosystem, helps manage the bulk amount of data generated in the

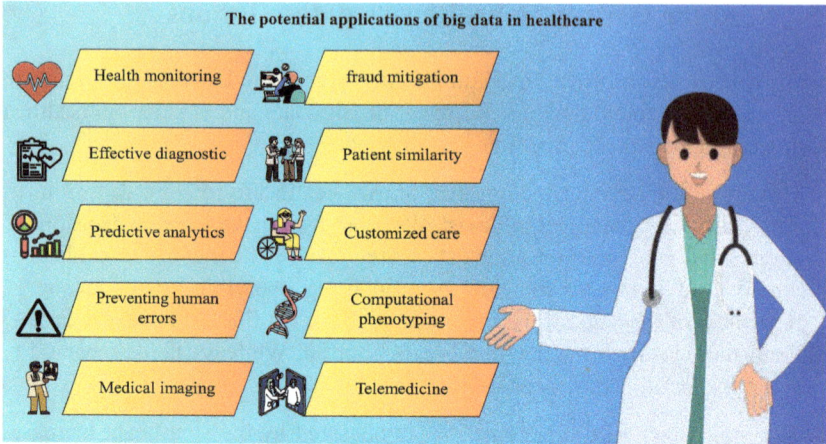

Figure 1.6   Applications of big data in healthcare

health sector. Traditional systems are unable to handle this bulk amount of healthcare data, while the Hadoop ecosystem provides a scalable solution for collecting, storing, managing, and processing such bulk amounts of datasets (Jee and Kim, 2013; Sharma *et al.*, 2024; Sharma and Kumar, 2022). Here are some applications of Hadoop-based ecosystems in the health sector.

1.8.1   Hadoop technology in advancing cancer treatments and genomics

Hadoop incorporation has seen particular growth in the sector of health innovation, significantly for the upcoming generation following and Foundational read-to-inference process as said by Deepak Singh, Principal Product Manager at Amazon Web Services. He elaborated on all the obstacles faced by using Hadoop mentioning that Hadoop is an efficient tool for research. About 300 million base pairs make human DNA, making research on a large amount of data important for defeating diseases like cancer. The toughest step in curing cancer is understanding the patterns of DNA and how it reacts based on different genetics. This leads the study of cancer to perceive specific treatment according to the person, which leads to every patient's distinct deoxyribonucleic acid profile. Hadoop with its simultaneous processing abilities and MapReduce programs, plays an important role in matching those 300 million base pairs effectively. The Global Director of Health and Life Sciences at Intel, Ketan Paranjape, mentioned the importance of Hadoop in healthcare analytics. The technology facilitates the gathering and evaluating the data on a large scale from tracing health trends for millions of people to recognizing accurate treatment options for individual cancer patients.

In 2014, then-UK Prime Minister, David Cameron, announced the government investment of 26 billion Indian rupees as a four-year project that processes about 1,00,000 genomes in collaboration with American biotech company Illumina and Genomics England. The project's basic need is to utilize big data in healthcare

(Kim *et al.*, 2014; Sharma *et al.*, 2022) to develop personalized cancer treatments. In the same way, Complex Adaptive Systems Initiative (CASI) at Arizona State University is building a genomic data lake having petabytes of genetic data. This attempt can help in recognizing cancer-related genes and place the base for life-saving treatments through big data analysis.

To grasp the difficulties of genomics data and understand how Hadoop addresses them, reflecting on cancer drugs declared to be 40% efficient. This number means that the drug is working proficiently for patients with a particular genetic environment, but without these conditions for others, it does not affect them at all. The difficulty of healthcare data from the reality that a single genome can have 20,000 genes. Storing this data in a Relational Database, which has a million variable DNA elements, can result in 20 billion rows of data per person. Legacy items that are simply unable to manage these large data and difficult to handle these sizes of data. Hadoop provides a much-needed solution for manipulating and examining such large datasets (Pramanik *et al.*, 2019).

1.8.2 Hadoop for real-time patient vital monitoring

Nowadays, big healthcare organizations are using Hadoop for continuous monitoring of patients. Healthcare organizations are utilizing Hadoop to educate healthcare professionals to work better and effectively. These types of organizations use sensors around patients' beds which collect and store patients' activities and behavior like cholesterol, blood pressure, etc. It records the issues that patients are experiencing and report. These sensors produce a bulk amount of data which a

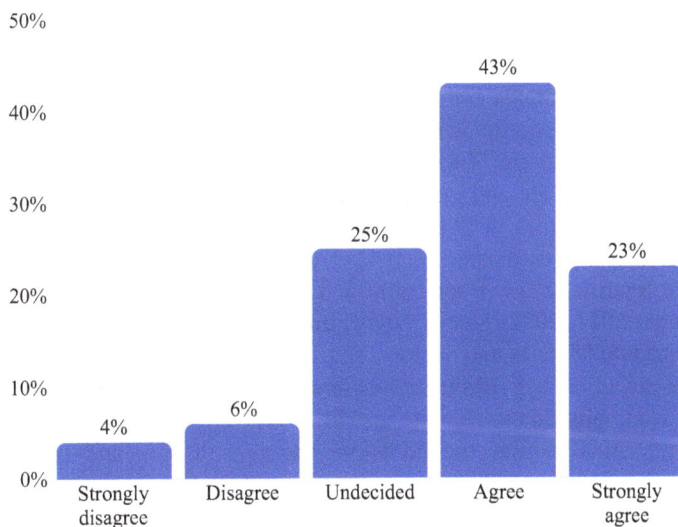

Figure 1.7 *This graph shows the survey report of people about the high cost of healthcare insurance because of the huge number of fraudulent claims.*

traditional database management can't manage for longer, and here there is a need for big data Hadoop for storing and processing the huge data (Shafqat *et al.*, 2020). These collected data will be then processed and analyzed to enhance treatment for patients and improve services for patients.

1.8.3   Enhancing fraud prevention and detection

Today's society is full of fraudulent claims, which is nothing new, but the fraud claims are increasing day by day. The issue we are looking up to be simple, but when we sum up all the fraudulent claims, it will show a greater impact that will be in billions. In a survey, nearly 40% of people agree that the cost of healthcare insurance is high because of the huge number of fraudulent claims.

Many healthcare insurance companies are utilizing big data Hadoop to diminish these false claims. These companies use both real-time data as well as previous historical data such as medical claims, details of wages, voice recording, demographic data, and call center notes.

1.8.4   Reducing hospital readmissions

Hospitals are facing a very big problem which is readmissions, those people who return in 30 days of treatment. These hospitals want to keep these types of cases away at any cost. Texas Hospital utilized Hadoop in electronic medical records and detected that patients who came back within 30 days should require further supervision and enhanced treatment. By following this, by using Hadoop in Electronic medical records, these hospitals can be able to minimize readmission by 5%.

1.8.5   Electronic health records

EHRs are broadly used in big data applications in healthcare. In developed countries like the United States and the United Kingdom, every patient will maintain their digital health records which include their medical history, demographic details, health records, etc. These records will be available to healthcare service providers in both private and public sectors. Every record contains a single modifiable file, which allows doctors to make updates whenever needed without any paperwork, this will eradicate the risk of data redundancy (Gupta *et al.*, 2022; Kumar and Sharma, 2023; Raghupathi and Raghupathi, 2014; Satti *et al.*, 2024; Sharma *et al.*, 2021; 2022; Sharma and Kumar, 2022). These EHRs will also send alerts to schedule lab tests and doctor visits, ensuring that they prioritize healthcare needs. The execution of EHRs in all health centers is an enduring task. The United States has accomplished this in more than 96% of hospitals. However, in places like India or the Asia-Pacific, the adoption rate is still low and presents more challenges.

1.8.6   Real-time alerting system

In hospitals, there is a system called Clinical Decision Support (CDS) systems which is used to analyze medical information in real-time and offer guidance to healthcare professionals as they make choices about the patient's treatment. Generally, doctors favor patients to stay out of the hospital because of the high living costs and patients themselves do not want to bear heavy treatment costs unless it is an emergency case. With the advantage of wearable technology, patients can now be able to send their health information to the cloud easily by using these wearable devices. With this technology, doctors can access and respond to this data in real-time. These types of innovations have sustainable enhanced real-time alerting capabilities. For example, if any patient's blood pressure or sugar level spikes, then doctors can be able to respond immediately by reviewing their historical records and taking immediate necessary actions. In addition, the information is helpful to analyze and improve healthcare service delivery with the help of Big Data technologies like Hadoop. By exploring these kinds of technologies (Hong *et al.*, 2018), healthcare organizations can identify patterns, forecast possible health issues, and make more useful decisions, ultimately improving patient results.

1.9   Hadoop's tools and techniques for big data

To manage data that does not have any structure and does not fit into any database, different types of tools are required. Information Technology uses the Apache Hadoop platform for different methods that have been implemented to record, organize, and analyze this type of unstructured data. We must use more efficient tools to produce an efficient output from big data. Many of the tools are applied in the Apache Hadoop Architecture which includes MapReduce and Hive. Here are the various tools used to process healthcare big datasets.

- **Apache Hadoop:**
  It is a framework that is used to store and process datasets of bulk size of information across a network of computers. It provides a robust and flexible

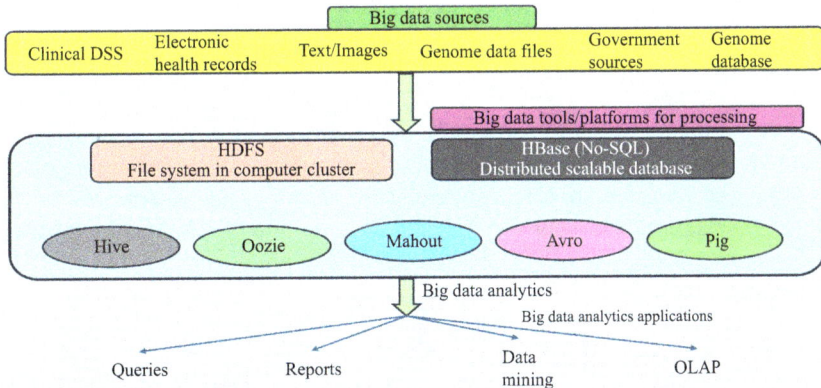

Figure 1.8   Conceptual architecture of big data analytics for health informatics.

infrastructure that supports the development and application of scattered computation of large-size data (Guo and Chen, 2023). This Apache Hadoop works on the principles of error resilience, expandability, and data locality. It became the main infrastructure of multiple organization data techniques. In other words, big data talks about collecting data from multiple sources such as Images on the internet, and data collected from sensors of both structured and unstructured data forms to be computed. Hadoop Topology contains MapReduce, HBase, and HDFS.

- **Hadoop Distributed File System:**
  HDFS is used for aggregating big data. Of course, multiple users can use Hadoop at a time, but the HDFS is not used for that kind of environment. Instead, its configuration is thought of as single instance write/multiple access that reduces the overload and manages the multi-tasking and integrity needs of a Parallel Data Management System. HDFS is created for reading a large amount of data from a disk. The block size of the HDFS is either 64 or 124 megabytes. Nodes are divided into two types, name node and data node (Najafabadi *et al.*, 2015). There are many data nodes in the HDFS. The name node oversees all the catalog information that is used to store and give the data present from the data nodes. There is no data present in the name node. Files are stored in blocks and a structured manner, and all these blocks are in same size. The main properties of the HDFS are its networked structure and its trustworthiness. The disk space of metadata and file data is different and divided. The file system information is stored in the name node and the data of the application is stored in the data node (Luo *et al.*, 2016).

- **MapReduce:**
  Apache Hadoop is combined with MapReduce processing. MapReduce processing is used in Health Application systems, and it is a simplified than most users realize. MapReduce Architecture is very easy to understand. There are two phases in MapReduce, i.e., mapper phase and the reducer phase. In the mapper phase, mapping is applied to the input and the reducer phase is applied when counting is completed. The mapper phase takes the input in key-value pairs and provides the output in key-value pairs. The data segments are divided into equal sizes, which is also known as input splits in terms of Hadoop. MapReduce is also a programming tool that can be written in languages like Java and C.

- **Apache Hive:**
  A data warehousing layer is built on top of Hadoop, in which it evaluates, and queries can be performed in procedure language like Structured Query Language. Apache Hive is also applicable to unplanned queries, recap, and data evaluation. Hive can handle data of large sizes such as (Aggarwal and

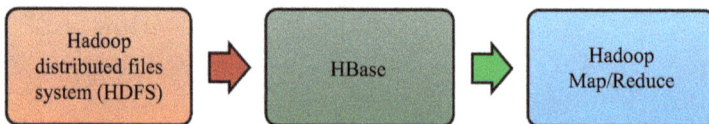

Figure 1.9   Hadoop system architecture

Wang, 2010) petabytes of data, and provides features like data retrieval, data modification, and access to the Hadoop Distributed File System consisting of data files or other storage systems like HBase.

- **Apache Pig:**
  It is used for evaluating big data. Pig is used as another optional tool instead of MapReduce Programming Tool. Apache Pig was created by Yahoo. Pig was a research project Pig that allows users to create our user-defined functions and supports many high-level operations like Joins, sort, filter, etc.

- **Apache HBase:**
  It is a column-based NoSQL database that is used in Hadoop in which users can handle large sizes of data. It has features like read/write operations. It also has a feature to store concurrent data storage by the essential parallel file systems on multiple servers. If we need a fast response (Wang *et al.*, 2018) to view data of large-size data stored in Hadoop, HBase is the best option to use. HBase is also capable of handling petabytes of data.

- **Apache Oozie:**
  If we have a sophisticated system or if there are many interconnected stations within the data, then we can use Apache Oozie. Oozie is capable of running many jobs. Oozie has two phases. It has workflow engines and a coordinator engine. The workflow engine stores and executes Hadoop-based jobs, coordinator engine processes workflow jobs based on their design in the processing timetable. The workflow jobs are produced in the form of Directed Acyclic graphs of Actions (Zhang and Huang, 2016).

- **Apache Avro:**
  Avro is a data encoding format that can be written in any language. It can be used to connect Flume data flows. It is schema based, which is used for read and write operations.

- **Apache Zookeeper:**
  Zookeeper is a single-point architecture used by many applications to sustain the healthcare system. It maintains common objects required in a large server environment, including metadata and hierarchical naming space. Apache Zookeeper is a reliable tool. If the master gets suspended or disconnected, the Zookeeper creates a new application master to resume those tasks.

- **Apache Yarn:**
  Apache Yarn is a command-line tool and an example of a Hadoop non-MapReduce tool built on top of YARN. Yarn has two elements, A Resource Manager and a Node Manager (Kuo *et al.*, 2014). Resource Manager is used to operate all the resources in a server and operates the available resources on the independent host. Both elements are used to handle and manage the scheduling of jobs, operating the containers, managing the memory, throughput of the CPU, and input–output systems that run the given application code.

- **Apache Sqoop:**
  Sqoop is a powerful tool that processes the collection of data from a Relational Database Management System and inputs data into Hadoop Architecture for processing the queries. We can use MapReduce programming or Hive.

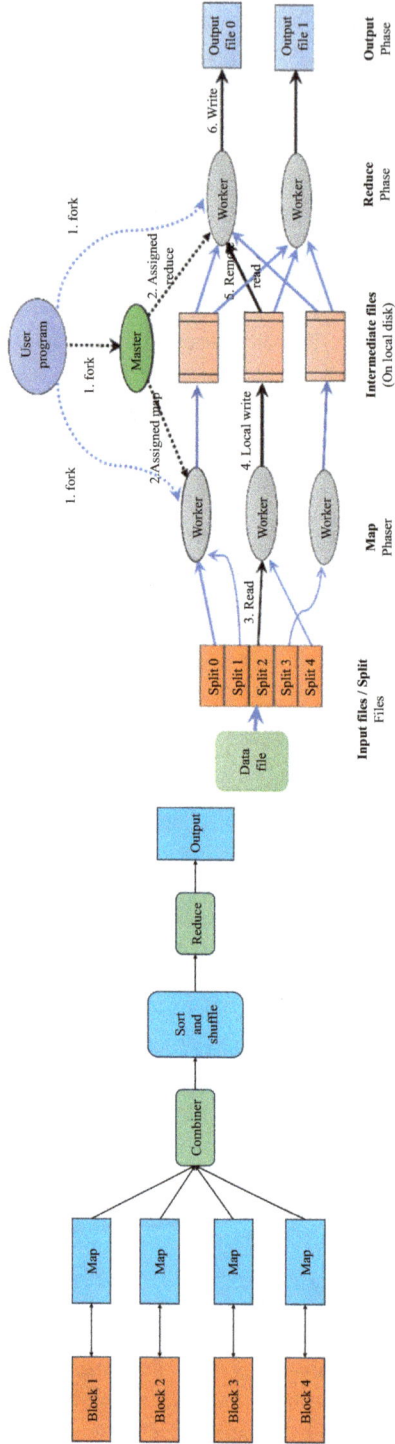

Figure 1.10   Architecture of MapReduce used in the big data for processing the large file containing different kinds of data.

- **Apache Flume:**
  Apache Flume is a highly stable service for collecting data efficiently and transforming huge data from an independent system to HDFS. Apache Flume is used for log files and data produced by email messages and social media.

1.10 Conclusion

The incorporation of big data analytics in healthcare is a ground-breaking transformation in how we approach healthcare delivery, research, and decision making. Several key conclusions emerge from this comprehensive exploration of health data analytics and big data applications. The healthcare sector is experiencing tremendous data growth, which information is being produced from different sources in structured, unstructured, and semi-structured formats. The capability to efficiently manage and analyze this information has become important for enhancing healthcare delivery and results.

The deployment of Hadoop-based systems confirmed fundamental in addressing the problems introduced by healthcare big data. These systems offer scalable, cost-friendly solutions that can be able to handle the volume, variety, velocity, and veracity of healthcare information while retaining data security and integrity.

The integration of various tech innovations including AI, ML, cloud computing, blockchain, and NLP with big data analytics has enabled powerful tools for healthcare professionals. These incorporated results created more accurate diagnoses, personalized care plans, and enhanced patient outcomes. The incorporation of big data analytics has proven real-world benefits in various healthcare applications, from EHRs management to real-time patient monitoring, clinical decision support, and fraud detection systems.

1.11 Future scope

The future of healthcare data analytics and big data applications introduces fascinating potentials and difficulties. Advanced data analytics and AI are set to revolutionize healthcare with more advanced models that can handle various data types. The incorporation of quantum computing will enable complex research, and predictive analytics which allows for enhancing the management of disease outbreaks and population health.

The integration of genomic and clinical data is set to take personalized medicine to new heights. This intersection will enable highly personalized treatments and interventions, optimized to the unique genetic makeup and medical history of each patient. By using this comprehensive data, healthcare professionals can be able to deliver more accurate diagnoses, predict disease tendencies, and develop personalized treatment plans that significantly improve patient outcomes.

The prospects of health data analytics will depend on balancing technological innovations with ethical considerations, data privacy, and accessibility. Tackling issues such as standardization, interoperability, and workforce development is needed for the full utilization of big data in reshaping the healthcare delivery and medical research. These efforts will guarantee that data from various sources can be efficiently integrated, utilized, and shared, which enables better-informed decision making, improved patient outcomes, and advancements.

References

Aggarwal, C. C., and Wang, H. (2010). *Managing and Mining Graph Data*. 610 New York: Springer, 40.

Andreu-Perez, J., Poon, C. C. Y., Merrifield, R. D., Wong, S. T. C., and Yang, G. Z. (2015). Big Data for health. *IEEE Journal of Biomedical and Health Informatics*, *19*(4), 1193–1208. https://doi.org/10.1109/JBHI.2015.2450362

Bouhriz, M., and Chaoui, H. (2015). Big data privacy in healthcare Moroccan context. *Procedia Computer Science*, *63*, 575–580. https://doi.org/10.1016/J.PROCS.2015.08.387

Dash, S., Shakyawar, S. K., Sharma, M., and Kaushik, S. (2019). Big data in healthcare: management, analysis, and prospects. *Journal of Big Data*, *6*(1), 1–25. https://doi.org/10.1186/S40537-019-0217-0/FIGURES/6

Dubey, R., Gunasekaran, A., Childe, S. J., Blome, C., and Papadopoulos, T. (2019). Big Data and predictive analytics and manufacturing performance: integrating institutional theory, resource-based view and Big Data culture. *British Journal of Management*, *30*(2), 341–361. https://doi.org/10.1111/1467-8551.12355

Fisher, E. S., Wennberg, D. E., Stukel, T. A., Gottlieb, D. J., Lucas, F. L., and Pinder, É. L. (2003). The implications of regional variations in Medicare spending. Part 2: health outcomes and satisfaction with care. *Annals of Internal Medicine*, *138*(4), 288–298. https://doi.org/10.7326/0003-4819-138-4-200302180-00007

Galetsi, P., Katsaliaki, K., and Kumar, S. (2020). Big data analytics in the health sector: theoretical framework, techniques, and prospects. *International Journal of Information Management*, *50*, 206–216. https://doi.org/10.1016/J.IJINFOMGT.2019.05.003

Graham, S., Depp, C., Lee, E. E., Nebeker, C., Tu, X., Kim, H. C., and Jeste, D. V. (2019). Artificial Intelligence for mental health and mental illnesses: an overview. *Current Psychiatry Reports*, *21*(11). https://doi.org/10.1007/S11920-019-1094-0

Guleria, V., Sharma, A., and Gupta, G. (n.d.). 2022. Biosensors in bioinformatics, biotechnology, and healthcare. *Researchgate. NetA Sharma, V Guleriaresearchgate.Net.* https://doi.org/10.52305/LDXT8191

Guo, C., and Chen, J. (2023). *Big Data Analytics in Healthcare*. 27–70. https://doi.org/10.1007/978-981-99-1075-5_2

Gupta, M., Sharma, S., Sakshi, and Sharma, C. (2022). Security and privacy issues in blockchain IoT: principles, challenges and counteracting actions. *Blockchain*

Technology: Exploring Opportunities, Challenges, and Applications, 27–56. https://doi.org/10.1201/9781003138082-3/SECURITY-PRIVACY-ISSUES-BLOCKCHAINED-IOT-MANIK-GUPTA-SHAMNEESH-SHARMA-SAK-SHI-CHETAN-SHARMA

Hong, L., Luo, M., Wang, R., Lu, P., Lu, W., and Lu, L. (2018). Big Data in health care: applications and challenges. *Data and Information Management*, *2*(3), 175–197. https://doi.org/10.2478/DIM-2018-0014

Jee, K., and Kim, G. H. (2013). Potentiality of big data in the medical sector: focus on how to reshape the healthcare system. *Healthcare Informatics Research*, *19*(2), 79–85. https://doi.org/10.4258/HIR.2013.19.2.79

Karatas, M., Eriskin, L., Deveci, M., Pamucar, D., and Garg, H. (2022). Big Data for Healthcare Industry 4.0: applications, challenges and future perspectives. *Expert Systems with Applications*, *200*, 116912. https://doi.org/10.1016/J.ESWA.2022.116912

Kim, G. H., Trimi, S., and Chung, J. H. (2014). Big-data applications in the government sector. *Communications of the ACM*, *57*(3), 78–85. https://doi.org/10.1145/2500873

Kumar, R., and Sharma, A. (2023). Computational strategies and tools for protein tertiary structure prediction. *Basic Biotechniques for Bioprocess and Bioentrepreneurship*, 225–242. https://doi.org/10.1016/B978-0-12-816109-8.00015-5

Kumar, S., and Singh, M. (2019). *Big Data Analytics for Healthcare Industry: Impact, Applications, and Tools*. *2*(1). https://doi.org/10.26599/BDMA.2018.9020031

Kuo, M. H., Sahama, T., Kushniruk, A. W., Borycki, E. M., and Grunwell, D. K. (2014). Health big data analytics: current perspectives, challenges, and potential solutions. *International Journal of Big Data Intelligence*, *1*(1/2), 114. https://doi.org/10.1504/IJBDI.2014.063835

Lee, C., Luo, Z., Ngiam, K. Y., Zhang, M., Zheng, K., Chen, G., Ooi, B. C., and Yip, W. L. J. (2017). *Big Healthcare Data Analytics: Challenges and Applications*. 11–41. https://doi.org/10.1007/978-3-319-58280-1_2

Luo, J., Wu, M., Gopukumar, D., and Zhao, Y. (2016). Big data application in biomedical *research and health care: a literature review. Biomedical Informatics Insights*, 19(8), 1–10. *Https://Doi.Org/10.4137/BII.S31559, 8,* BII.S31559. https://doi.org/10.4137/BII.S31559

Lv, Z., and Qiao, L. (2020). Analysis of healthcare big data. *Future Generation Computer Systems*, *109*, 103–110. https://doi.org/10.1016/j.future.2020.03.039

Mehta, N., and Pandit, A. (2018). Concurrence of big data analytics and healthcare: a systematic review. *International Journal of Medical Informatics*, *114*, 57–65. https://doi.org/10.1016/J.IJMEDINF.2018.03.013

Mikalef, P., Boura, M., Lekakos, G., and Krogstie, J. (2019). Big data analytics capabilities and innovation: the mediating role of dynamic capabilities and moderating effect of the environment. *British Journal of Management*, *30*(2), 272–298. https://doi.org/10.1111/1467-8551.12343

Najafabadi, M. M., Villanustre, F., Khoshgoftaar, T. M., Seliya, N., Wald, R., and Muharemagic, E. (2015). Deep learning applications and challenges in big data analytics. *Journal of Big Data*, 2(1). https://doi.org/10.1186/S40537-014-0007-7

Pramanik, P. K. D., Pal, S., and Mukhopadhyay, M. (2019). Healthcare big data: A comprehensive overview. *Intelligent systems for healthcare management and delivery*, 72–100. *Https://Services.Igi-Global.Com/Resolvedoi/Resolve.Aspx? Doi=10.4018/978-1-6684-3662-2. Ch006, 1*, 119–147. https://doi.org/10.4018/978-1-6684-3662-2.CH006

Raghupathi, W., and Raghupathi, V. (2014). Big data analytics in healthcare: promise and potential. *Health Information Science and Systems*, 2(1), 1–10. https://doi.org/10.1186/2047-2501-2-3/TABLES/2

Rehman, A., Naz, S., and Razzak, I. (2021). Leveraging big data analytics in healthcare enhancement: trends, challenges, and opportunities. *Multimedia Systems 2021 28:4, 28*(4), 1339–1371. https://doi.org/10.1007/S00530-020-00736-8

Rumsfeld, J. S., Joynt, K. E., and Maddox, T. M. (2016). Big data analytics to improve cardiovascular care: promise and challenges. *Nature Reviews Cardiology*, 13(6), 350–359. https://doi.org/10.1038/nrcardio.2016.42

Satti, S. R., Lankadasu, J. S. K., Sharma, A., Sharma, S., and Gochhait, S. (2024). Deep learning in medical image diagnosis for COVID-19. *2024 ASU International Conference in Emerging Technologies for Sustainability and Intelligent Systems, ICETSIS 2024*, 1858–1865. https://doi.org/10.1109/ICETSIS61505.2024.10459430

Shafqat, S., Kishwer, S., Rasool, R. U., Qadir, J., Amjad, T., and Ahmad, H. F. (2020). Big data analytics enhanced healthcare systems: a review. *Journal of Supercomputing*, 76(3), 1754–1799. https://doi.org/10.1007/S11227-017-2222-4/METRICS

Sharma, A., Guleria, V., and Jaiswal, V. (2022a). The future of blockchain technology, recent advancement and challenges. *Studies in Big Data, 105*, 329–349. https://doi.org/10.1007/978-3-030-95419-2_15

Sharma, A., Gupta, A., and Jaiswal, V. (2021). Solving image processing critical problems using machine learning. *Studies in Big Data, 82*, 213–248. https://doi.org/10.1007/978-981-15-9492-2_11

Sharma, A., Kala, S., Kumar, A., Sharma, S., Gupta, G., and Jaiswal, V. (2024). Deep learning in genomics, personalized medicine, and neurodevelopmental disorders. *Intelligent Data Analytics for Bioinformatics and Biomedical Systems*, 235–264. https://doi.org/10.1002/9781394270910.CH10

Sharma, A., Kaur, D., Gupta, A., and Jaiswal, V. (2019). Application and Analysis of Hyperspectal Imaging. *Proceedings of IEEE International Conference on Signal Processing, Computing and Control, 2019-October*, 30–35. https://doi.org/10.1109/ISPCC48220.2019.8988436

Sharma, A., and Kumar, R. (2022). Recent advancement and challenges in deep learning, big data in bioinformatics. *Studies in Big Data, 105*, 251–284. https://doi.org/10.1007/978-3-030-95419-2_12

Sharma, A., Kumar, R., and Jaiswal, V. (2021). Classification of Heart Disease from MRI Images Using Convolutional Neural Network. *Proceedings of IEEE International Conference on Signal Processing, Computing and Control, 2021, October*, 358–363. https://doi.org/10.1109/ISPCC53510.2021.9609408

Sharma, A., Pal, T., and Jaiswal, V. (2022). Heart disease prediction using convolutional neural network. *Cardiovascular and Coronary Artery Imaging: Volume 1*, 245–272. https://doi.org/10.1016/B978-0-12-822706-0.00012-3

Sharma, A., Pal, T., Naithani, U., Gupta, G., and Jaiswal, V. (2024). Emerging trends of big data in bioinformatics and challenges. *Intelligent Data Analytics for Bioinformatics and Biomedical Systems*, 265–290. https://doi.org/10.1002/9781394270910.CH11

Vassakis, K., Petrakis, E., and Kopanakis, I. (2018). Big data analytics: applications, prospects and challenges. *Lecture Notes on Data Engineering and Communications Technologies*, *10*, 3–20. https://doi.org/10.1007/978-3-319-67925-9_1

Wang, Y., Kung, L. A., and Byrd, T. A. (2018). Big data analytics: understanding its capabilities and potential benefits for healthcare organizations. *Technological Forecasting and Social Change*, *126*, 3–13. https://doi.org/10.1016/J.TECHFORE.2015.12.019

Wang, Y., Kung, L. A., Wang, W. Y. C., and Cegielski, C. G. (2018). An integrated big data analytics-enabled transformation model: Application to health care. *Information & Management*, *55*(1), 64–79. https://doi.org/10.1016/J.IM.2017.04.001

Zhang, J., and Huang, M. L. (2016). Data behaviours model for Big Data visual analytics. *International Journal of Big Data Intelligence*, *3*(1), 1. https://doi.org/10.1504/IJBDI.2016.073899

Chapter 2

Advances in brain tumor detection: role of feature extraction and machine learning

Kuldeep Chaurasia[1], Aryan Aryan[1], Ayush Dikshit[1], Chirag Sanskrityayan[1] and Vikas Kumar Jain[1]

Abstract

The brain is widely regarded as the most vital organ in the human body, as it governs and regulates all essential bodily functions. In contrast, a tumor is an abnormal mass of cells that grows uncontrollably. Any tumor that originates in the brain or spreads to it is classified as a brain tumor. Currently, the primary cause of brain tumor development remains unknown. While brain tumors are relatively rare, their detection and management are critical for patient outcomes.

To address the challenge of identifying brain tumors, this study explores various models for detecting them in MRI scans. In this chapter, multiple Artificial Intelligence algorithms have been implemented to determine the most accurate approach for brain tumor detection from image data. The techniques employed include Watershed, Particle Swarm Optimization (PSO), Lazy IBK, Support Vector Machines (SVM), and Convolutional Neural Networks (CNN). Among these, the combined approach of Watershed and CNN attained an accuracy of 98%, demonstrating its efficacy. The findings of this chapter have the potential to significantly benefit clinical practice and can be further refined for deployment in real-world medical applications.

Keywords: Brain tumor; MRI scans; Watershed algorithm; PSO algorithm; CNN; Classification

2.1 Introduction

2.1.1 Background

The humans are made up of many cells. Uncontrolled cell proliferation turns the excess mass of cells into a tumor. To identify the tumor, magnetic resonance imaging (MRI) and CT scans are utilized. This study's objective is to find a more accurate and

[1]School of Computer Science, Engineering, and Technology, Bennett University, India

efficient method of identifying brain tumor using a variety of methods that include computer vision, pattern analysis, and medical image processing for the improvement, segmentation, and classification of brain diagnoses. Medical professionals can utilize the findings of the study to improve brain tumor identifications by analyzing MRI scans collected from the dataset. This chapter evaluates and compares different classification and segmentation techniques. The goal is to identify a better and more accurate method for detecting brain tumor, which would ultimately enhance the diagnosis of patients by reducing unnecessary procedures.

2.1.2   Problem definition

The aim of this chapter is to find an accurate and efficient method of identifying a brain tumor. MRI scans are widely used techniques for investigating brain anatomy. In this chapter, algorithms such as Watershed are evaluated to detect if a brain tumor is present in the provided MRI scan as well as if the spot has a tumor or not.

2.1.3   Motivation

This chapter aims to assist professionals in establishing how different algorithms accurately and efficaciously diagnose brain tumors on MRI scans. The goal of this study is to make brain tumor detection more accurate.

2.1.4   Scope

This chapter aims to evaluate and compare different algorithms to detect the brain tumor. This chapter outcomes can be of benefit to medical professionals. The conclusions can potentially improve the accuracy and the speed of brain tumor detection via MRI image processing, pattern analysis, and computer vision. Such tools to support early diagnosis and detection monitoring of these algorithms can be developed through approaching them properly.

2.2   Related work

Mishra (2010): The paper introduces the application of an expert system designed for the diagnosis of brain tumors using artificial neural networks (ANNs) and wavelet packet analysis of MRI data. Detailed feature extraction by using wavelet packets present a richer representation of MRI spectra than the standard transforms of wavelets. The approach is to use the back-propagation training algorithm and optimize the neural networks for minimal prediction errors by adjusting weights. The analysis concludes that wavelets are less effective than wavelet packets in capturing the subtle patterns of MRI data, thereby enhancing the diagnostic accuracy. This is a revolutionary combination of ANN and wavelet packets, a novel approach for the accurate diagnosis of brain tumors and certainly holds great promise for many clinical applications.

Dubey and Mushrif (2012): The paper reviews advanced fuzzy c-means (FCM) clustering algorithms for segmenting brain MR images, addressing problems such as noise and intensity inhomogeneity. Methods, including Bias-Corrected FCM (BCFCM), spatially constrained FCM, and multiscale fuzzy clustering, were utilized to improve robustness and accuracy. Validation was conducted using the McGill BrainWeb MR image dataset, with synthetic images and controlled noise and intensity variations. All three metrics, namely, accuracy of segmentation, Jaccard similarity, and Dice coefficient, showed significant improvement, indicating better performance over regular approaches.

Gowthaman *et al.* (2022): The study by Gowthaman focuses on classification and detection of brain tumors using MRI images by means of various machine learning and image processing techniques. It mainly includes methodologies like pre-processing, segmentation, feature extraction, and classification. For the purpose of analysis, T-1 and T-2 weighted MRI scans are used. Comparison is made between the conventional techniques such as support vector machine (SVM) with more sophisticated classifiers like ANNs and convolutional neural networks (CNNs). It is discovered that the latter one gives better accuracy. The authors concluded that the earlier it can detect, the better will be the treatment. The article hopes to improve the accuracy in the classification of tumors and thus the clinical outcome in the patient.

Mohsen *et al.* (2012): A new image segmentation approach as proposed in this paper: the "partially supervised watermark optimization" (PSWO) technique method. The learning procedure of combining supervised and unsupervised learning improved segmentation performance was applied. The training and testing dataset contained a diverse set of images, both labeled and unlabeled data. It enables the model to learn from explicit annotations as well as inherent patterns present in images. The overall conclusion of the study is that the proposed PSWO method improved segmentation accuracy significantly more than traditional techniques of segmentation. The results showed that partially supervised learning frameworks are effective tools, and thus, by using a combination of labeled and unlabeled datasets, the performance in image processing tasks can be improved. This provides avenues for further research in other applications where accurate image segmentation is required.

Khiyal *et al.* (2009): The paper discusses problems in image segmentation, specially focusing on the watershed algorithm, which generally suffers over-segmentation and under-segmentation in the case of low-contrast images. For this, the authors suggested a preprocessing step using the Random Walk method, enhancing the image contrast before the application of the watershed algorithm. Several techniques for better watershed segmentation have been known. These include wavelet-based denoising and marker-controlled watershed. However, many focus only on the issue of over-segmentation, hardly giving attention to low-contrast problems. The results from the proposed method show good improvements in the quality of the segmentation, as errors from low-contrast images are reduced. This work contributes to the field of image processing by giving a more effective solution to segmenting challenging images and thus hints at further research on extending these techniques to three-dimensional images.

Saini *et al.* (2017): The paper presents an overview of various techniques for detecting brain tumors using medical imaging, especially MRI and CT scans. It emphasizes the challenge of early detection because of the subtlety of initial symptoms and the importance of timely diagnosis for effective treatment options. The authors discussed several image processing techniques, namely segmentation, thresholding, fuzzy clustering, and ANNs, used in the identification of tumors during the recent past. Further, they classify brain tumors into benign and malignant types, highlighting their features and grades. These characteristics and grades play a key role in the treatment approach. In this regard, the study clearly highlights the fact that progress in image processing techniques leads to the efficient and precise detection of brain tumors in medical practice.

Bahadure *et al.* (2017): This paper evaluates the enhancement of MRI brain tumor segmentation and classification using a combination of Berkeley Wavelet Transformation along with Support Vector Machine techniques. MRI images from 15 patients were selected for the study, all containing T1, T2, and FLAIR sequences. The technique comprised a series of preprocessing: skull stripping, segmentation, and then feature extraction steps that generated texture-based features from the images themselves. The results are shown as significant improvements, where accuracy attained 96.51%, specificity at 94.2%, and sensitivity at 97.72%. The paper concludes that the proposed technique provides effective and timely brain tumor detection, with strong potential for integration into clinical decision support systems for initial diagnosis and screening. Future work will explore methods to further enhance classification accuracy using feature selection and combined approaches.

Kazi *et al.* (2017): This paper proposes a better kernelized rough-fuzzy c-means (KRFCM) approach for segmenting brain tissue in MRI data. The authors use the machine learning methods: fuzzy set theory and rough set theory, to improve clustering that is vulnerable to problems such as noise, uncertainty, and the overlapping nature of regions in complex brain tissues. A total of 20 images with a size of 256×256 pixels, derived from a public repository of T1-weighted MRI brain images, were used in the evaluation process. Results showed that KRFCM outperformed the classical segmentation techniques and obtained a higher accuracy and efficiency when segmenting the brain tissues, especially with noisy environments. This research study concluded that the proposed method considerably improves the accuracy of brain tissue segmentation, which is vital for diagnosing neurological disorders. Future work can be directed towards applying this technique to other MRI datasets and types of noise.

Jabbar *et al.* (2023): The study that this article presents deals with the segmentation of brain tissue based on an improved KRFCM technique applied to MRI data. The authors utilized a number of machine learning techniques, particularly the FCM algorithm combined with kernel functions and spatial contextual information to improve segmentation accuracy. The dataset used for evaluation consists of T1-weighted MRI brain images from a public repository, including 20 images of size 256×256 pixels. In a nutshell, the proposed KRFCM method has achieved better accuracy and efficiency than traditional methods such as FCM and RFCM. The experimental results showed marked improvements in segmentation metrics, namely sensitivity, dice similarity coefficient, and specificity, validating the efficacy of the KRFCM technique in brain tissue region identification, a crucial aspect of diagnosing neurological disorders.

Leo *et al.*, (2019): This paper by Venkatesh and M. Judith Leo titled "MRI Brain Image Segmentation and Detection Using KNN Classification" is discussing the importance of early detection of brain tumors through MRI. The authors highlight that the tumors in the brain are either benign or malignant and, therefore, proper segmentation of MRI images is important for diagnosis. The methodology involves four major steps: preprocessing using a median filter to eliminate noise, segmentation using the K-means clustering algorithm, feature extraction using the gray level co-occurrence matrix (GLCM), and finally, classification using the k-nearest neighbour (k-NN) algorithm. The proposed approach reached a high accuracy of 85% in the classification process for distinguishing between different tumor types. The authors emphasize the potential for broader applications of their method in detecting various cancers and leave the scope for future work toward improving the performance of the algorithm by using larger datasets and maybe incorporating neural networks to obtain a better diagnosis.

Balasooriya and Nawarathna (2017): A paper proposed a CNN model, classifying the brain MRI images into five classes, including astrocytoma, glioblastoma multiforme, oligodendroglioma, normal healthy tissue of the brain, and other unknown tumors. This CNN architecture comprises four convolutional layers, two max pooling layers, one dropout layer, and two fully connected layers. It used the cross-validation method, and particularly the 6-fold cross validation was adopted to train the CNN. The dataset consists of 96,115 MRI images obtained from open databases, such as REMBRANDT (with different tumor types) and BRAINS ImageBank (healthy brain images). The model achieved an average F1-score of 99.46% and an accuracy of 99.68%, showing very high classification performance, especially in the case of healthy brain images. Misclassifications occurred most frequently in the astrocytoma category, with the complexity of the images being the main reason. Furthermore, another widely recognized CNN-based deep learning architecture (UNet) was also utilized by Chaurasia *et al.* (2021) to segment the landcover features in complex images into different classes.

Wadhai and Kawathekar (2021): The system developed is based on MRI data and ANNs to focus on the automated detection of a brain tumor. It makes use of wavelet packet analysis for extracting visual features from MRI scans, providing information more detailed than that of standard wavelet transforms. Such features are used in training an ANN that can classify MRI scans as normal or as a possible case of a brain tumor. The neural network is trained and optimized using the error backpropagation algorithm. The proposed approach has high accuracy in detecting brain tumors from MRI data with significant potential for clinical applications. However, the specific dataset used for training and evaluation is not provided in the excerpts. It thus presents a new neural network-based approach for automated brain tumor detection by using advanced signal processing techniques, such as wavelet packets, in the development of a dependable computer-aided diagnosis system.

Liu (2015): The authors describe the implementation of a face recognition system using convolutional neural networks, specifically detailing their CNN architecture such that it consists of layers for feature extraction, convolution, and sampling. The method is tested with the Yale Face Database, comprising 136 images, and optimized with a parallel computing strategy to enhance training efficiency. Experimental results showed the effectiveness of face recognition of the CNN. The theoretical and experimental analysis confirmed improvements in processing speeds through parallelization. Future improvement will contain more refinement of the algorithm for better granularity in parallel execution.

Mohan and Subashini (2018): The research article discusses a new approach for brain tumor classification in MRI images using CNN combined with data augmentation techniques. The work points to the growing number of patients suffering from brain tumors, which are difficult to detect early and emphasizes the importance of MRI for diagnosis. The authors have created a simple CNN model, which was trained using only a small dataset of 253 MRI images. They reached an accuracy of 100% compared to the accuracies achieved by pre-trained models. It used fewer computations; hence, it outperformed all those models in terms of both accuracy and training time. Their method might be very helpful in the early identification of brain tumor, and they have postulated possible future applications in other areas of medical imaging.

Vishwanathan and Murty (2002): The document describes a study on an automated system for the detection of brain tumors using MRI data and neural networks. It uses wavelet packet analysis for detailed feature extraction from MRI scans, with an ANN trained to differentiate between normal and tumor-affected scans. The training is done using the error backpropagation algorithm for optimal performance. The details of the dataset are not discussed, but the paper is aimed at improving clinical diagnosis by having a reliable computer-aided detection system.

Pal and Foody (2012): The paper compares the performance of SVM, relevance vector machine (RVM), and sparse multinomial logistic regression (SMLR) for image classification with limited ground data. It shows that RVM and SMLR can obtain similar accuracy to SVM at a fraction of the number of training cases. For example, RVM achieved 93.75% accuracy using only 7.33% of available training data, compared to SVM's 92.50% using 4.5 times more data. The study discusses the specific characteristics of valuable training cases for each type of classifier, which may motivate training data acquisition strategies optimized for each classifier. In sum, RVM and SMLR are viable alternatives for SVM in settings where available ground data is limited. Machine learning methods have been applied in the medical domain for predicting communicable diseases such as H1N1.

Ayachit *et al.* (2020) Inampudi *et al.* (2021) and dengue Babu *et al.* Additionally, advanced deep learning approaches have been extensively utilized in the identification and classification of conditions like skin cancer Ranjan *et al.* (2023), showcasing their effectiveness in medical research and diagnostics.

2.3　Material and methods

2.3.1　*Methodology*

In this chapter, seven AI-based algorithms have been used to predict brain tumor from image data. The overall methodology of the chapter is shown in Figure 2.1.

2.3.1.1　Dataset used for analysis

The dataset that was chosen for the current research work was obtained via Kaggle.

There are 800 MRI pictures in it. This dataset contained two distinct folders, one containing images of brain MRIs which contain both brain tumors as well as healthy brains (as shown in Figure 2.2). The models are then trained. Hundred images were used for testing and 200 for validation (as shown in Table 2.1). The collection contains MRI pictures with varying diameters. This dataset was selected due to the difficulty of obtaining hospital datasets.

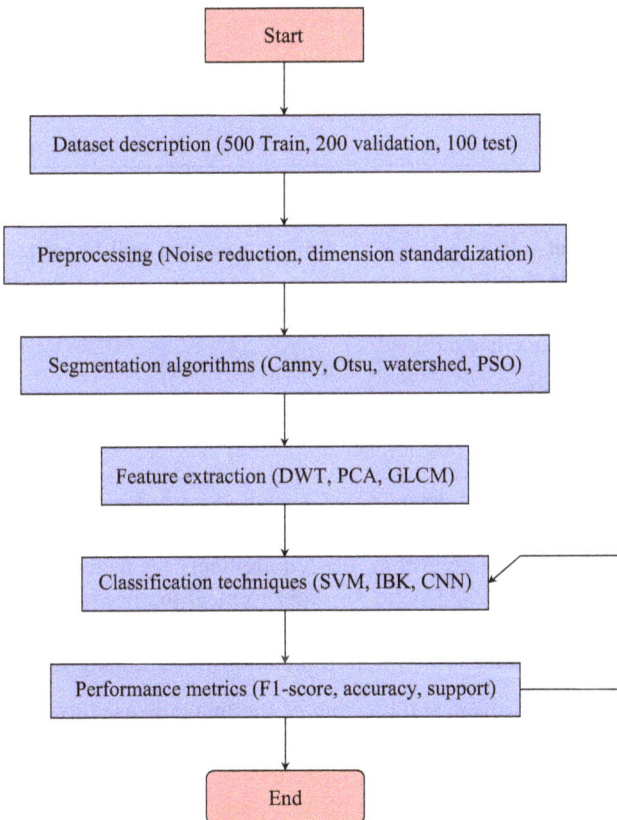

Figure 2.1　Methodology for brain tumor detection

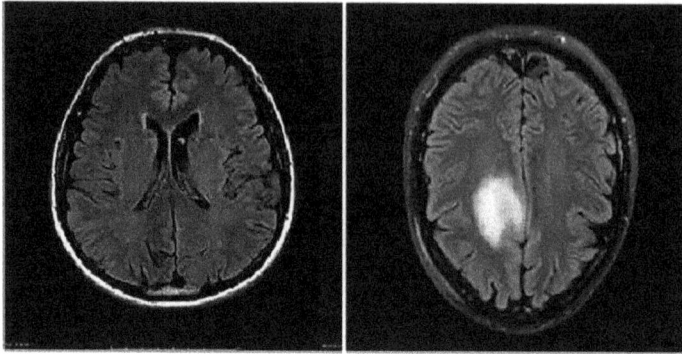

Figure 2.2   MRI images of the brain without tumor and with tumor

Table 2.1   Set of images

Folder	No. of images
Train	500
Test	100
Validate	200

- **Validation set** – It is the collection of pictures that will be utilized to modify the parameters during training.
- **Testing set** – This collection of photos won't be used until the model's ultimate performance is examined.

2.3.1.2   Segmentation

- **Canny algorithm**

The Canny algorithm (edge detection) is a method for reducing the quality of data that needs to be worked on by extracting significant features from various vision objects. According to the algorithm, the requirements for using edge detection on images are comparatively the same. In order to meet the needs, an edge detection method can be applied in a variety of scenarios (Sekehravani *et al.*, 2020).

The canny edge detection algorithm includes the following steps:

1. Smoothing of the image and blurring of the image is used to reduce noise.
2. Finding derivatives, magnitude, and orientation of the gradient.
3. To eliminate the fake response to edge detection, use non-max suppression.
4. To identify possible edges in the pictures, apply a double threshold.
5. By overwhelming every other edge that isn't attached to a strong edge, it's utilized to finish the edges.

- **Otsu algorithm**

This algorithm is designed for images with bimodal histograms, where two distinct pixel classes, typically foreground and background, are present. Otsu's method determines the ideal threshold point that maximizes the interclass variance between these classes or, conversely, decreases the overall intraclass variance (Huang *et al.*, 2021).

The algorithm works like this:

1. Generate the histogram and probability distribution.
2. Set initial class probabilities and mean values.
3. Traverse all possible thresholds.
4. Compute the between-class variance.
5. Identify the optimal threshold.

- **Watershed algorithm**

A grayscale image can be viewed as a topographic map where valleys are denoted by low intensity and hills represent high intensity. The initial step is to fill each isolated local minima with differently colored water (label). As the level rises, it mixes between valleys of different colors, influenced by the gradients of neighboring peaks. Barriers are placed where the water mixes to prevent merging. This continues until all peaks are submerged, defining the segmentation by the barriers (Ng *et al.*, 2006).

However, this method may lead to excessive segmentation due to noise or visual artifacts. To mitigate this, a marker-based watershed method was devised, enabling users to specify which valleys should merge and which should not. The segmentation process is interactive. Foreground or object regions are marked with one color (or intensity), background or non-object regions with another color, and uncertain regions with zero. The watershed algorithm updates object boundaries to −1 and uses the given labels.

Marker-controlled watershed segmentation follows this step:

1. Develop an image segmentation function.
2. Determine foreground markers.
3. Determine background markers.
4. Modify the segmentation function and retain local minima only at the background and foreground marker points.
5. Run the watershed transform on the adjusted segmentation function.

- **Particle swarm optimization**

The particle swarm optimization (PSO) algorithm (Marini and Walczak, 2015) starts in the initialization zone, $\Theta' \subseteq \Theta$, by placing particles randomly. To prevent particles from straying outside the search space during initial iterations, they can set their velocities to zero or tiny, arbitrary quantities, although these velocities are typically initialized within Θ'. The main loop iteratively updates the positions and velocities of the particles until the stopping condition is met. The rules for these updates are shown in equations 2.1 and 2.2:

$$v_i \rightarrow^{t+1} = wv_i \rightarrow^t + \phi 1 U_1 \rightarrow^t b_i \rightarrow^t - x_i \rightarrow^t + \phi 2 U_2 \rightarrow^t l_i \rightarrow^t - x_i \rightarrow^t \tag{2.1}$$

$$x_i \rightarrow t^{+1} = x_i \rightarrow^t + v_i \rightarrow t^{+1} \tag{2.2}$$

where $U_1 \rightarrow t$ and $U_2 \rightarrow t$ are two n diagonal matrices with random entries uniformly distributed within the range [0, 1] on their main diagonals, w is a parameter known as the inertia weight, and ϕ_1 and ϕ_2 are two parameters referred to as acceleration coefficients. These matrices are regenerated in each iteration. The optimal location that any particle in the vicinity finds of particle p_i, or $f(l_i \rightarrow t) \leq f(b_j \rightarrow t) \; \forall p_j \in N_i$, is commonly denoted as $l_i \rightarrow t$, known as the neighborhood best. If w, ϕ_1, and ϕ_2 are chosen appropriately, the particle velocities will not diverge to infinity.

2.3.1.3   Feature extraction

- wavelet transform and principal component analysis combined with the gray-level co-occurrence matrix method (Fayaz *et al.*, 2021)

The discrete wavelet transform (DWT) is a powerful application of the wavelet transform (WT) that utilizes dyadic scales and positions. An overview of the basics of DWT is provided below. For a given wavelet, $\theta(t)$, the continuous WT of $x(t)$ is defined as follows when $x(t)$ is a square integrable function as shown in (2.3) and (2.4):

$$W_\theta(a,b) = \int_{-\infty}^{\infty} x(t)\theta_{a,b}(t)dt \tag{2.3}$$

$$\theta_{a,b}(t) = \frac{1}{\sqrt{a}}\theta\left(\frac{t-b}{a}\right) \tag{2.4}$$

The wavelet $\psi_{a,b}(t)$ is calculated from the mother wavelet $\psi(t)$ using translation and dilation, where a is the translation and b is the dilation (both being positive real numbers). Various wavelets have been developed as wavelet analysis has progressed, with the Haar wavelet being among the simplest and most used types.

GLCM is a very frequently used method for extracting the textural features from an image. It has gained the popularity not only in medical imaging but also in a complex form of images such as satellite images with multiple bands (Kuldeep and Garg, 2014; Kuldeep *et al.*, 2021). GLCM is also a very popular technique for classifying and analyzing medical images (Hall-Beyer, 2000). It evaluates the relative positions of pixel pairs in an image by constructing a matrix based on their frequency at specific distances. A distance vector, $d = (x, y)$, is used to compute the GLCM for an image $f(i, j)$. The entry $P(i, j)$ in the matrix indicates the probability of pixels with values i, j appearing at distance d and angle θ. The GLCM is formed using distances (1, 2, 3, 4) and angles (0°, 45°, 90°, 135°) as shown in Figure 2.3. There are multiple GLCM features which can be extracted from the image (Chaurasia and Garg, 2013) to be used for classification.

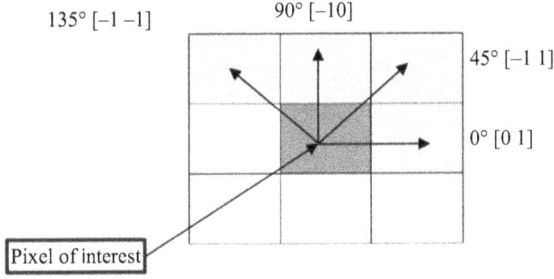

Figure 2.3   Calculation of co-occurrence matrix in GLCM

- **Correlation**
  Quantifies the degree of linear association between pixel intensities in an image. The mathematical formulation for correlation is given in (2.5).

$$\text{Correlation} = \sum_{m,n=0}^{N-1} \frac{P_{mn}(m-\mu)(n-\mu)}{\sigma^2} \tag{2.5}$$

- **Contrast**
  Measures the pixel intensity variation between neighboring areas, with high contrast showing more distinct differences. The mathematical formulation for contrast is given in (2.6).

$$\text{Contrast} = \sum_{m,n=0}^{N-1} P_{mn}(m-n)^2 \tag{2.6}$$

- **Energy**
  The uniformity of pixel distribution in an image, computed as the sum of squared GLCM entries. The mathematical formulation for energy is given in (2.7).

$$\text{Energy} = \sum_{m,n=0}^{N-1} (P_{mn})^2 \tag{2.7}$$

- **Homogeneity**
  Reflects how similar the pairs of pixels are based on GLCM proximity to the diagonal. The homogeneity equation is given in (2.8).

$$\text{Homogeneity} = \sum_{m,n=0}^{N-1} \frac{P_{mn}}{1+(m-n)^2} \tag{2.8}$$

- **Mean**
  The average value of all pixels in an image, calculated as the sum of pixel values divided by their total. The mathematical formulation of mean is given in (2.9).

$$\text{Mean} = \frac{1}{N^2} \sum_{m=0}^{N-1} \sum_{n=0}^{N-1} P_{m,n} \tag{2.9}$$

- Standard deviation
  Indicates how pixel intensities deviate from the mean, with a higher value showing greater variability. The equation for standard deviation is given in (2.10) and (2.11).

$$\text{Standard deviation} = \sqrt{\sigma^2} \tag{2.10}$$

$$\sigma^2 = \frac{1}{N^2} \sum_{m=0}^{N-1} \sum_{n=0}^{N-1} (P_{mn} - \mu)^2 \tag{2.11}$$

- **Entropy**
  A metric of texture complexity, with higher values suggesting detailed and varied structures. The mathematical formulation for entropy is given in (2.12).

$$\text{Entropy} = \sum_{m,n=0}^{N-1} -\ln(P_{mn}) P_{mn} \tag{2.12}$$

- **Root mean square**
  The average magnitude of pixel intensities is calculated as the square root of the mean squared pixel values. The mathematical formulation for root mean square (RMS) is given in (2.13).

$$\text{RMS noise} = \sqrt{\frac{\sum_{i=1}^{n} \left(X_i - \frac{\sum_{i=1}^{n} (X_i)}{n} \right)^2}{n}} \tag{2.13}$$

- **Variance**
  Shows the extent of pixel intensity spread around the mean. The mathematical formulation for variation is given in (2.14).

$$\text{Variance} = \frac{\sum_{i=1}^{n} \left(X_i - \frac{\sum_{i=1}^{n} (X_i)}{n} \right)^2}{n} \tag{2.14}$$

- **Inverse difference movement**
  A feature that quantifies the uniformity of an image based on pixel similarity. The mathematical formulation for inverse difference movement (IDM) is shown in (2.15).

$$\text{IDM} = \sum_{m=0}^{N_g-1} \sum_{n=0}^{N_g-1} \frac{P_{m,n}}{1 + (m+n)^2} \tag{2.15}$$

2.3.1.4   Classification

- **Support vector machine**

Support vector machines (SVMs) are a form of supervised learning algorithm that helps in analyzing data and recognizing patterns, primarily for use in classification (Vishwanathan and Murty, 2002). An SVM processes input data and assigns each data point to one of two categories, such as malignant or benign, functioning as a deterministic binary linear classification model. Once labeled training data is available, an SVM algorithm builds a model that can classify new data points. The model visualizes data points in a space, ensuring a wide gap that divides the classes. New data points are then assigned to a class based on their position relative to the margin, with a larger margin generally leading to better generalization and performance.

- **Lazy IBK**

Instance-based learning algorithms focus on comparing new data instances with stored training examples, bypassing explicit generalizations. This methodology, termed instance-based, constructs hypotheses directly from training data, leading to increasing complexity with hypothesis size. The computational complexity of classifying an instance is $O(n)$, with the worst-case hypothesis containing all n training examples.

Earlier modeling approaches involved creating a model during the training phase and applying it for classification or prediction, a practice referred to as eager learning, due to its emphasis on early model construction (Chellam *et al.*, 2018)

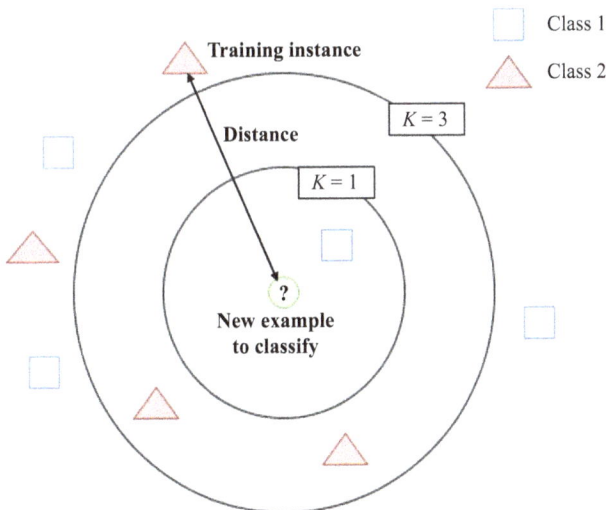

Figure 2.4   Instance-based KNN

Instance-based learning postpones generalization until the classification or prediction stage. These methods rely on local subsets of training data to create models rather than employing the entire dataset. Commonly used for regression and classification.

- **Convolutional neural networks**

Traditional neural networks are made up of neurons with learnable weights and biases, and they are comparable to CNNs. After receiving several inputs, each neuron processes the weighted total of those inputs, passes it through an activation function, and then produces an output (Chauhan *et al.*, 2018).

A multilayer perceptron, the CNN method, is specifically tailored for extracting information from two-dimensional images. It comprises input, convolutional, sampling, and output layers as its primary components. Weight sharing, which involves using the same convolution kernels for image deconvolution, serves as a key parameter for each neuron. The process followed in a CNN is illustrated in Figures 2.5 and 2.6 (Chauhan *et al.*, 2018).

The CNN algorithm features two primary components: convolution and sampling.

In the convolution phase, the input image undergoes processing to produce a feature map for each layer. This is achieved by applying a trainable filter F_x, followed by adding a bias b_x, forming the convolution layer C_x.

In the sampling phase, a reduced feature map S_{x+1} is created. This step involves pooling n neighboring pixels into one, scaling them with weights W_{x+1}, and applying a bias b_{x+1} through an activation function.

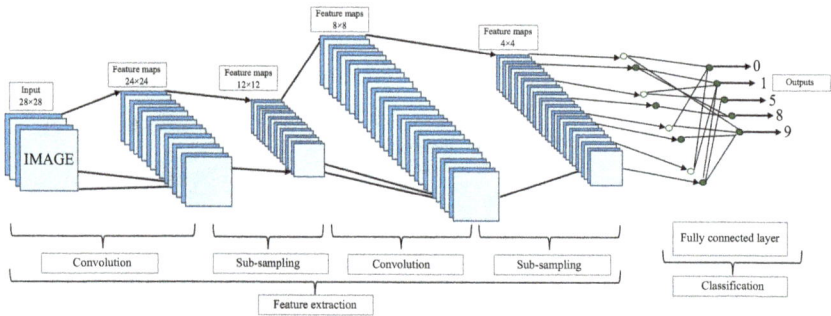

Figure 2.5   CNN block diagram

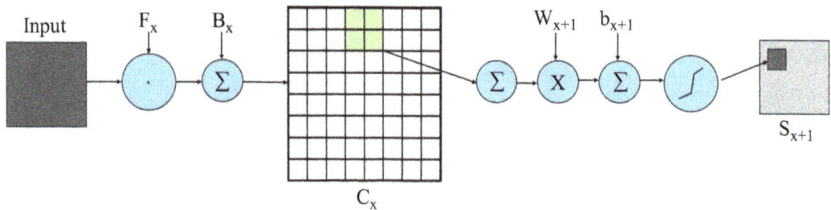

Figure 2.6   Working of CNN

The CNN algorithm emphasizes the use of weight sharing, spatial and temporal subsampling, and local receptive fields to streamline feature extraction and limit training parameter usage. This methodology allows CNNs to learn directly from training datasets without requiring explicit feature extraction techniques.

Uniform neuron weights across feature maps ensure that learning occurs in parallel, thereby reducing overall network complexity. Additionally, spatial or temporal subsampling introduces robustness to the model and enables it to adapt to changes in scale or displacements caused by deformation.

2.3.2 Programming language used for implementation

Multiple programming languages are supported by the free, open-source web-based interactive computing environment known as the Jupyter Notebook. It makes it possible to create and share interactive documents with narrative text, live code, mathematics, and visualizations using Project Jupyter. Data analysis, machine learning, scientific computing, and teaching all make extensive use of it. It offers real-time data exploration, works well with existing libraries, and facilitates user collaboration with shareable.ipynb files.

2.3.3 Tools used

- Neural network toolbox.
- **Languages**: Python.
- **Documentation**: Microsoft Word, Google docx.
- **Software**: Jupyter.

2.3.4 Testing approach

The models are tested using a different tests to make sure the model operates correctly and to gauge its strengths and weaknesses. Below is a quick description of the suggested test types that will be carried out.

1. Unit testing:
   - Focuses on verifying the smallest units of software design, known as modules.
   - Uses the component-level design as a reference to analyze key control paths for detecting defects within the module's boundaries.
   - Unit testing follows a white-box methodology and can be performed simultaneously across multiple modules in parallel.
2. Integration testing:
   - The interfacing of multiple modules can be challenging.
   - Data loss can occur at the interface, with one module affecting another, and minor inaccuracies acceptable at the module level becoming amplified when interconnected.
   - Integration testing is used to design the software structure and execute tests aimed at detecting interface-related faults.

3. Stress testing:
    - Stress testing is conducted to push programs through abnormal situations.
    - This type of testing forces the model to operate in ways that require excessive resources in terms of volume, frequency, or quantity, enabling an evaluation of the model's limits.
4. Performance testing:
    - Performance testing is done at each step of the testing procedure to assess the software's runtime performance within an integrated model setting.
5. Security testing:
    - This model handles confidential patient information, which may make it susceptible to potential threats or unauthorized access attempts.
    - Security testing aims to ensure that the model's security measures can effectively protect it from unauthorized intrusions.
    - During security testing, the tester plays the part of an intruder attempting to breach the model.
    - With adequate resources and time, effective security testing will eventually find ways to compromise the model. The designer's objective is to make the cost of gaining access higher than the value of the data obtained.
6. Recovery testing:
    - Many computer models are designed to recover from failures and maintain operations within a predetermined timeframe. In some cases, models need to be fault-tolerant, ensuring that problems do not cause the entire model to fail. In other situations, failures must be resolved within a specific period to mitigate potential economic consequences.
    - The testing strategy employed in the project started at an early stage and expanded outward to fully integrate the model.
    - This approach to testing was a continuous process throughout the model's development. Each module was tested upon completion before moving to the next component.
    - After selecting a dataset for kidney tumor analysis and determining effective network combinations, testing was conducted, and the output was evaluated against to the expected results.
    - Following the creation of the Graphical User Interface (GUI), integration testing was performed to assess the network's functionality with the GUI.
    - The testing approach involved various forms, including unit testing and integration testing.

2.3.5   Testing plan

2.3.5.1   Unit testing

- Each module will be tested individually to ensure that it generates the desired results. The output images will be reviewed for proper tumor segmentation after the segmentation algorithms run.
- During feature extraction, it will be ensured that features are stored in an Excel document without producing any NaN values.

- The datasets will be used to train the classification algorithms, and their accuracy will be assessed using the training data.

2.3.5.2 Integration testing

- Every possible combination of component modules will be tested to ensure the output is correct. Each segmentation module will be tested with a feature extraction module and then with a classification module.

2.4 Results and discussion

2.4.1 Visual results

The visual outcomes of brain tumor prediction using different machine learning algorithms are shown in Figure 2.7.

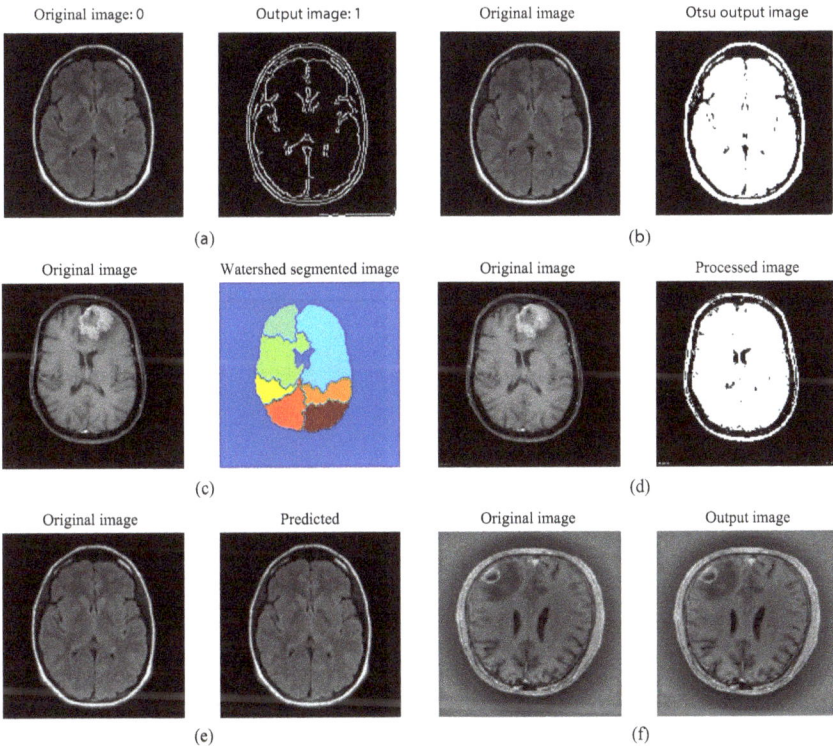

Figure 2.7 Outputs from different image processing algorithms. (a) Output from Canny algorithm. (b) Output from Otsu algorithm. (c) Output from watershed algorithm (malignant). (d) Output from PSO algorithm. (e) Output from lazy IBK algorithm. (f) Output from SVM algorithm.

2.4.2   *Performance analysis*

2.4.2.1   Confusion matrix

The computation results from various classification algorithms are mapped using a confusion matrix. This matrix provides a summary of the true and predicted classes from the classification model, which is used to evaluate the model's performance. Table 2.2 displays a two-class classifier's confusion matrix. It can be leveraged to assess the accuracy of a dataset. Contextually, the information in the shown confusion matrix means the following:

- m – number of true negative predictions,
- n – count of false positive predictions,
- o – count of false negative predictions,
- p – number of true positive prediction.

Here are the findings of a single run of the aforementioned testing, shown as confusion matrices for each of the possibilities in Table 2.2.

Table 2.2   Confusion matrix

	Predicted	
Actual	**Negative**	**Positive**
Negative	m	n
Positive	o	p

2.4.2.2   Test cases

Unit test cases

Table 2.3   Unit test case accuracies

Test case	**Average accuracy (%)**
SVM accuracy	97
Lazy IBK accuracy	90
CNN accuracy	96

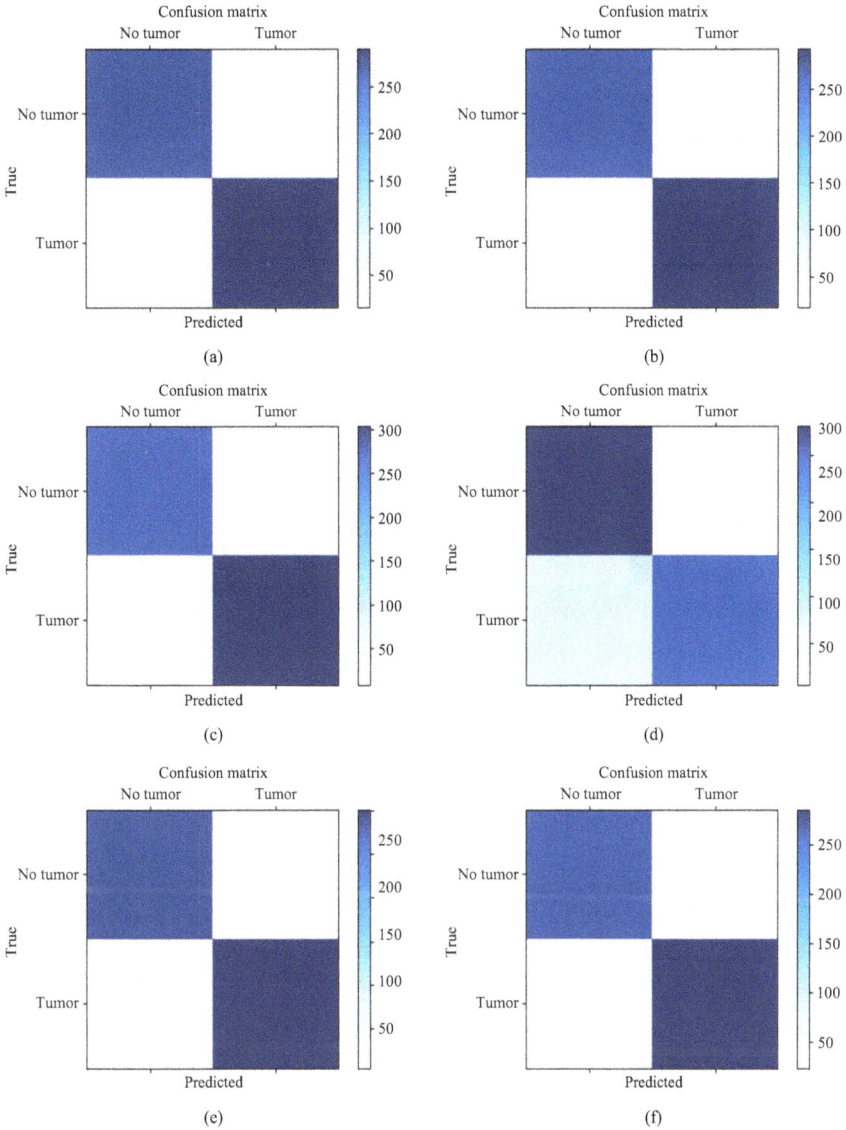

Figure 2.8   *Confusion matrices from integration tests for different methods (a) to (f). (a) OTSU thresholding combined with DWT, PCA, and SVM for classification. (b) OTSU thresholding with DWT, PCA, and lazy IBK classifier. (c) OTSU thresholding with CNN for feature extraction and classification. (d) Particle swarm optimization (PSO) enhanced CNN-based method. (e) CANNY edge detection combined with DWT, PCA, and lazy IBK classifier. (f) CANNY edge detection combined with CNN for feature extraction and classification.*

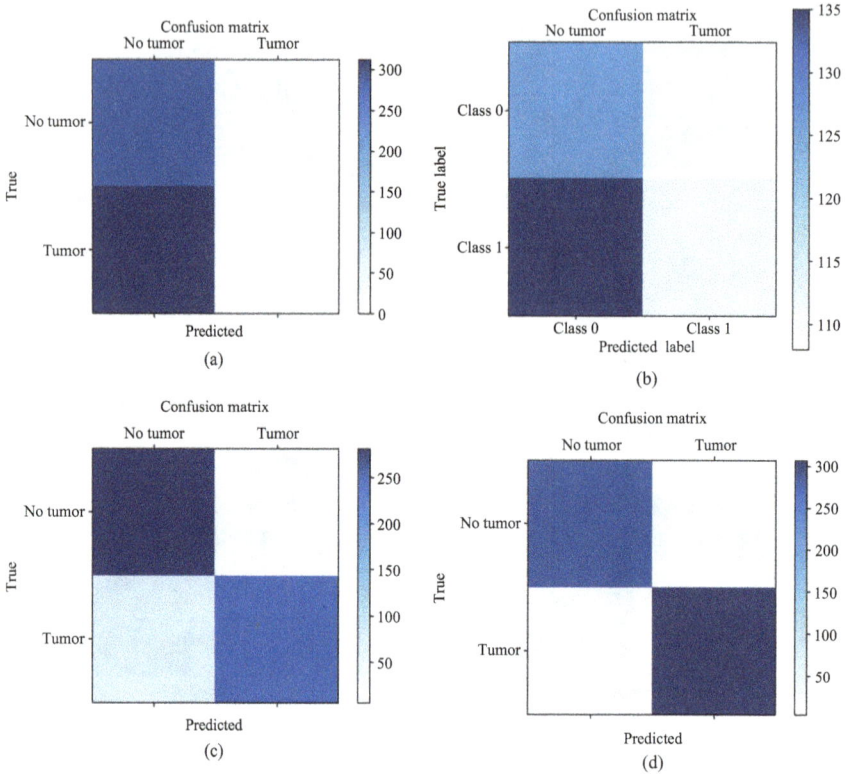

Figure 2.9   Confusion matrices from integration tests for methods described in subcaptions (a)–(d). (a) PSO thresholding combined with DWT, PCA, and SVM for classification. (b) PSO thresholding with DWT, PCA, and Lazy IBK classifier. (c) Watershed thresholding with DWT, PCA, and Lazy IBK. (d) Watershed combined with CNN.

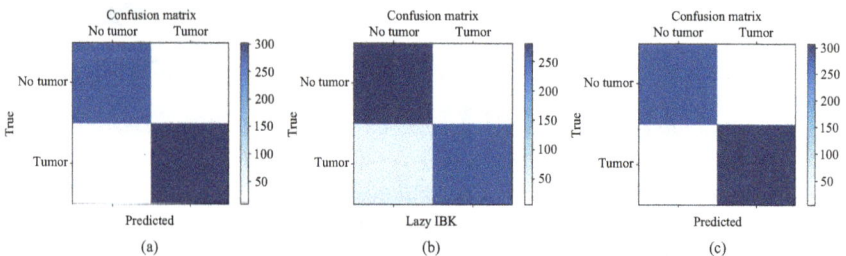

Figure 2.10   Confusion matrices for individual classifiers described in subcaptions (a)–(c). (a) SVM, (b) lazy IBK, and (c) CNN.

Integration test cases

Table 2.4   Integration test cases accuracies

Test case	Accuracy (%)
OTSU + (DWT + PCA) + SVM	94
OTSU + (DWT + PCA) + Lazy IBK	94
OTSU + CNN	96
CANNY + (DWT + PCA) + Lazy IBK	88
CANNY + CNN	92
PSO + (DWT + PCA) + SVM	48
PSO + (DWT + PCA) + Lazy IBK	49.38
PSO + CNN	97
Watershed + (DWT + PCA) + Lazy IBK	89
Watershed + CNN	98

2.5   Limitations

The study has a few shortcomings that might influence its output. For example, its dataset is small, with only 800 MRI images and might not be large enough to offer real- world variety for the performances of the models. Besides, the data was sourced online rather than collected from clinical settings, so it may fail to represent the complexity of real clinical data. This study has limited algorithms like SVM, CNN, and PSO; hence, there are several other methods that can prove to be better. The area of security testing is touched but not elaborated about the kind of privacy maintenance for patient data. Further research in this area will make the datasets bigger and more comprehensive, as well as try more algorithms on to ensure greater data privacy protection.

2.6   Conclusion

An abnormal tissue growth in the brain that disrupts normal day to day brain functions is known as a brain tumor. The main goal of medical image processing is to extract precise and valuable information through algorithms that minimize errors. Identifying and classifying these brain tumors from MRI images include four main stages: pre- processing the data, segmentation, feature extraction, and classification. This study explores various segmentation methods, with the algorithms and parameters chosen for the proposed model aimed at enhancing performance and optimizing results.

While border-based and edge-based segmentation techniques are commonly applied, region-growing methods tend to yield superior results. The findings indicated that PSO achieved the most accurate tumor segmentation. Additionally, feature extraction through the GLCM method proved beneficial by capturing fine details of the tumor, contributing to improved model performance. Among the different classification methods tested, empirical evidence demonstrated that convolutional neural networks provided the highest accuracy. Since accuracy and reliability are crucial for patient diagnoses, the proposed approach effectively enhances accuracy and helps achieve the desired outcomes. The best-performing algorithm(s) in the unit testing was SVM, with an accuracy of 97%, and integration testing and Watershed + CNN achieved an accuracy of 98%.

These iterations maintain the core meaning and intention of the original text, presenting it with varied phrasing and structure.

2.7   Future scope

Building on these results, future research will aim to enhance classification performance and overall accuracy. With the availability of more extensive data, the range of output classes can be expanded, potentially boosting classification accuracy through a larger and more diverse dataset.

Another approach to improving results involves increasing the hidden layers in the neural network. This expansion can lead to better weight adjustments, thereby enhancing classification performance. Furthermore, by utilizing insights from previously trained models, the model can be further optimized through the use of fine-tuning and transfer learning approaches, making the learning process more efficient.

References

Sai Sanjay Ayachit, Tanmay Kumar, Shriya Deshpande, Nayan Sharma, Kuldeep Chaurasia, and Mayank Dixit. Predicting H1N1 and seasonal flu: vaccine cases using ensemble learning approach. In *2020 2nd International Conference on Advances in Computing, Communication Control and Networking (ICACCCN)*, pages 172–176. IEEE, 2020.

Babu, P. D., Samyuktha, T., Krishnaveni, U., Kavya, T., and Priya, M. (2023). Time Series Analysis-based Prediction of Dengue Spread using Climate Data. *Journal of Cardiovascular Disease Research*, 14(7). https://jcdronline.org/index.php/JCDR/article/view/8055.

Nilesh Bhaskarrao Bahadure, Arun Kumar Ray, and Har Pal Thethi. Image analysis for MRI based brain tumor detection and feature extraction using biologically inspired BWT and SVM. *International Journal of Biomedical Imaging*, 2017(1):9749108, 2017.

Narmada M Balasooriya and Ruwan D Nawarathna. A sophisticated convolutional neural network model for brain tumor classification. In *2017 IEEE International Conference on Industrial and Information Systems (ICIIS)*, pages 1–5. IEEE, 2017.

Rahul Chauhan, Kamal Kumar Ghanshala, and RC Joshi. Convolutional neural network (CNN) for image detection and recognition. In *2018 First*

International Conference on Secure Cyber Computing and Communication (ICSCCC), pages 278–282. IEEE, 2018.

Kuldeep Chaurasia and PK Garg. "A Brief Review on Texture Analysis Methods", *Studies in Surveying and Mapping Science*, 1(2):28–36, 2013.

Kuldeep Chaurasia, Rijul Nandy, Omkar Pawar, Ravi Ranjan Singh, and Meghana Ahire. Semantic segmentation of high-resolution satellite images using deep learning. *Earth Science Informatics*, 14(4):2161–2170, 2021.

Aditya Chellam, L Ramanathan, and S Ramani. Intrusion detection in computer networks using lazy learning algorithm. *Procedia Computer Science*, 132:928–936, 2018.

Yogita K Dubey and Milind M Mushrif. Segmentation of brain MR images using intuitionistic fuzzy clustering algorithm. In *Proceedings of the Eighth Indian Conference on Computer Vision, Graphics and Image Processing*, pages 1–6, 2012.

Muhammad Fayaz, Nurlan Torokeldiev, Samat Turdumamatov, Muhammad Shuaib Qureshi, Muhammad Bilal Qureshi, and Jeonghwan Gwak. An efficient methodology for brain MRI classification based on DWT and convolutional neural network. *Sensors*, 21(22):7480, 2021.

R Gowthaman, K Jayavignesh, V Jeevanantham, *et al.* Brain neoplasm classification and detection of accuracy on MRI images. *International Journal on Orange Technologies*, 4(5):25–44, 2022.

Mryka Hall-Beyer. GLCM texture: a tutorial. *National Council on Geographic Information and Analysis Remote Sensing Core Curriculum*, 3(1):75, 2000.

Chunyan Huang, Xiaorui Li, and Yunliang Wen. An OTSU image segmentation based on fruit fly optimization algorithm. *Alexandria Engineering Journal*, 60 (1):183–188, 2021.

Srividya Inampudi, Greshma Johnson, Jay Jhaveri, S Niranjan, Kuldeep Chaurasia, and Mayank Dixit. Machine learning based prediction of H1N1 and seasonal flu vaccination. In *Advanced Computing: 10th International Conference, IACC 2020, Panaji, Goa, India, December 5–6, 2020, Revised Selected Papers, Part I 10*, pages 139–150. Springer, 2021.

Hiyam Hatem Jabbar, Raed Majeed Muttasher, and Ali Fattah Dakhil. Segmentation of brain tissue using improved kernelized rough-fuzzy c-means technique. *Indonesian Journal of Electrical Engineering and Computer Science*, 32(1):216–226, 2023.

IM Kazi, SS Chowhan, and UV Kulkarni. MRI brain image segmentation using adaptive thresholding and k-means algorithm. *International Journal of Computer Applications*, 167(8):11–15, 2017.

Malik Sikandar Hayat Khiyal, Aihab Khan, and Amna Bibi. Modified watershed algorithm for segmentation of 2D images. *Issues in Informing Science & Information Technology*, 6, 2009.

Kuldeep and PK Garg. Texture based information extraction from high resolution images using object based classification approach. In *2014 Third International Workshop on Earth Observation and Remote Sensing Applications (EORSA)*, pages 299–303, 2014. doi: 10.1109/EORSA.2014.6927899.

Kuldeep, PK Garg, and RD Garg. Texture-based riverine feature extraction and flood mapping using satellite images. *Advances in Remote Sensing for Natural Resource Monitoring*, pages 405–430, 2021.

Leo, M. Judith *et al.* MRI brain image segmentation and detection using K-NN classification. *Journal of Physics: Conference Series*, Vol. 1362. No. 1. IOP Publishing, 2019.

T Liu. *Implementation of training convolutional neural networks. arXiv preprint arXiv:1506.01195*, 2015.

Federico Marini and Beata Walczak. Particle swarm optimization (PSO). A tutorial. *Chemometrics and Intelligent Laboratory Systems*, 149:153–165, 2015.

Richa Mishra. MRI based brain tumor detection using wavelet packet feature and artificial neural networks. In *Proceedings of the International Conference and Workshop on Emerging Trends in Technology*, pages 656–659, 2010.

Geethu Mohan and M Monica Subashini. MRI based medical image analysis: survey on brain tumor grade classification. *Biomedical Signal Processing and Control*, 39:139–161, 2018.

Fahd Mohsen, Mohiy M Hadhoud, Kamel Moustafa, and Khalid Ameen. A new image segmentation method based on particle swarm optimization. *The International Arab Journal of Information Technology*, 9(5):487–493, 2012.

HP Ng, SH Ong, KWC Foong, Poh-Sun Goh, and WL Nowinski. Medical image segmentation using k-means clustering and improved watershed algorithm. In *2006 IEEE Southwest Symposium on Image Analysis and Interpretation*, pages 61–65. IEEE, 2006.

Mahesh Pal and Giles M Foody. Evaluation of SVM, RVM and SMLR for accurate image classification with limited ground data. *IEEE Journal of Selected Topics in Applied Earth Observations and Remote Sensing*, 5(5):1344–1355, 2012.

Vibhav Ranjan, Kuldeep Chaurasia, and Jagendra Singh. Multi-class skin lesion classification using intelligent techniques. In *International Conference on Advanced Computing and Intelligent Technologies*, pages 597–605. Springer, 2023.

Mehak Saini, Sanju Saini, Priyanshu Tripathi, KK Saini, and Madhwendra Nath. A survey on brain tumor identification through medical images. *International Journal of Advanced Research in Computer Science*, 8(7), 2017.

Ehsan Akbari Sekehravani, Eduard Babulak, and Mehdi Masoodi. Implementing canny edge detection algorithm for noisy image. *Bulletin of Electrical Engineering and Informatics*, 9(4):1404–1410, 2020.

SVM Vishwanathan and M Narasimha Murty. SSVM: a simple SVM algorithm. In *Proceedings of the 2002 International Joint Conference on Neural Networks. IJCNN'02 (Cat. No. 02CH37290)*, volume 3, pages 2393–2398. IEEE, 2002.

Sheetal Ashokrao Wadhai and Seema S Kawathekar. A segmentation of brain tumor detection from MRI images transform information using algorithms in CBMIR. *Turkish Online Journal of Qualitative Inquiry*, 12(7), 2021.

Chapter 3

Understanding of Parkinson's disease: pathology, diagnosis, therapies, and importance of biomarkers

Srishti Seth[1], Muniraj Gupta[2], Aman Verma[1], Md. Zubbair Malik[3], Saurabh Kumar Sharma[2] and Nidhi Verma[4]

Abstract

Being a complex and progressive neurodegenerative disorder, Parkinson's disease (PD) is manifested to loss of dopaminergic neurons from substantia nigra pars compacta region of the brain which is the accumulation of Lewy bodies or neurites. Tremors, fatigue, bradykinesia, postural dysfunction, and uncontrollable muscle movement are the main symptoms found in PD. The diagnosis is sometimes difficult pertaining to late appearance of motor as well as non-motor symptoms and often misdiagnosis. The current drugs and therapies available only alleviate the symptoms of the disease in the early stages but cannot modify or cease the disease progression. The need of the hour still remains the identification of such biomarkers that can help early diagnosis of the disease to develop more effective drugs and therapies.

Keywords: Lewy bodies; alpha-synuclein; levodopa; LRRK2; dopaminergic neurons; biomarkers

3.1 Introduction

Parkinson's disease (PD) refers to disorders of the group of motor system which cause uncontrollable and unwanted body movements. Following Alzheimer's disease, it holds the second largest position of cause of neurodegeneration in human beings. Some cases of PD are suggested to be inherited (familial) while the others are known to be triggered by environmental factors (sporadic), but the precise

[1]School of Computational and Integrative Sciences, Jawaharlal Nehru University, India
[2]School of Computer and Systems Sciences, Jawaharlal Nehru University, India
[3]Department of Genetics and Bioinformatics, Dasman Diabetes Institute, Kuwait
[4]Department of Microbiology, Ram Lal Anand College, University of Delhi, India

cause of the disease is yet unclear [1]. Death or damage of the cells of the dopamine-producing part of the brain has been firmly established as the reason behind PD. Parkinson's is physiologically related to depletion of dopaminergic neurons (DNs) in substantia nigra pars compacta (SNpc), which is the result of several mechanisms among which α-synuclein aggregation plays the central role while others being mitochondrial dysfunction, oxidative stress, neuroinflammation, and malfunction in autophagy. It is the result of protein build-up into Lewy bodies in this brain area. The major symptoms observed in PD include tremors, fatigue, bradykinesia, postural abnormalities, and uncontrollable muscle movements [2]. Along with being chronic, PD is also progressive, that is, it worsens with time [3]. Males are affected over females by the ratio 3:2. It most likely affects individuals above 60 years of age (late onset PD), though there are cases where individuals below 50 years have been inflicted with PD (early onset PD). The life expectancy is 7 to 15 years after the diagnosis [4].

Most of the cases of PD are sporadic, the etiology of which is contributed to by environmental and genetic factors. Exposure to pesticides such as rotenone, welding, etc., increase the risk of sporadic PD [5]. Rotenone has been found to inhibit complex 1 involved in the mitochondrial respiratory chain of dopaminergic cells and reproduce features unique to PD such as inclusion of alpha-synuclein as seen in rat models [6]. The remaining cases of PD belong to the rare familial category that is caused due to mutations in genes such as *Parkin, LRRK2, SNCA, DJ-1, PINK1* and many more [7].

A levodopa–carbidopa combination is given as the first line of therapy to PD patients. The role of carbidopa is to delay the conversion of levodopa into dopamine till the time it reaches the brain [8]. The presence of levodopa in the brain allows the neurons to produce dopamine and thus replenish the diminishing supply of the brain. Although levodopa serves as a relief to majority of the PD patients, the response to the drug is not similar in all cases [9]. In cases where any of the medications fail to show a positive respond, surgery is an alternative option. Deep brain simulations (DBS) have also shown promising results in patients who cannot respond well to drugs [10].

3.2   Epidemiology

At any given time, 1–2 per 1000 individuals in a population are affected by PD [11]. The rise in prevalence is directly proportional to age, affecting 1% and 4% of the global population above 60 years and 80 years of age, respectively [12]. On the basis of age of symptoms, it can be categorized as an early onset PD (EOPD) with pre-40 year-old features comprising 3–5% of all cases of PD, pre-age 21 years-old symptoms causing juvenile PD, and young onset PD (YOPD) 21–40 year-old symptoms [13]. Males are more affected from PD than females in the ratio of 3:2. The predominance in male can be explained by gender inclined genetic or environmental risk factors [14]. The life expectancy following the diagnosis is 7–15 years [4]. It is less prevalent in individuals with Asian and African ancestry. PD's prevalence is high in Native Americans. This can be attributed to persistent organic pollutant and genetically modified exposure [15].

In India, prevalence rate in Kashmir, northern part of India, and in Bangalore, southern part of India, as high as 247/100,000 and as low as 27/100,000, respectively, was reported, [16,17]. In rural Bengal, eastern part of India, prevalence of 16.1/100,000 was reported [18]. A relatively high rate of 328.3/100,000 individuals in a 14,010 Parsi population living in Mumbai, western India, was reported [19].

3.2.1 Environmental factors

In 1983, it was discovered that several people who were injected with a drug tainted with MPTP (1-methyl-4-phenyl-1,2,3,6-tetrahydropyridine) developed symptoms of PD. MPTP is a neurotoxin that inhibits mitochondria complex-1 leading to decline of DN [20]. Several studies also show relation of pesticide and late onset PD. Paraquat and rotenone also selectively inhibit complex-1 of mitochondria leading to cell death [21]. There are many other studies that suggest how the PD can be associated with rural residence, well drinking water, welding, and heavy metal exposure [22].

3.2.2 Aging

Studies suggest that with increasing age, the dopamine synthesizing neurons in SNpc region of the brain begin to deplete. It can be a risk factor, but its exact pathology in PD is still elusive [23].

3.2.3 Genetics factors

Only 5–10% among all the PD cases have been reported to have been affected from genetic form of the disease or have familial PD while the rest are sporadic [24]. Genetically stating, many monogenetic types have been identified, along with genetic factors that increase the risk of disease. The 30% of the rare monogenic forms are familial while 3–5% are sporadic. The single mutation in a gene inherited either dominantly or recessively causes monogenic heritable form of PD [25]. Mutations in more than 20 genes are reported to be linked with both autosomal dominant and recessive form of the disease [25]. The major genes involved are *PARK1* and *PARK4/SNCA*, *PARK8/LRRK2*, *PARK2/Parkin*, *PARK9/ATP13A2*, *PARK7/DJ-1*, *PARK6/PINK1*, *PARK5/UCHL1*, *PARK11/GIGYF2*, and *PARK13/Omi/Htra2* [26,27] (Table 3.1; Figure 3.2).

Mutations in *PARK1* and *PARK4/SNCA*, *PARK8/LRRK2*, *PARK5/UCHL1*, *PARK11/GIGYF2*, and *PARK13/Omi/Htra2* give rise to autosomal dominant forms of PD, while mutations in *PARK2/Parkin*, *PARK9/ATP13A2*, *PARK7/DJ-1*, and *PARK6/PINK1* give rise to autosomal recessive forms [25] (Table 3.1).

3.2.3.1 *PARK1* and *PARK4/SNCA*

SNCA (α-Synuclein) was the first gene to have reported autosomal dominant PD causing mutations. It encodes SNCA protein with Lewy bodies being its chief component. *SNCA* mutations are infrequent and until now three missense

Table 3.1   Major genes in PD with their symbol, chromosomal loci, inheritance pattern, form, and type of mutation [25,7].

PARK loci symbol	Gene	Chromosomal loci	Inheritance	Form	Mutations
PARK1/ PARK4	*SNCA*	4q21	Autosomal Dominant	EOPD	A30P, A53T, E46K, Genomic duplications or triplications
PARK8	*LRRK2*	12q12	Autosomal Dominant	LOPD	>50 Missense and non-sense variants, 16 pathogenic, including common G2019S
PARK2	*Parkin*	6q25-q27	Autosomal Recessive	EOPD and juvenile	>100 Mutations (frameshift, non-sense, missense mutations)
PARK6	*PINK1*	1p35-p36	Autosomal Recessive	EOPD	>40 Point mutations
PARK7	*DJ-1*	1p36	Autosomal Recessive	EOPD	>10 Point and deletion mutations
PARK9	*APT123*	1p36	Autosomal Recessive	Juvenile, Kufor-Rakeb syndrome and EOPD	>5 Point mutations
PARK5	*UCHL1*	4p14	Autosomal Dominant	LOPD	One mutation in a PD sibling pair
PARK11	*GIGYF2*	2p36-q37	Autosomal Dominant	LOPD	Missense variants (seven)
PARK13	*Omi/ HTRA2*	2p13	Autosomal Dominant	LOPD	Missense variants (two)

(substitution) mutations (A53T, E46K, and A30P) along with its duplications and triplications have been reported [25]. These mutations usually cause EOPD. A53T mutation is the most frequent of all and is common in Greek population and was found in eight Greek families. Along with it, it was also found in one Italian, one Swedish, and two Korean families [28–33]. Each A30P and E46K missense mutations have also been reported in one Greek, Korean, Italian, and Swedish family [34,35]. The three missense mutations disable the amino terminal domain of *SNCA*. This impairment leads to α–helical folding of natively unfolded α-Synuclein and the three missense mutations results in formation of β-sheets which aggravate the formation of toxic fibrils and protofibrils. Duplications and triplications of the *SNCA* copies result in its overexpression and is thus toxic and show acute disease progression [7]. Gain of function of *SNCA* results in aggregation of Lewy bodies in neural cells is one of the many reasons of familial PD cases. Overproduction of Lewy bodies can also be the cause of sporadic PD pathogenesis [36].

3.2.3.2 *PARK8/LRRK2*

LRRK2 (leucine-rich repeat kinase 2) consists of 51 exons spanning 144 kb. Its protein is 2527 amino acids long. LOPD in autosomal dominant and sporadic form is most frequently caused by *LRRK2* mutations at PARK8 chromosomal locus [7]. The protein consists of several conserved domains flanked by a leucine-rich repeat (LRR) at the amino terminus and a kinase domain at the carboxyl terminus. The conserved motifs/domains present in *LRRK2* from N- to C-terminal are ARM (Armadillo), followed by ANK (Ankyrin), followed by LRR (leucine rich repeat), which is again followed by Roc (Ras of complex proteins), then COR (C-terminal of Roc), then Kinase, and finally WD-40 [37]. The gene has been recorded for more than 50 mutations, most of whose exact cause is still not clear while only 16 of them have proven to be pathogenic [38]. Out of these, six established dominant mutations present on the conserved domains of *LRRK2* are R114C, R1441G, R1441H (ROC domain), Y1699C (COR domain), G2019S, and I2020T (Kinase) [7](Figure 3.1). The most frequent of these mutations is G2019S, accounting for 1–2% and 2–5% of sporadic and familial cases, respectively. Some studies report that the frequency of G2019S mutations depends upon ethnicity. G2019S mutations have been reported in 30% of Arab Berber [39] and 13% of Ashkenazi Jewish PD populations [40], but its prevalence is very low in Asian populations [41].

3.2.3.3 *PARK2/Parkin*

Parkin was the first identified gene for the early onset as well as juvenile autosomal recessive form of PD [7] and the second most common gene causing Parkinson [42]. It has 12 exons encoding 465 amino acid long protein [43]. It comprises of ubiquitin-like domain in the amino- terminal while its carboxy-terminal consists of

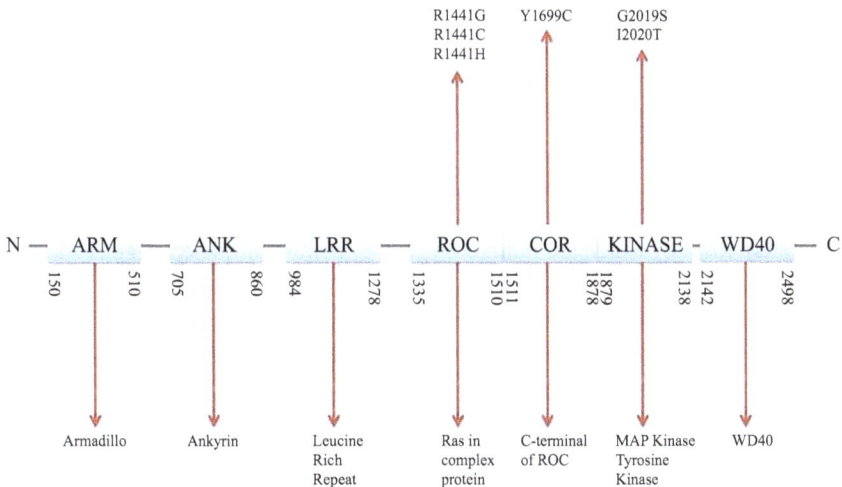

Figure 3.1 *LRRK2: pathogenic mutations, functional domains along with their boundaries [25,37].*

three RING domains and one IBR (in-between-ring) domain [7,43]. While the amino-terminal controls the expression level of *Parkin*, the carboxy-terminal contributes to Parkin's functioning as ubiquitin E3 ligase [7]. Parkin plays role in monoubiquitination and polyubiquitination through lysine-48 and lysine-63 during post translational modifications. Loss of ubiquitin properties is significant in pathogenic mechanism of familial and sporadic PD [42]. Mutation in the sixth chromosome of *Parkin* usually causes early onset PD (30–40 years of age) but its homologous variants can cause juvenile PD. It also destroys the dysfunctional mitochondria to maintain a healthy cellular network [43].

3.2.3.4   *PARK6/PINK1*

Mutations in *PINK1* (PTEN-induced kinase 1) gene can induce early onset, autosomal recessive form of PD [44,43]. This mutation is located at the *PARK6* locus on chromosome 1p36. This was first reported in an Italian family. About 581 amino acids protein PINK1 is encoded by six exons possessing *PARK6* [45]. It has a 34 amino acid long mitochondrial targeting motif at amino-terminal which is a conserved kinase domain and an auto-regulatory domain at carboxy-terminal [46]. Almost two-thirds of the *PINK1* mutations are reported to be of loss of function which degrade its kinase domain activity, thus, demonstrating the relevance of enzymatic role of *PINK1* in the PD pathogenesis [7]. Cell culture studies have shown *PINK1* to be located on mitochondria imparting protective function to the cell, failing of which due to mutation results in cellular stress and thus contribute to pathogenic mechanism of PD [45]. It has been reported in recent studies that *PINK1* and *Parkin* functions in a specific way, in which impaired mitochondria are sensed and selectively removed from the mitochondrial system [7].

3.2.3.5   *PARK7*/DJ-1

DJ-1(protein deglycase), a cysteine protease, is encoded by *PARK7*. It is a protein spanning 189 amino acids which are encoded by 7 exons. It was identified to be involved in 1–2% of early onset autosomal recessive PD cases and its oxidized form is found in the sporadic PD [47]. Ubiquitous DJ-1 protein function as an oxidative stress cellular sensor [48]. Protein DJ-1 functions as a chaperon particle and assists the three-dimensional folding of newly formed protein and their delivery to proteasome. *DJ-1* accounts for about 10 different mutations (deletions/point mutations), most of which are in homozygous or compound-heterozygous state [7]. It also forms dimeric structure under physiological conditions [49]. The disease-causing mutants that form heterodimer with wild type *DJ-1* are E64D, L166P, D149A, and M26I [50]. When mutated, DJ-1 generally loses the protein-folding function and is degraded by the proteasome. Furthermore, its antioxidant and neuroprotective activities are also minimized [51,52].

3.2.3.6   *PARK9/ATP13A2*

ATP13A2 mutation causes early onset and autosomal recessive form of PD. It is a comparatively larger gene with 29 exons, encoding a 1180 amino acid long protein that is present in the membrane of lysosome and comprises of 10 transmembrane domain an ATPase domain [53]. Its mutation causes Kufor-Rakeb syndrome, an autosomal recessive atypical form of PD [54]. It is a juvenile onset syndrome linked with a secondary clinical trial of dementia, supranuclear gaze palsy, spasticity and has rapid progression. Until now, 11 cases of compound heterozygous and homozygous *ATP13A2* mutations have been identified [55].

3.3   Pathophysiology

Parkinson is physiologically related to dopaminergic neurons' loss in SNpc which is the result of several mechanisms among which α-synuclein aggregation plays the central role while others being mitochondrial dysfunction, oxidative stress, neuroinflammation, and malfunction in autophagy [5] (Figure 3.2).

3.3.1   α-Synuclein aggregation (Lewy bodies)

One of the most important neuropathology of PD is α-synuclein irregular deposition known as Lewy bodies in the cytoplasmic regions of neurons of several regions of the brain [56]. These aggregates are initially present in olfactory bulb, as well as in cholinergic and monoaminergic brainstem, corresponding to the non-symptomatic or presymptomatic stage of the disease. With the disease progression, they are formed in the limbic areas of forebrain and midbrain, SNpc as well as in the neocortical regions of the brain which manifest to clear symptoms of Parkinson like gait problems and dementia [57].

α-Synuclein is a polypeptide chain of 140 amino acid, present in the presynaptic axon terminals regulating vesicle recycling [58]. It is also known to have

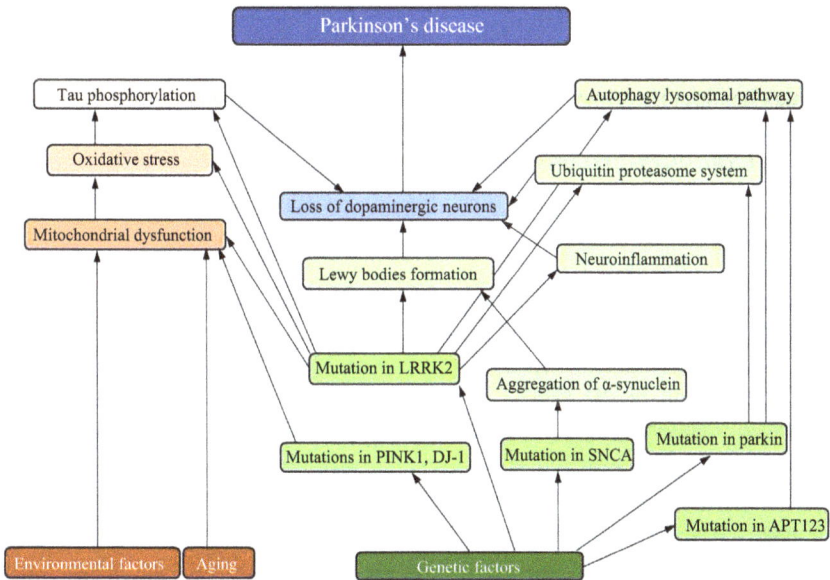

Figure 3.2   Schematic representation of mechanism of Parkinson's disease [89].

directly or indirectly involved in DNA repair mechanism by activating a major DNA repair signaling enzyme ATM (ataxia telangiectasia mutated) that codes protein kinase functioning in cell cycle, DNA double stranded break repair, apoptosis, etc., which is compromised in Lewy bearing neurons, thus triggering the cell death [58]. This provides for the association between α-synuclein aggregated Lewy bodies and PD [36].

3.3.2   Mitochondrial dysfunction

Disruption in mitochondrial function leading to cell death is one of the major causes of both idiopathic and familial PD [59]. A vital component of the electron transport chain, mitochondrial complex-1, was reported to be deficient in the postmortem studies of SNpc region in the brains of PD patients [60]. Complex-1 deficiency leads to cell death as a result energy depletion [61], which was found to be more in blood platelets and skeletal muscles of PD subjects over healthy individuals [62]. It was also further reported that, MPTP (1-methyl-4-phenyl-1,2,3,6-tetrahydropyridine) [63], a pro-drug to neurotoxin MPP+ (1-methyl-4-phenylpyridinium), caused the PD symptoms by destroying the dopaminergic neurons in SNpc [64]. Rotenone and paraquat are the toxins which disrupt the mitochondrial complex-1 activity and lead to dopamine loss in cells of animals as well as potentially in humans [6].

Parkin and *PINK1* genes [46] regulate the pathways of mitophagy (removal of dysfunctional mitochondria) [65]. Loss-of-function mutation in any of these genes results in mitophagy impairment and leads to autosomal dominant PD [66,67]. α-Synuclein can also show interaction with and accumulate in mitochondrial membrane, thus interrupting its function [68]. α-Synuclein damages complex-1 activity and leads to mitochondrial dysfunction and oxidative stress [69].

Reactive oxygen species (ROS), which are oxygen-containing highly reactive molecules disrupt the function of mitochondria. These ROS are regularly removed by mitochondria. α-Synuclein interacts with mitochondrial receptor TOMO interferes with the mitochondrial machinery to remove ROS and thus mitochondrial dysfunction [70].

3.3.3   Ubiquitin-proteasome system

Protein-clearing machinery, known as ubiquitin proteasome system (UPS), is manifested to removal of dysfunctional proteins. Dysfunctional proteins are first tagged with ubiquitin and then removed via proteasome machinery [71]. This UPS was found to be functionally weakened in SNpc regions of brain of PD patients [72]. Activities of 20S proteasome α-subunit [73] and of other molecules like PA700 and PA28 (proteasome activators), which are a part of normal UPS machinery were found to be diminishing in PD subjects as compared to healthy individuals [74]. This diminished function of UPS machinery can lead to cell death in SNpc region.

3.3.4   Autophagy-lysosome system

Autophagy-lysosome system is also a protein-clearing system that degrades the dysfunctional cytosolic protein by firstly engulfing them with autophagosomes, later fusing with lysosome where it breaks down into small fragments [75]. Level of

LC3-II, which is the central protein in the autophagy pathway, was found to be increased in SNpc neurons of brain leading to an agglomeration of autophagic vacuoles [76]. Also, in samples of PD patients few lysosomal membrane proteins such as LAMP1 and LAMP2A had also been found to be diminished [77,78]. Point mutation in the lysosomal gene, *APT123*, leads to an autosomal recessive atypical Parkinsonian Kufor-Rakeb syndrome [54]. In addition, *GBA1* mutations also induce lysosome autophagy system dysfunction. All these contribute to the evidence of autophagy-lysosome pathogenesis of PD.

3.3.5   Neuroinflammation

Parkinson patients' samples have confirmed to have higher concentration of pro-inflammatory cytokines in SNpc than in healthy individuals [79]. Microglial studies on the PD patients' samples show that microglial cells which are usually in the M2 state (resting state) in normal case, enter the M1 state (active, pro-inflammatory factors secretion state) rapidly in the PD patients due to the presence of α-synuclein aggregates. The pro-inflammatory factors can lead to the depletion of motor nerves and the dying cells form the positive feedback loop by activating M1 microglia cells [80].

According to genome wide association studies (GWAS) [81], genes involved in immune regulation are found to be associated with Parkinson such as *LRRK2* [82,83]. This is suggested by close relation between certain genes, α-synuclein aggregates and neuroinflammation [84]. α-Synuclein aggregation is responsible for innate and adaptive immunity in PD as having evidences of association of human leukocyte antigen (HLA) class II antigen with PD [79]. Nonsteroidal anti-inflammatory drugs consuming people have been studied to have lesser risk of PD [85].

3.3.6   Oxidative stress

One of the effects of mitochondrial dysfunction recorded in the PD patients' brain tissue is oxidative stress [86]. Mutation in gene *DJ-1*, which is suggested to encode antioxidant, is reported to have caused autosomal PD due to oxidative stress in nigral DN [87]. In a mice model, knocking out of *DJ-1* gene increases oxidation in already stressed nigral dopaminergic neurons [88].

3.4   Clinical features

PD symptoms can be grouped into motor and non-motor. The four fundamental motor features which come under an acronym TRAP are Tremors, Rigidity, Akinesia, and Postural instability [3]. Rest tumors are the most common and easily recognizable of all. Tremor or shaking usually begins in limb in which the limbs shake while at rest [90]. Hand tremor is referred to as supination-pronation ("pill-rolling"). Rigidity refers to the stiffness in any part of the body. It is characterized by increased resistance and can limit the range of movements. Bradykinesia implies slowed down movements. It manifests to slowing down of simple tasks, difficulty in planning, and executing any movement, shortening of steps while walking, inability to do multiple tasks at the same time [91]. Also, spontaneous movements, gesturing, facial expressions, blinking, arm movements while waling are affected. Although the pathophysiology of bradykinesia is still not known, reduced dopaminergic function disrupting the normal motor neuron activity is the hypothesized cause of bradykinesia [3]. Rigidity refers to the stiffness in any part of the body. It is characterized by increased resistance and can limit the range of movements. Postural instability is the stooping of the body posture which occurs at the later stages of PD. It occurs due to loss of reflexes controlling body postures. This also leads to falling, increasing risk of hip fractures.

Akinesia, which is known as freezing, is also one of the many debilitating symptoms of PD. Also called motor blocks, it mostly affect legs but can also rarely affect arms and eyelids [92,93]. Other motor symptoms include speech problems (dysarthria) and writing changes, emotional changes, difficulty in thinking, chewing, and swallowing problems(dysphagia) [3].

The non-motor symptoms of PD include, difficulty in urinating due to bladder problems, sleeping disorders, constipation, blood pressure changes, fatigue, smell dysfunction (hyposmia), pain, sexual dysfunction, obsessive-compulsive, and impulsive behavior [94]. Autonomic dysfunctions include orthostatic hypotension, sweating dysfunction. Patients with PD also have six times higher risk of dementia and it also has no treatment [95].

3.5 Diagnosis

PD has no definitive diagnosis test. Parkinsonism is the primary diagnostic criteria of PD. It relies on the four basic motor symptoms, tremor at rest, rigidity, postural instability, and akinesia or bradykinesia [3]. Of these, bradykinesia is the most common along with any of the other one. Along with motor symptoms, non-motor symptoms are also of importance because of its prominence in the early stage of PD [59]. However, on autopsy, the presence of Lewy bodies is the standard criteria for pathological diagnosis of the disease [3]. The misdiagnosis of the disease is also common during the early phase which can be checked by observing patient's response to Parkinson's medications [96]. History of any other parkinsonism disorder or any other neurological signs upon examination can also avoid misdiagnosis of the disease [97]. However, neuroimaging techniques such as magnetic resonance imaging [98], fluorodopa positron emission tomography [98], raclopride imaging of dopamine D2 receptors [99], and single photon emission topography [100] can also be used as for disease diagnosis and also preventing misdiagnosis.

3.6 Management

There is still no definitive cure of this disease but only disease modifying medication or therapy which helps in slowing down the progression of the disease. Medical therapies such as levodopa–carbidopa combination medication, dopamine agonists (apomorphine), N-methyl-D-aspartate (NMDA) receptor inhibitors, monoamine oxidase-B (MAO-B) inhibitors, catechol-O-methyltransferase (COMT) inhibitors, are available against the early stage motor symptoms [101].

Levodopa is a drug used by patients to replace striatal dopamine, which was lost due to dopamine neuron loss in the SNpc [102]. Levodopa, which as a substitute of dopamine, can and acts as a substrate for DOPA decarboxylase to convert into dopamine [103]. Levodopa is usually given in combination with carbidopa. Carbidopa assists levodopa in delaying the conversion into dopamine until reaching the brain. Although levodopa can control most of the motor symptoms but only marginally controls tremor. Also, its prolonged usage can result in some complications such as dyskinesias [8] which can be minimized by using it in combination with DOPA decarboxylase inhibitors [102]. For the patients who don't react well to levodopa/carbidopa, add-on medications of safinamide tablets [104], istradefylline tablets [105], and opicapone [106] are given.

Dopaminergic agonists are the drugs which activate the dopaminergic receptor family. They are grouped into two categories namely, ergoline and non-ergoline derivatives. Ergoline derivatives are not a safe recommendation because of having several side effects whereas non-ergoline derivatives are safer to use, especially for younger patients [107]. Several dopamine agonists, such as bromocriptine, riponirole, apomorphine, pramipexole, can be used to manage the early year symptoms of PD [59]. An antiviral Amantadine, which is an antiviral drug, can also be used to reduce PD symptoms [108]. Tolcapone, an inhibitor of dopamine-degrading enzyme COMT(catechol-*O*-methyltransferase), is given to patients to prolong the effects of levodopa by preventing the breakdown of dopamine in nerve cells [109]. Similarly, inhibitors of MAO-B(monoamine oxidase type B) such as selegiline and rasagiline decline the activity of MAO-B enzyme of dopamine breaking activity [110,111,9].

In the modern era of rapidly advancing technology, the integration of sophisticated computational approaches plays a crucial role in addressing challenges across diverse fields, including healthcare and data management. This research investigates innovative methodologies that harness machine learning (ML) and hybrid intelligent systems to enhance public health outcomes, strengthen data security, and facilitate real-time processing. A robust ML-based platform [112] demonstrates remarkable efficacy in early disease diagnosis, achieving 98% accuracy in breast cancer prediction through logistic regression, 85% in heart disease detection using Naïve Bayes, and 75% in diabetes prediction with SVM. Furthermore, a hybrid intelligent intrusion detection system (HIIDS) [113] designed for IoT-based healthcare applications ensures the protection of sensitive electronic health records (EHRs). Employing GA-DT algorithms, the proposed HIIDS achieves 99.88% accuracy in detecting DoS attacks, outperforming contemporary techniques. Additionally, the adoption of edge computing, powered by 5G technology, enhances efficiency by processing data near its source, thereby reducing latency. This capability enables applications such as IoT and AR/VR to deliver real-time functionalities, exemplified by AR-based gaming platforms.

These advancements highlight the transformative impact of technology in revolutionizing healthcare, fortifying security, and enriching user experiences [114].

In certain cases where disease does not respond well to the drugs, surgery is an alternative. DBS is one such method in which electrodes are inserted into the brain and the signals are stimulated by a device called pulse generator which suppress the signals that trigger many of the motor symptoms of PD [10]. However, neither DBS nor medication can cease the disease progression.

3.7   Biomarkers

To combat the ongoing problem of diseases diagnosis, early diagnosis of the disease, prior to the onset of symptoms can be done, for which, identification of biomarkers can be useful. Network theory has its applications in various fields to study the complex dynamic systems in economics, computer science, statistics, engineering, social sciences, and biological sciences. Biologists apply this theory to study the gene regulatory networks, metabolic networks, cell-signaling networks, etc. The systemic network approach helps to identify and study the genes/proteins playing important roles in complex disease outcome [115]. It is, therefore, important to understand complex diseases such as Parkinson's through systemic network approach to identify the novel key regulators in the disease system and help in developing novel therapeutic drugs and the disease-modifying therapies in the future.

Jiang *et al.* at Department of Neurology, Dongying People's Hospital, China performed one such study to identify PD biomarkers. Using microarray database, they extracted differentially expressed genes (DEGs) by comparing PD patients and normal control groups and identified nine potential biomarkers namely, *PTGDS* (upregulated), *GPX3* (upregulated), *SLC25A20* (upregulated), *CACNA1D* (upregulated), *LRRN3* (downregulated), *POLR1D* (downregulated), *ARHGAP26* (upregulated), *TNFSF14* (upregulated), and *VPS11* (downregulated). The protein–protein interaction (PPI) network was constructed containing 138 nodes and 133 edges, *VPS11* having the highest (44) degree. Their expression levels of all the genes in the groups were also analyzed. *PTGDS* codes an enzyme that is responsible for the regulation of sleep/wake cycle. In this study, *PTGDS* is found to be upregulated suggesting that it plays a major role in sleeping problems in PD patients. *GPX3* which was also found to be upregulated may also have the possible role in PD development as it is present in pathways associated with oxidative stress, hydrogen peroxide catabolic process. Similarly, all others except *LRRN3*, *POLR1D*, and *VPS11* were upregulated when comparison was made between PD and control [116].

While many studies have been done on microarray datasets, there is study which explains the significance of study of PD at transcriptomic level because of the high inconsistency and concordance in the microarray studies [117]. One such study employed genetic data from genome-wide association studies and gene expression microarray data to reposition drugs for usage in PD. Using drug-protein regulatory relationship databases, drug effect sum score was developed and evaluated all the candidate drugs of which 12 were reported to have highest drug effect sum score scores among which 5 had the potential for PD treatment [118].

3.8 Conclusion

PD is a prevalent, neurodegenerative disorder, the exact mechanism of which is known imprecisely. A small population of patients has monogenic form of the disease but the majority population is said to have sporadic form of the disease. The available drugs and therapies can only partially attenuate the disease but cannot treat it completely. Understanding the pathological processes and identifying PD biomarkers can help in developing drugs/small molecules against them to open new dimensions of Parkinson's disease study and disease interventions in future.

Author contributions

S.K.S., N.V. and M.Z.M. conceptualized the work. S.S., NV., AV., M.G., S.K.S., and M.K.H. did the preparation of the figures. S.S., A.V., N.V., M.G., M.Z.M., and S.K.S wrote the manuscript. All the authors read and approved the manuscript.

Acknowledgments

Nidhi Verma and Saurabh Kumar Sharma are financially supported by Indian Council of Medical Research (ICMR), New Delhi, India.

Competing interests

The authors declare that they have no competing interests.

References

[1]   L. V. Kalia and A. E. Lang, "Parkinson's disease," *The Lancet*, vol. 386, no. 9996. Lancet Publishing Group, pp. 896–912, 2015, doi: 10.1016/S0140-6736(14)61393-3.

[2]   A. J. Lees, J. Hardy, and T. Revesz, "Parkinson's disease," *The Lancet*, vol. 373, no. 9680, pp. 2055–2066, 2009, doi: 10.1016/S0140-6736(09)60492-X.

[3]   J. Jankovic, "Parkinson's disease: clinical features and diagnosis," *Journal of Neurology, Neurosurgery and Psychiatry*, vol. 79, no. 4. BMJ Publishing Group, pp. 368–376, 2008, doi: 10.1136/jnnp.2007.131045.

[4]   L. I. Golbe and C. E. Leyton, "Editorial life expectancy in Parkinson disease," *Neurology*, vol. 91, no. 22, pp. 991–992, 2018, doi: 10.1212/WNL. 0000000000006560.

[5]   A. Kouli, K. M. Torsney, and W.-L. Kuan, "Parkinson's disease: etiology, neuropathology, and pathogenesis," *in Parkinson's Disease: Pathogenesis and Clinical Aspects, Codon Publications*, 2018, pp. 3–26.

[6]   C. M. Tanner, "Rotenone, paraquat, and Parkinson's disease," *Environmental Health Perspectives*, vol. 119, no. 6, pp. 866–872, 2011, doi: 10.1289/ehp.1002839.

[7]   C. Klein and A. Westenberger, "Genetics of Parkinson's disease," *Cold Spring Harbor Perspectives in Medicine*, vol. 2, no. 1, 2012, doi: 10.1101/cshperspect.a008888.

[8]   J. Jankovic and M. Stacy, "Medical management of levodopa-associated motor complications in patients with Parkinson's disease," *CNS Drugs*, vol. 21, no. 8. CNS Drugs, pp. 677–692, 2007, doi: 10.2165/00023210-200721080-00005.

[9]   J. Jankovic and W. Poewe, "Therapies in Parkinson's disease," *Current Opinion in Neurology*, vol. 25, no. 4. pp. 433–447, 2012, doi: 10.1097/WCO.0b013e3283542fc2.

[10] Pedrosa, David J. and Lars Timmermann, "Management of Parkinson's disease." *Neuropsychiatric disease and treatment*, 2013, pp. 321–340, doi: 10.2147/NDT.S32302.

[11] O. B. Tysnes and A. Storstein, "Epidemiology of Parkinson's disease," *Journal of Neural Transmission*, vol. 124, no. 8, pp. 901–905, 2017, doi: 10.1007/s00702-017-1686-y.

[12] C. M. Tanner and S. M. Goldman, "Epidemiology of Parkinson's disease," *Neuroepidemiology*, vol. 14, no. 2, pp. 317–335, 1996.

[13] A. Schrag and J. M. Schott, "Epidemiological, clinical, and genetic characteristics of early-onset parkinsonism," *Lancet Neurology*, vol. 5, no. 4, pp. 355–363, 2006, doi: 10.1016/S1474-4422(06)70411-2.

[14] T. Pringsheim, N. Jette, A. Frolkis, and T. D. L. Steeves, "The prevalence of Parkinson's disease: a systematic review and meta-analysis," *Movement Disorders*, vol. 29, no. 13, pp. 1583–1590, 2014, doi: 10.1002/mds.25945.

[15] P. Mehta, W. Kaye, J. Raymond, R. Punjani, T. Larson, and J. Cohen, "Prevalence of Amyotrophic Lateral Sclerosis – United States, 2015," *MMWR. Morbidity and mortality weekly report*, vol. 67, no. 46, 2018.

[16] S. Razdan, R. L. Kaul, A. Motta, S. Kaul, and R. K. Bhatt, "Prevalence and pattern of major neurological disorders in rural Kashmir (India) in 1986," *Neuroepidemiology*, vol. 13, no. 3, pp. 113–119, 1994, doi: 10.1159/000110368.

[17] M. Gourie-Devi, G. Gururaj, P. Satishchandra, and D. K. Subbakrishna, "Prevalence of neurological disorders in Bangalore, India: a community-based study with a comparison between urban and rural areas," *Neuroepidemiology*, vol. 23, no. 6, pp. 261–268, 2004, doi: 10.1159/000080090.

[18] S. K. Das and K. Sanyal, "Neuroepidemiology of major neurological disorders in rural Bengal," *Neurology India*, vol. 44, no. 2, pp. 47–58, 1996.

[19] N. E. Bharucha, E. P. Bharucha, A. E. Bharucha, A. V. Bhise, and B. S. Schoenberg, "Prevalence of Parkinson's disease in the Parsi community of

Bombay, India," *Archives of Neurology*, vol. 45, no. 12, pp. 1321–1323, 1988, doi: 10.1001/archneur.1988.00520360039008.

[20] K. Xu., C. Xu, J. Xu, and J. Schwarzschild. "Estrogen prevents neuroprotection by caffeine in the mouse 1-methyl-4-phenyl-1,2,3,6-tetra-hydropyridine model of Parkinson's disease," *The Journal of Neuroscience*, vol. 26, no. 2, pp. 535–541, 2006, doi: 10.1523/JNEUROSCI.3008-05.2006.

[21] R. Betarbet, T. B. Sherer, G. MacKenzie, M. Garcia-Osuna, A. V. Panov, and J. T. Greenamyre, "Chronic systemic pesticide exposure reproduces features of Parkinson's disease," *Nature Neuroscience*, vol. 3, no. 12, pp. 1301–1306, 2000, doi: 10.1038/81834.

[22] C. M. Tanner, "Epidemiology of Parkinson's disease," *Neurologic Clinics*, vol. 10, no. 2, pp. 317–329, 1992, doi: 10.1016/s0733-8619(18)30212-3.

[23] R. Mayeux, Richard, Jean Denaro, *et al.*, "A population-based investigation of Parkinson's disease with and without dementia: relationship to age and gender," *Archives of Neurology*, vol. 49, no. 5, pp. 492–497, 1992, doi: 10. 1001/archneur.1992.00530290076015.

[24] T. T. Warner, T. Thomas, Anthony HV Schapira, *et al.*, "Genetic and environmental factors in the cause of Parkinson's disease," *Annals of Neurology*, vol. 53, no. suppl. 3, 2003, doi: 10.1002/ana.10487.

[25] S. Lesage and A. Brice, "Parkinson's disease: from monogenic forms to genetic susceptibility factors," *Human Molecular Genetics*, vol. 18, no. R1, pp. 48–59, 2009, doi: 10.1093/hmg/ddp012.

[26] M. R. Cookson, G. Xiromerisiou, and A. Singleton, "How genetics research in Parkinson's disease is enhancing understanding of the common idiopathic forms of the disease," *Current Opinion in Neurology*, vol. 18, no. 6, pp. 706–711, 2005, doi: 10.1097/01.wco.0000186841.43505.e6.

[27] D. M. Radhakrishnan and V. Goyal, "Parkinson's disease: a review," *Neurology India*, vol. 66, no. 7, pp. S26–S35, 2018, doi: 10.4103/0028-3886.226451.

[28] M. H. Polymeropoulos, C. Lavedan, E. Leroy, *et al.*, "Mutation in the α-synuclein gene identified in families with Parkinson's disease," *Science (80).*, vol. 276, no. 5321, pp. 2045–2047, 1997, doi: 10.1126/science. 276.5321.2045.

[29] A. Athanassiadou, G. Voutsinas, L. Psiouri, *et al.*, "Genetic analysis of families with Parkinson disease that carry the Ala53Thr mutation in the gene encoding α-synudein [1]," *American Journal of Human Genetics*, vol. 65, no. 2. pp. 555–558, 1999, doi: 10.1086/302486.

[30] P. J. Spira, D. M. Sharpe, G. Halliday, J. Cavanagh, and G. A. Nicholson, "Clinical and pathological features of a parkinsonian syndrome in a family with an Ala53Thrα-synuclein mutation," *Annals of Neurology*, vol. 49, no. 3, pp. 313–319, 2001, doi: 10.1002/ana.67.

[31] C. S. Ki , E. F. Stavrou, N. Davanos, *et al.*, "The Ala53Thr mutation in the α-synuclein gene in a Korean family with Parkinson disease [2]," *Clinical Genetics*, vol. 71, no. 5, pp. 471–473, 2007, doi: 10.1111/j.1399-0004.2007. 00781.x.

[32] J. M. Choi, M. S. Woo, H. I. Ma, *et al.*, "Analysis of PARK genes in a Korean cohort of early-onset Parkinson disease," *Neurogenetics*, vol. 9, no. 4, pp. 263–269, 2008, doi: 10.1007/s10048-008-0138-0.

[33] A. Puschmann, O. A. Ross, C. Vilariño-Güell, *et al.*, "A Swedish family with de novo α-synuclein A53T mutation: evidence for early cortical dysfunction," *Parkinsonism & related disorders*, vol. 15, no. 9, pp. 627–632, 2009, doi:10.1016/j.parkreldis.2009.06.007.

[34] R. Krüger, W. Kuhn, T. Müller, *et al.*, "Ala30Pro mutation in the gene encoding α-synuclein in Parkinson's disease," *Nature Genetics*, vol. 18, no. 2, pp. 106–108, 1998, doi: 10.1038/ng0298-106.

[35] J. J. Zarranz, J. Alegre, J. C. Gómez-Esteban, *et al.*, "The new mutation, E46K, of α-synuclein causes Parkinson and Lewy body dementia," *Annals of Neurology*, vol. 55, no. 2, pp. 164–173, 2004, doi: 10.1002/ana.10795.

[36] N. Alkanli and A. Ay, "The relationship between alpha-synuclein (SNCA) gene polymorphisms and development risk of Parkinson's disease," Chapter 3, 96 in *Synucleins – Biochemistry and Role in Diseases*, *IntechOpen*, 2020, ISBN: 9781-78984-5655.

[37] J. Q. Li, L. Tan, and J. T. Yu, "The role of the LRRK2 gene in Parkinsonism," *Molecular Neurodegeneration*, vol. 9, p. 47, 2014, doi: 10.1186/1750-1326-9-47.

[38] K. Nuytemans, J. Theuns, M. Cruts, and C. Van Broeckhoven, "Genetic etiology of Parkinson disease associated with mutations in the SNCA, PARK2, PINK1, PARK7, and LRRK2 genes: a mutation update," *Human Mutation*, vol. 31, no. 7, pp. 763–780, 2010, doi: 10.1002/humu.21277.

[39] S. Lesage, A. Dürr, M. Tazir, *et al.*, "LRRK2 G2019S as a cause of Parkinson's disease in North African Arabs [13]," *New England Journal of Medicine*, vol. 354, no. 4, pp. 422–423, 2006, doi: 10.1056/NEJMc055540.

[40] L. J. Ozelius, G. Senthil, R. Saunders-Pullman, *et al.*, "LRRK2 G2019S as a cause of Parkinson's disease in Ashkenazi Jews [14]," *New England Journal of Medicine*, vol. 354, no. 4, pp. 424–425, 2006, doi: 10.1056/NEJMc055509.

[41] E. K. Tan, H. Shen, L. C. S. Tan, *et al.*, "The G2019S LRRK2 mutation is uncommon in an Asian cohort of Parkinson's disease patients," *Neuroscience Letters*, vol. 384, no. 3, pp. 327–329, 2005, doi: 10.1016/j.neulet.2005.04.103.

[42] T. M. Dawson and V. L. Dawson, "The role of Parkin in familial and sporadic Parkinson's disease," *Movement disorders*, vol. 25, no. S1, pp. S32–S39, 2010, doi: 10.1002/mds.22798.

[43] S. Selvaraj and S. Piramanayagam, "Impact of gene mutation in the development of Parkinson's disease," *Genes and Diseases*, vol. 6, no. 2, pp. 120–128, 2019, doi: 10.1016/j.gendis.2019.01.004.

[44] "PINK1 Gene – GeneCards | PINK1 Protein | PINK1 Antibody." [Online]. Available: https://www.genecards.org/cgi-bin/carddisp.pl?gene=PINK1. [Accessed: 08-Sep-2020].

[45] E. M. Valente, P. M. Abou-Sleiman, V. Caputo, *et al.*, "Hereditary early-onset Parkinson's disease caused by mutations in PINK1," *Science (80)*, vol. 304, no. 5674, pp. 1158–1160, 2004, doi: 10.1126/science.1096284.

[46]   D. Truban, X. Hou, T. R. Caulfield, F. C. Fiesel, and W. Springer, "PINK1, Parkin, and mitochondrial quality control: What can we learn about Parkinson's disease pathobiology?," *Journal of Parkinson's Disease*, vol. 7, pp. 13–29, 2017, doi: 10.3233/JPD-160989.

[47]   M. Repici and F. Giorgini, "DJ-1 in Parkinson's disease: clinical insights and therapeutic perspectives," *Journal of Clinical Medicine*, vol. 8, no. 9, p. 1377, 2019, doi: 10.3390/jcm8091377.

[48]   R. M. Canet-Avilés, M. A. Wilson, D. W. Miller, *et al.*, "The Parkinson's disease DJ-1 is neuroprotective due to cysteine-sulfinic acid-driven mitochondrial localization," *Proceedings of National Academy of Sciences of United States of America*, vol. 101, no. 24, pp. 9103–9108, 2004, doi: 10.1073/pnas.0402959101.

[49]   M. G. Macedo, Burcu Anar, Iraad F. Bronner, *et al.*, "The DJ-1 L166P mutant protein associated with early onset Parkinson's disease is unstable and forms higher-order protein complexes," *Human molecular genetics*, vol. 12, no. 21, pp. 2807–2816, 2003, doi: 10.1093/hmg/ddg304.

[50]   K. Takahashi-Niki, T. Niki, T. Taira, S. M. M. Iguchi-Ariga, and H. Ariga, "Reduced anti-oxidative stress activities of DJ-1 mutants found in Parkinson's disease patients," *Biochemical and Biophysical Research Communications*, vol. 320, no. 2, pp. 389–397, 2004, doi: 10.1016/j.bbrc.2004.05.187.

[51]   P. C. Anderson and V. Daggett, "Molecular basis for the structural instability of human DJ-1 induced by the L166P mutation associated with Parkinson's disease," *Biochemistry*, vol. 47, no. 36, pp. 9380–9393, 2008, doi: 10.1021/bi800677k.

[52]   G. Malgieri and D. Eliezer, "Structural effects of Parkinson's disease linked DJ-1 mutations," *Protein Science*, vol. 17, no. 5, pp. 855–868, 2008, doi: 10.1110/ps.073411608.

[53]   A. Ramirez, A. Heimbach, J. Gründemann, *et al.*, "Hereditary parkinsonism with dementia is caused by mutations in ATP13A2, encoding a lysosomal type 5 P-type ATPase," *Nature Genetics*, vol. 38, no. 10, pp. 1184–1191, 2006, doi: 10.1038/ng1884.

[54]   D. R. Williams, A. Hadeed, A. S. Najim al-Din, A. L. Wreikat, and A. J. Lees, "Kufor Rakeb disease: autosomal recessive, levodopa-responsive Parkinsonism with pyramidal degeneration, supranuclear gaze palsy, and dementia," *Movement Disorders*, vol. 20, no. 10, pp. 1264–1271, 2005, doi: 10.1002/mds.20511.

[55]   J. S. Park, N. F. Blair, and C. M. Sue, "The role of ATP13A2 in Parkinson's disease: clinical phenotypes and molecular mechanisms," *Movement Disorders*, vol. 30, no. 6, pp. 770–779, 2015, doi: 10.1002/mds.26243.

[56]   H. Braak, K. Del Tredici, U. Rüb, R. A. I. De Vos, E. N. H. Jansen Steur, and E. Braak, "Staging of brain pathology related to sporadic Parkinson's disease," *Neurobiology of Aging*, vol. 24, no. 2, pp. 197–211, 2003, doi: 10.1016/S0197-4580(02)00065-9.

[57]   G. M. Halliday and H. McCann, "The progression of pathology in Parkinson's disease," *Annals of the New York Academy of Sciences*, vol. 1184, pp. 188–195, 2010, doi: 10.1111/j.1749-6632.2009.05118.x.

[58]   A. J. Schaser, V. R. Osterberg, S. E. Dent, *et al.*, "Alpha-synuclein is a DNA binding protein that modulates DNA repair with implications for Lewy body disorders," *Scientific Reports*, vol. 9, no. 1, pp. 1–19, 2019, doi: 10.1038/s41598-019-47227-z.

[59]   T. B. Stoker, M. BChir, and J. C. Greenland, *Parkinson's Disease (book) Pathogenesis and Clinical Aspects*. Brisbane (AU): Codon Publications, 2018, doi: 10.15586/codonpublications.parkinsonsdisease.2018.

[60]   A. H. V. Schapira, J. M. Cooper, D. Dexter, J. B. Clark, P. Jenner, and C. D. Marsden, "Mitochondrial complex I deficiency in Parkinson's disease," *The Journal of Neurochemistry*, vol. 54, no. 3, pp. 823–827, 1990, doi: 10.1111/j.1471-4159.1990.tb02325.x.

[61]   L. A. Bindoff, M. A. Birch-Machin, N. E. F. Cartlidge, W. D. Parker, and D. M. Turnbull, "Respiratory chain abnormalities in skeletal muscle from patients with Parkinson's disease," *Journal of the Neurological Sciences*, vol. 104, no. 2, pp. 203–208, 1991, doi: 10.1016/0022-510X(91)90311-T.

[62]   D. Krige, M. T. Carroll, J. M. Cooper, C. D. Marsden, and A. H. V. Schapira, "Platelet mitochondria function in Parkinson's disease," *Annals of Neurology*, vol. 32, no. 6, pp. 782–788, 1992, doi: 10.1002/ana.410320612.

[63]   J. Sian, M. Youdim, P. Riederer, and M. Gerlach, "MPTP-Induced Parkinsonian Syndrome," *Molecular, Cellular and Medical Aspects*, 1999.

[64]   W. m. Do. J. Nicklas, I. Vyas, and R. E. Heikkila, "Inhibition of NADH-linked oxidation in brain mitochondria by 1-methyl-4-phenyl-pyridine, a metabolite of the neurotoxin, 1-methyl-4-phenyl-1,2,5,6-tetrahydropyridine," *Life Sci.*, vol. 36, no. 26, pp. 2503–2508, 1985, doi: 10.1016/0024-3205(85)90146-8.

[65]   A. M. Pickrell and R. J. Youle, "The roles of PINK1, Parkin, and mitochondrial fidelity in Parkinson's disease," *Neuron*, vol. 85, no. 2, pp. 257–273, 2015, doi: 10.1016/j.neuron.2014.12.007.

[66]   T. Kitada, S. Asakawa, N. Hattori, *et al.*, "Mutations in the parkin gene cause autosomal recessive juvenile parkinsonism," *Nature*, vol. 392, no. 6676, pp. 605–608, 1998, doi: 10.1038/33416.

[67]   E. M. Valente, A. R. Bentivoglio, P. H. Dixon, *et al.*, "Localization of a novel locus for autosomal recessive early-onset parkinsonism, PARK6, on human chromosome 1p35-p36," *The American Journal of Human Genetics*, vol. 68, no. 4, pp. 895–900, 2001, doi: 10.1086/319522.

[68]   L. Devi, V. Raghavendran, B. M. Prabhu, N. G. Avadhani, and H. K. Anandatheerthavarada, "Mitochondrial import and accumulation of α-synuclein impair complex I in human dopaminergic neuronal cultures and Parkinson disease brain," *Journal of Biological Chemistry*, vol. 283, no. 14, pp. 9089–9100, 2008, doi: 10.1074/jbc.M710012200.

[69]   E. S. Luth, I. G. Stavrovskaya, T. Bartels, B. S. Kristal, and D. J. Selkoe, "Soluble, prefibrillar α-synuclein oligomers promote complex I-dependent, $Ca2+$-induced mitochondrial dysfunction," *Journal of Biological Chemistry*, vol. 289, no. 31, pp. 21490–21507, 2014, doi: 10.1074/jbc.M113.545749.

[70]   R. Di Maio, P. J. Barrett, E. K. Hoffman, *et al.*, "α-Synuclein binds to TOM20 and inhibits mitochondrial protein import in Parkinson's disease,"

Science Translational Medicine, vol. 8, no. 342, 2016, doi: 10.1126/scitranslmed.aaf3634.

[71] C. McKinnon and S. J. Tabrizi, "The ubiquitin-proteasome system in neurodegeneration," *Antioxidants and Redox Signaling,* vol. 21, no. 17, pp. 2302–2321, 2014, doi: 10.1089/ars.2013.5802.

[72] K. S. P. McNaught and P. Jenner, "Proteasomal function is impaired in substantia nigra in Parkinson's disease," *Neuroscience Letters*, vol. 297, no. 3, pp. 191–194, 2001, doi: 10.1016/S0304-3940(00)01701-8.

[73] K. S. P. McNaught, R. Belizaire, P. Jenner, C. W. Olanow, and O. Isacson, "Selective loss of 20S proteasome α-subunits in the substantia nigra pars compacta in Parkinson's disease," *Neuroscience Letters*, vol. 326, no. 3, pp. 155–158, 2002, doi: 10.1016/S0304-3940(02)00296-3.

[74] K. S. P. McNaught, R. Belizaire, O. Isacson, P. Jenner, and C. W. Olanow, "Altered proteasomal function in sporadic Parkinson's disease," *Experimental Neurology*, vol. 179, no. 1, pp. 38–46, 2003, doi: 10.1006/exnr.2002.8050.

[75] B. Dehay, J. Bové, N. Rodríguez-Muela, *et al.*, "Pathogenic lysosomal depletion in Parkinson's disease," *The Journal of Neuroscience*, vol. 30, no. 37, pp. 12535–12544, 2010, doi: 10.1523/JNEUROSCI.1920-10.2010.

[76] K. Tanji, F. Mori, A. Kakita, H. Takahashi, and K. Wakabayashi, "Alteration of autophagosomal proteins (LC3, GABARAP and GATE-16) in Lewy body disease," *Neurobiology of Disease*, vol. 43, no. 3, pp. 690–697, 2011, doi: 10.1016/j.nbd.2011.05.022.

[77] Y. Chu, H. Dodiya, P. Aebischer, C. W. Olanow, and J. H. Kordower, "Alterations in lysosomal and proteasomal markers in Parkinson's disease: relationship to alpha-synuclein inclusions," *Neurobiology of Disease*, vol. 35, no. 3, pp. 385–398, 2009, doi: 10.1016/j.nbd.2009.05.023.

[78] L. Alvarez-Erviti, M. C. Rodriguez-Oroz, J. M Cooper, *et al.*, "Chaperone-mediated autophagy markers in Parkinson disease brains," *Archives of Neurology*, vol. 67, no. 12, pp. 1464–1472, 2010, doi: 10.1001/archneurol.2010.198.

[79] E. C. Hirsch and S. Hunot, "Neuroinflammation in Parkinson's disease: a target for neuroprotection?," *The Lancet Neurology*, vol. 8, no. 4, pp. 382–397, 2009, doi: 10.1016/S1474-4422(09)70062-6.

[80] M. S. Moehle and A. B. West, "M1 and M2 immune activation in Parkinson's Disease: Foe and ally?," *Neuroscience*, vol. 302, pp. 59–73, 2015, doi: 10.1016/j.neuroscience.2014.11.018.

[81] M. A. Nalls, N. Pankratz, C. M. Lill, *et al.*, "Large-scale meta-analysis of genome-wide association data identifies six new risk loci for Parkinson's disease," *Nature Genetics*, vol. 46, no. 9, pp. 989–993, 2014, doi: 10.1038/ng.3043.

[82] B. Ma, L. Xu, X. Pan, *et al.*, "LRRK2 modulates microglial activity through regulation of chemokine (C-X3-C) receptor 1-mediated signalling pathways," *Human Molecular Genetics*, vol. 25, no. 16, pp. 3515–3523, 2016, doi: 10.1093/hmg/ddw194.

[83] J. Schapansky, J. D. Nardozzi, F. Felizia, and M. J. LaVoie, "Membrane recruitment of endogenous LRRK2 precedes its potent regulation of

autophagy," *Human Molecular Genetics*, vol. 23, no. 16, pp. 4201–4214, 2014, doi: 10.1093/hmg/ddu138.

[84]   R. M. Ransohoff, "How neuroinflammation contributes to neurodegeneration," *Science*, vol. 353, no. 6301, pp. 777–783, 2016, doi: 10.1126/science.aag2590.

[85]   E. Esposito, V. Di Matteo, A. Benigno, M. Pierucci, G. Crescimanno, and G. Di Giovanni, "Non-steroidal anti-inflammatory drugs in Parkinson's disease," *Experimental Neurology*, vol. 205, no. 2, pp. 295–312, 2007, doi: 10.1016/j.expneurol.2007.02.008.

[86]   V. Dias, E. Junn, and M. M. Mouradian, "The role of oxidative stress in Parkinson's disease," *Journal of Parkinson's Disease*, vol. 3, no. 4, pp. 461–491, 2013, doi: 10.3233/JPD-130230.

[87]   V. Bonifati, P. Rizzu, Van Baren, *et al.*, "Mutations in the DJ-1 gene associated with autosomal recessive early-onset parkinsonism," *Science (80)*, vol. 299, no. 5604, pp. 256–259, 2003, doi: 10.1126/science.1077209.

[88]   M. Di Nottia, M. Masciullo, D. Verrigni, *et al.*, "DJ-1 modulates mitochondrial response to oxidative stress: clues from a novel diagnosis of PARK7," *Clinical Genetics*, vol. 92, no. 1, pp. 18–25, 2017, doi: 10.1111/cge.12841.

[89]   P. Maiti, J. Manna, G. L. Dunbar, P. Maiti, and G. L. Dunbar, "Current understanding of the molecular mechanisms in Parkinson's disease: targets for potential treatments," *Translational Neurodegeneration*, vol. 6, no. 1, pp. 1–35, 2017, doi: 10.1186/s40035-017-0099-z.

[90]   "Parkinson's disease – Symptoms and causes – Mayo Clinic." [Online]. Available: https://www.mayoclinic.org/diseases-conditions/parkinsons-disease/symptoms-causes/syc-20376055. [Accessed: 29-Aug-2020].

[91]   A. Berardelli, J. C. Rothwell, P. D. Thompson, and M. Hallett, "Pathophysiology of bradykinesia in Parkinson's disease," *Brain*, vol. 124, no. 11, pp. 2131–2146, 2001, doi: 10.1093/brain/124.11.2131.

[92]   N. Giladi, M. P. McDermott, S. Fahn, *et al.*, "Freezing of gait in PD: prospective assessment in the DATATOP cohort," *Neurology*, vol. 56, no. 12, pp. 1712–1721, 2001, doi: 10.1212/WNL.56.12.1712.

[93]   F. Rocha Cabrero and O. De Jesus, *Apraxia of Lid Opening*. StatPearls Publishing. 2024.

[94]   R. D. Palmiter, "Is dopamine a physiologically relevant mediator of feeding behavior?," *Trends Neuroscience*, vol. 30, no. 8, pp. 375–381, 2007, doi: 10.1016/j.tins.2007.06.004.

[95]   D. Aarsland, K. Andersen, J. P. Larsen, A. Lolk, H. Nielsen, and P. Kragh-Sørensen, "Risk of dementia in Parkinson's disease: a community-based, prospective study," *Neurology*, vol. 56, no. 6, pp. 730–736, 2001, doi: 10.1212/WNL.56.6.730.

[96]   L. Marsili, G. Rizzo, and C. Colosimo, "Diagnostic criteria for Parkinson's disease: from James Parkinson to the concept of prodromal disease," *Frontiers in Neurology*, vol. 9, no. MAR, 2018, doi: 10.3389/fneur.2018.00156.

[97]   "Parkinson's Disease – Symptoms, Diagnosis and Treatment." [Online]. Available: https://www.aans.org/en/Patients/Neurosurgical-Conditions-and-Treatments/Parkinsons-Disease. [Accessed: 29-Aug-2020].

[98] P. Piccini and D. J. Brooks, "New developments of brain imaging for Parkinson's disease and related disorders," *Movement Disorders*, vol. 21, no. 12, pp. 2035–2041, 2006, doi: 10.1002/mds.20845.

[99] D. J. Brooks, V. Ibanez, G. V. Sawle, *et al.*, "Striatal D2 receptor status in patients with Parkinson's disease, striatonigral degeneration, and progressive supranuclear palsy, measured with 11C-raclopride and positron emission tomography," *Annals of Neurology*, vol. 31, no. 2, pp. 184–192, 1992, doi: 10.1002/ana.410310209.

[100] K. L. Marek, J. P. Seiby, S. S. Zoghbi, *et al.*, "[123I] β-CIT/SPECT imaging demonstrates bilateral loss of dopamine transporters in hemi-Parkinson's disease," *Neurology*, vol. 46, no. 1, pp. 231–237, 1996, doi: 10.1212/WNL.46.1.231.

[101] J. Zhang and L. Chew-Seng Tan, "Revisiting the medical management of Parkinson's disease: levodopa versus dopamine agonist," *Current Neuropharmacology*, vol. 14, no. 4, pp. 356–363, 2016, doi: 10.2174/1570159x14666151208114634.

[102] P. A. L. Witt and S. Fahn, "Levodopa therapy for Parkinson disease: a look backward and forward," *Neurology*, vol. 86, no. 14, pp. S3–S12, 2016, doi: 10.1212/WNL.0000000000002509.

[103] National Collaborating Centre for Chronic Conditions (Great Britain). "Parkinson's disease: national clinical guideline for diagnosis and management in primary and secondary care." *Royal College of Physicians*, 2006.

[104] R. M. deSouza and A. Schapira, "Safinamide for the treatment of Parkinson's disease," *Expert Opinion on Pharmacotherapy*, vol. 18, no. 9, pp. 937–943, 2017, doi: 10.1080/14656566.2017.1329819.

[105] "NOURIANZ (istradefylline) for the Treatment of Parkinson's Disease." [Online]. Available: https://www.clinicaltrialsarena.com/projects/nourianz-istradefylline/. [Accessed: 07-Sep-2020].

[106] J. J. Ferreira, A. Lees, J. F. Rocha, W. Poewe, O. Rascol, and P. Soares-da-Silva, "Opicapone as an adjunct to levodopa in patients with Parkinson's disease and end-of-dose motor fluctuations: a randomised, double-blind, controlled trial," *The Lancet Neurology*, vol. 15, no. 2, pp. 154–165, 2016, doi: 10.1016/S1474-4422(15)00336-1.

[107] S. H. Fox, R. Katzenschlager, S. Y. Lim, *et al.*, "The movement disorder society evidence-based medicine review update: treatments for the motor symptoms of Parkinson's disease," *Movement Disorders*, vol. 26, no. SUPPL. 2011, doi: 10.1002/mds.23829.

[108] A. Thomas, L. Bonanni, F. Gambi, A. Di Iorio, and M. Onofrj, "Pathological gambling in Parkinson disease is reduced by amantadine," *Annals of Neurology*, vol. 68, no. 3, pp. 400–404, 2010, doi: 10.1002/ana.22029.

[109] T. Müller, "Catechol-O-methyltransferase inhibitors in Parkinson's disease," *Drugs*, vol. 75, no. 2. pp. 157–174, 2015, doi: 10.1007/s40265-014-0343-0.

[110] W. Birkmayer, P. Riederer, L. Ambrozi, and M. B. H. Youdim, "Implications of combined treatment with 'madopar' and l-deprenil in Parkinson's

disease a long-term study," *Lancet*, vol. 309, no. 8009, pp. 439–443, 1977, doi: 10.1016/S0140-6736(77)91940-7.

[111]   S. E. Pålhagen and E. Heinonen, "Use of selegiline as monotherapy and in combination with levodopa in the management of Parkinson's disease: perspectives from the MONOCOMB study," *Prog. Neurother. Neuropsychopharmacol.*, vol. 3, no. 1, pp. 49–71, 2008, doi: 10.1017/S174823210700002X.

[112]   A. Agarwal, A. Verma, and M. Khari, "Comparative assessment of machine learning methods for early prediction of diseases using health indicators." *Approaches to Human-Centered AI in Healthcare. IGI Global*, 160–186, 2024.

[113]   S. Saif, P. Das, S. Biswas, M. Khari and V. Shanmuganathan. "HIIDS: Hybrid intelligent intrusion detection system empowered with machine learning and metaheuristic algorithms for application in IoT based healthcare." *Microprocessors and Microsystems,* vol. 118, pp. 104622, 2022.

[114]   R. Satheeshkumar, K. Saini, A. Daniel and M. Khari. "5G—Communication in healthcare applications." *Advances in Computers*, vol. 127, pp.485–506, 2022.

[115]   Z. N. Barabasi and A.L. Oltvai, "Network biology: understanding the cell's functional organization," *Nature Review Genetics*, vol. 5, pp. 101–113, 2004.

[116]   F. Jiang, Q. Wu, S. Sun, G. Bi, and L. Guo, "Identification of potential diagnostic biomarkers for Parkinson's disease," vol. 9, pp. 1460–1468, 2019, doi: 10.1002/2211-5463.12687.

[117]   E. Oerton and A. Bender, "Concordance analysis of microarray studies identifies representative gene expression changes in Parkinson's disease: a comparison of 33 human and animal studies," pp. 1–14, 2017, doi: 10.1186/s12883-017-0838-x.

[118]   Z. Yue, I. Arora, E. Y. Zhang, V. Laufer, S. L. Bridges, and J. Y. Chen, "Repositioning drugs by targeting network modules: a Parkinson's disease case study," vol. 18, no. Suppl 14, pp. 17–30, 2017, doi: 10.1186/s12859-017-1889-0.

Chapter 4

Amyotrophic lateral sclerosis (ALS) disease: genes, molecular pathology and diagnostic treatment

Srishti Seth[1], Muniraj Gupta[2], Aman Verma[1],
Md. Zubbair Malik[3], Nidhi Verma[4] and
Saurabh Kumar Sharma[2]

Abstract

The debilitating neurodegenerative disease, amyotrophic lateral sclerosis (ALS) is peculiarized by the dysfunctioning of upper and lower motor neurons. It is an intensifying disease that becomes fatal after a few years of onset. There is no curative therapy or drug available which can stop or reverse the disease progression. The available drugs can only slow the disease progression. This chapter covers the studies made in the past in understanding the pathological mechanisms and molecular genetics as well as in diagnosing and managing the disease. It also encloses the recent growth and improvement in the field concerning genetics, pharmacological, neuroimaging, and stem cell therapy in finding the cure for the disease and how the identification of biomarkers is important to understand the signal propagation mechanism with its strength in the network.

Keywords: Amyotrophic lateral sclerosis (ALS); Neurodegenerative disorder; Network medicine; Biomarkers

4.1 Introduction

Amyotrophic lateral sclerosis (ALS), originally termed Lou Gehrig's disease (named after American baseball player Henry Lou Gehrig who was diagnosed with this during 1930), [1] is a fatal neurodegenerative disorder of the human motor system in which the motor neurons become incapable of sending signals to parts of

[1]School of Computational and Integrative Sciences, Jawaharlal Nehru University, India
[2]School of Computer and Systems Sciences, Jawaharlal Nehru University, India
[3]Department of Genetics and Bioinformatics, Dasman Diabetes Institute, Kuwait
[4]Department of Microbiology, Ram Lal Anand College, University of Delhi, India

the brain and the spinal cord that control the voluntary muscles which help in maintaining body posture, stabilizing and strengthening joints, maintaining normal body temperature, etc. ALS is a progressive disease [2], which means it gets worse with time. Motor neurons are the ones extending from the brain toward the spinal cord and then projecting outside of the spinal cord to the muscles and glands throughout the body. There are mainly two types of motor neurons: upper motor neurons and lower motor neurons. Upper motor neurons are the nerve cells originating in the motor region connected to the brain stem and carrying signal to the spine. They often transmit the signals to the lower motor neuron through a neurotransmitter glutamate, detected by glutamatergic receptors but they do not carry signals to the effector muscles. Lower motor neurons are the nerve cells having origin in the spinal column and carrying signals from the upper motor neuron and transmitting it to the effector muscles by connecting the spinal column to the brain [3]. Few of the cases (approximately 5%) of ALS are caused by the transfer of genes from one of the parents which follow Mendelian autosomal inheritance pattern. This type of ALS is called familial ALS (FALS). The other 90–95% approximate number of cases of sporadic ALS (SALS) have as such no definitive cause [4]. Mutations in genes like *SOD1* [5], *TARDBP* [6], *FUS* [7], *C9orf72* [8], and several others are linked to the disease. Although there is no permanent treatment of ALS, Riluzole (Rilutek) and Edaravone (Radicava) are the two drugs approved by U.S. Food and Drug administration for the slowing down the disease progression [3]. In this review, we have discussed the underlying mechanisms of ALS disease and the present advances and future possibilities in finding the cure of the disease.

4.2   Epidemiology

National ALS registry is an authorized population-based registry in United States established by US Congress in 2008 under the ALS Registry Act to gather information about the ALS cases throughout the country and work towards the cause and cure of the disease [9]. This registry is maintained by Centers for Disease Control and Prevention (CDC) which works with the aid of Agency for Toxic Substances and Disease Registry (ATSDR) [10]. This registry aims towards maintaining the occurrence and frequency record of the ALS patients within the country and understanding the demographics, environmental risk factors, and occupational risk factors associated with the patients [11,12]. ATSDR uses two approaches to identify ALS cases, firstly, an algorithm on the basis of three databases (Medicare [13], Veterans Health administration [14], Veterans Benefits Administration [15]), International Classification of Disease (ICD) code for ALS, frequency of visits to a neurologist, prescription drug use [16,17]. The ICD-10 code categorizes ALS cases as "definitive ALS", "possible ALS", and "not ALS". Just "definitive ALS" cases enter within the registry and the possible ones are evaluated for few subsequent years to become definite. Second approach is the online portal which facilitates individuals suffering from ALS to self-register themselves to help in the survey [16].

A total of 16,583 definite ALS cases, with a prevalence of 5.2 per 100,000 population, were identified in the 2015 ALS registry report which was prepared from January 1 to December 31, 2015. The annual incidence was found to be 2–3 per 100,000 person-years [18]. These findings were very similar to the previous year reports (5.0 per 100,000; 15,927 cases in 2014, and 5.0 per 100,000; 15,908 cases in 2013) [16]. The highest prevalence (20.0 per 100,000) was among people of 70–79 years of age while the lowest (0.5 per 100,000) was among the 18–39 years aged people. It was also 1.7 times more prevalent in males than females. The prevalence in whites (5.5 per 100,000) is double to that in blacks (2.3 per 100,000). Among the four census regions of United States, prevalence rate was in the order; Midwest (5.5), Northwest (5.1) followed by South (4.7) and West (4.1) [16].

In Europe, first registry was established in Scotland in 1989 aiming to study the aspects of motor neuron disease in populations. Later, similar ALS registries were built in Lancashire, England, Ireland, and three regions of Italy namely, Piemonte and Val D' Aosta, Lombardia, Puglia. These six registries worked on the 24 million European population dataset for the period of 2 years, 1998–1999. About 1,028 ALS cases were reported with the occurrence of 2.6 per 100,000 person-years with male and female incidence being 3.0 and 2.4 per 100,000 person-years, respectively. However, the occurrence was much lower in Asia with lowest being in East Asia and South Asia, 0.89 and 0.79 per 100,000 person-years, respectively [19].

4.3   Signs and symptoms

ALS causes motor neuron break down in parts of the brain and spinal column. Thus, the brain can no longer send messages to the body muscles which ultimately results in weakening of the muscles, a condition called atrophy [2]. Gradually, the muscles cease to work, and person loses control over his/her movements.

ALS symptoms can be categorized into lower and upper limb onset ALS and bulbar onset ALS [20]. In 75–80% of the patients, it occurs as limb onset in which the first signs appear in lower limb movements, which include stumbling and tripping while walking and running, foot drop (difficulty in lifting first part of the foot), and slapping gait (striking of heel first on the ground followed by foot). The upper limb onset includes complaints like cramping, stiffening of the fingers, weakening of finger dexterity, wasting of intrinsic hand muscles. In other 20–25% of the patients, it starts with bulbar onset in which patients initially notice problem in speech (dysarthria) and swallowing (dysphagia) [21]. They face the problem in chewing food which increases the risk of choking [3]. The calories in them are burnt at a faster rate than the normal people, as a result of which they lose weight rapidly resulting in their bodies becoming fragile [4].

Most of the ALS individuals have their mental ability intact and can perform mental processes such as thinking, reasoning, solving problems, etc. They are well aware of their disease progression which may lead them to anxiety and depression.

As the disease progresses, patients suffer through breathing problem (dyspnea) [3]. Eventually, diaphragm and chest muscles also become weak and they need

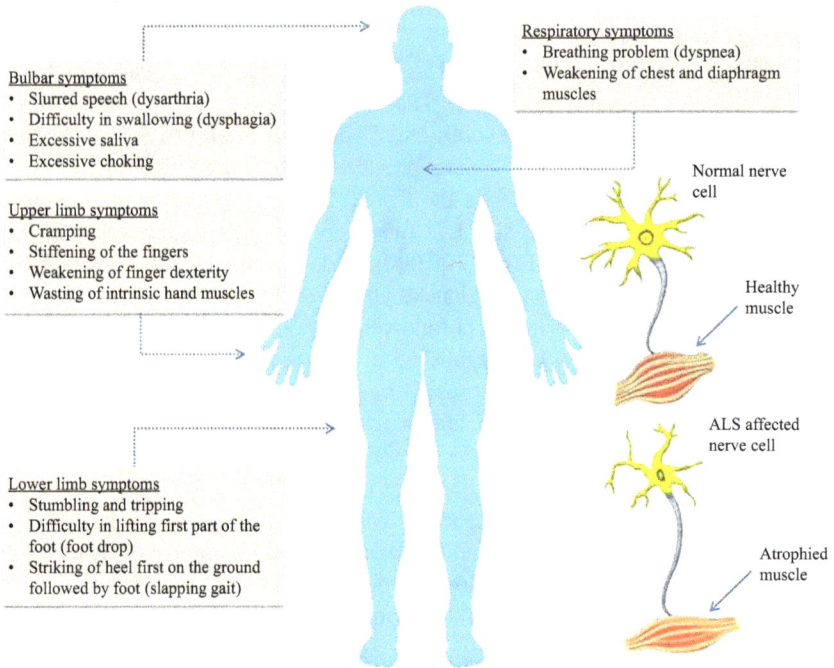

Figure 4.1   Diagrammatic representation of signs and symptoms of amyotrophic lateral sclerosis (ALS) affecting various levels of human system. Also, the comparison of a healthy nerve cell with the ALS affected nerve cell: showing the degeneracy in axon terminals and wasted muscle in ALS affected cells.

support of a ventilator to breathe. Loss in breathing results in death within 3–5 years of diagnosis on an average. But, some people may even survive for more than 10 years with the disease [2] (Figure 4.1).

4.4   Classification

ALS can be divided into sporadic and familial.

Sporadic ALS (SALS): 95% of ALS cases are sporadic [2]. Sporadic disease means occurring randomly without any clear cause and no family history. It is most widely hypothesized that sporadic ALS triggers due to the interactions between environmental, genetic, as well as risk-factors associated with old age. So far, the only established environmental risk factor is smoking. Some studies indicate that active smoker is twice as likely as non-smoker to develop ALS while former smoker is intermediately at risk [4]. Intensive sport and physical exertion [22] are some of the putative risk factors of ALS, though their exact reason and validation is yet unknown.

Familial ALS (FALS): Only about 5% of the cases of ALS are familial, i.e., inheriting the disease from either paternal or maternal gametes [3]. It often follows the Mendelian autosomal pattern of inheritance. Mutation in the genes is found to be responsible for familial ALS. ALSoD is a database website in which the curated list of genes having mutation, genotype–phenotype correlation, and analysis tools are maintained [4,23]. Some of the commonly found genes responsible for familial ALS are *SOD1*, *C9orf72*, and *TARDBP, FUS, AGN, OPTN* (Figure 4.2; Table 4.1).

This distinction does not hold very true now because of the absence of any clear differences between patients with and without genetic mutation [24]. In case of unexpected death of any family member or close relative, error in diagnosis or low gene penetrance, there is a lack of information of family history which can attribute for familial ALS [25]. It has also been found that the relatives of sporadic ALS have eight times higher chances of risking themselves to ALS than the general

Table 4.1 Chromosomal locus, protein function, inheritance pattern, and type of mutation of the most common genes-causing ALS

Gene	Protein	Chromosomal locus	Function	Inheritance	Mutation type
SOD1	SOD1 (superoxide dismutase 1)	21q22.11	Superoxide metabolism	Autosomal dominant, Autosomal recessive	Point mutations
TARDBP	TAR (Trans-active response element) DNA-binding protein 43	1p36.22	DNA/RNA binding Transcription and splicing regulation	Autosomal Dominant	Missense and nonsense
FUS	FUS (fusion in sarcoma) RNA-binding protein	16p11.2	DNA/RNA binding Transcription and splicing regulation	Autosomal dominant, Autosomal recessive	Missense
C9orf72	C9orf72 (Chromosome 9 open reading frame 72)	9p21.2	Guanine nucleotide exchange factor (GEF) activity	Autosomal Dominant	GGGGCC expansion
UBQLN2	Ubiquilin 2	Xp11.21	Proteasome-mediated protein degradation	Autosomal Dominant, X-linked	Missense
ANG	Angiogenin	14q11.2	rRNA synthesis stimulation tRNA modification vascularization	Autosomal Dominant	Missense
OPTN	Optineurin	10p13	Gogli complex maintenance Vesicular trafficking	Autosomal dominant, Autosomal recessive	Missense, nonsense, deletion

population. Siblings and offspring of apparent sporadic ALS patients have 0.5% and 1% chance, respectively, of developing ALS [26]. The mean age of sporadic ALS onset is 65 and that of familial ALS is 46–55 years [4]. But the incidence rapidly decreases after 80 years of age [27].

4.5   Molecular genetics and pathophysiological mechanisms

ALS is a multifactorial disease with different pathophysiologic processes, which commonly leads to progressive loss of motor neurons. Glutamate-induced excitotoxicity was considered to be one of the causes of ALS. The postsynaptic receptors N-methyl-D-aspartate (NMDA) and α-amino-3-hydroxy-5-methyl-4-isoxazolepropionic acid (AMPA) are induced by glutamate, an excitatory neurotransmitter in CNS on the postsynaptic membrane [28,29]. When these receptors are excited excessively, glutamate-induce excitotoxicity occurs which results in calcium-dependent enzymatic activation leading to neurodegeneration [30] (Regan *et al.*, 1995). It can also generate free radicals, which can not only damage intracellular organelles but also upregulate proinflammatory mediators, resulting in neurodegeneration [31].

Sometimes, microglial cells which provide immunity against germs and damaged cells, also destroy healthy motor neurons by secretion of neurotoxic factors and insufficient neurotrophic factors release [32]. This can be one of the reasons of ALS. Mitochondrial abnormalities [33], sodium–potassium pump dysfunction and axonal transport system disruption [34] are some of the other implications of ALS.

4.6   *SOD1* mutations

SOD1 gene is located on the human chromosome 21q22.11 [35], encoding for copper/zinc ion-binding superoxide dismutase (SOD1) [21] which is a homodimeric metalloenzyme [5]. It catalyzes the conversion of superoxide radical (O_2^-) to hydrogen peroxide, which further converts into water by glutathione peroxidase or catalase [5]. This loss of function was known to cause oxidative stress and give rise to ALS, but this was abandoned [36] and it was implicated that *SOD1* undergoes toxic gain of function mutation, with free radical generation [5] and thus leads into protein's failure and eventually cell death [35]. Approximately 180 mutations till date have been accounted for *SOD1* which include single point mutations, deletions, insertions [35]. Several human mutant *SOD1* transgenic strains of mice have been found to have overexpressed mutant *SOD1* activity leading to increased dismutase activity and thus toxic gain of function [37]. However, the exact pathophysiology is still unclear [21]. But some recent studies show that loss of function of SOD1 can also play a role in ALS [37]. *SOD1* mutations accounts for 10–20% of FALS [4] and 1–2.5% of total ALS cases (Figure 4.2) [38].

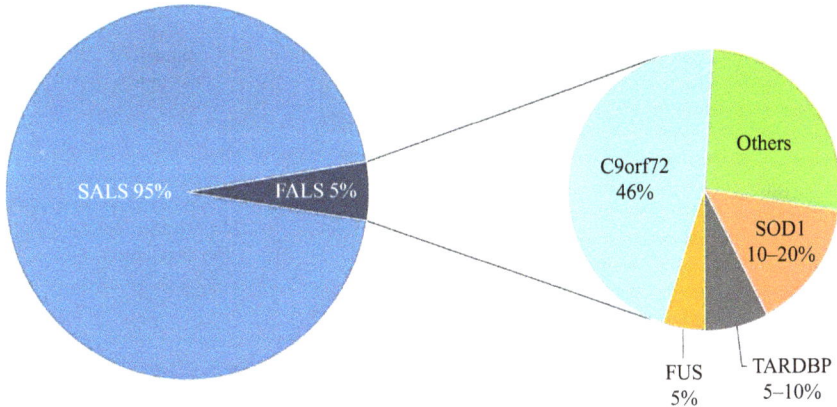

Figure 4.2   Chart showing the percentage of SALS (sporadic ALS) and FALS (familial ALS) cases among the total ALS cases. Also, the percentage of mutations in common genes like SOD1 (10–20%), TARDBP (5–10%), FUS (5%), and Coorf72 (46%) contributing to total FALS cases.

4.7   *TARDBP* and *FUS* mutations

The trans-active response element DNA binding protein *(TARDBP)* gene present on human chromosome 1p36.22 encodes the TAR DNA-binding protein-43 (TDP-43). This 141 amino acid long, highly conserved protein is ubiquitously present in variety of tissues including human brain [6]. It represses the transcription of human HIV-1 by binding to TAR–DNA [39], and is also a part of complex regulating splicing in cystic fibrosis transmembrane conductance regulator gene *(CFTR)*[40] and apoA-II gene [41]. Missense and deletion mutations have been found in most of the SALS and FALS patients. In SALS patients, TDP-43 was reported to be a significant component of ubiquitinated cytoplasmic protein aggregates which are normally present in the nucleus. Highly conserved sequences of *TARDBP* undergoing mutations were the evidence of pathogenicity in ALS patients as these mutations were not evident in controls [42]. It accounts for 5–10% mutations in FALS and 1–4% of total ALS cases (Figure 4.2).

FUS located on human chromosome 16p11.2 encodes fusion in sarcoma (FUS) protein (also known as TLS (translated in liposarcoma)) [43]. There have been around 50 mutations in FALS and SALS [44]. Mutations in *FUS* accounts 5% of FALS mutations (Figure 4.3) [21]. FUS regulates several RNA metabolism processes [44]. FUS is involved in regulating gene expression, transcription, splicing, transport as well as translation. The microRNA processing, and RNA maturation and splicing, is done with both FUS and TDP-43 [7,43]. It was reported that mutations in *FUS* occurred in an autosomal dominant form of ALS, ALS-6 [7,43].

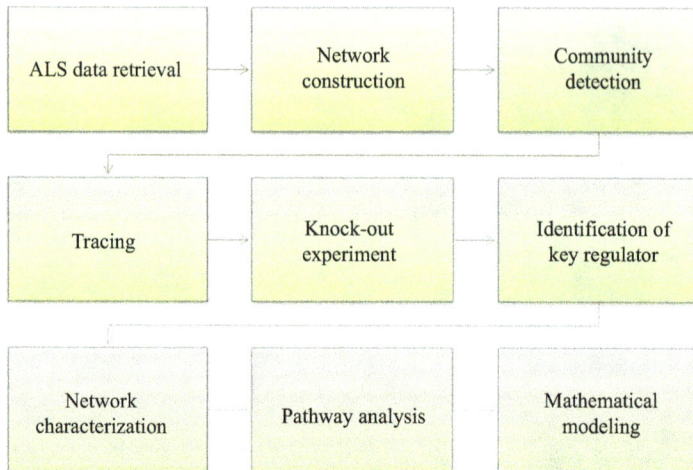

Figure 4.3   Flowchart of the proposed method of identifying novel key regulator.

4.8   *C9orf72* mutations

Chromosome 9 open reading frame 72 (C9orf72) is a protein encoded by *C9orf72*, which is located on the human chromosome 9p21.2 [45]. The first intron of hexanucleotide (GGGGCC) repeat expansion in the first intron of the *C9orf72* gene was found to be the cause of ALS [46,47]. Dipeptide repeat proteins (DPRs), associated with non-ATG, were found in 46% of FALS patients and 21.1% of SALS patient (Figure 4.2) [4]. But the function of this repeat expansion remains unknown.

4.9   Other genes

Mutations in *UBQLN2* gene is known to cause X-linked dominant FALS [48]. *ANG* encoding angiogenin protein, RNAse A is a hypoxia responsive gene regulating RNA transcription. Missense mutation in highly conserved residues of *ANG* results in the development of ALS [49]. Mutations in *ANG* account for at least 1% of FALS [21].

Mutations in *OPTN* gene encoding optineurin is the causative of the typical clinical phenotype of ALS [21]. The missense and nonsense mutations in this gene stop the activation of nuclear kappa factor (NFκB) which results in upregulation of *OPTN* and thus cell death [50].

4.10   Overlap with frontotemporal dementia

ALS and FTD both are the neurodegenerative disorders, deteriorating the brain and affecting of few common genes. Recent advances in neuropathology and genetics show that ALS and FTD are overlapping diseases because of the common molecular genetics. In most ALS cases and nearly half FTD cases, TDP-43 inclusions involve the typical neuropathological mechanism [51]. Expansion on the first intron of *C9orf72* gene was also discovered in both the diseases [52]. Neurophysiological assessments show that 20–25% of the ALS patients fulfill the criteria of probable or definitive FTD. Conversely, motor neuron disease tends to appear in about 15% of FTD patients [21].

Additionally, 2-fluoro-2-deoxy-D-glucose positron emission tomography (FDG-PET) [53] determines the frontotemporal hypometabolism in both ALS and FTD–ALS patients. Also, structural abnormalities such as frontotemporal atrophy are common in both ALS and FTD–ALS which are identified by voxel-based morphometry MRI [21].

4.11   Diagnosis

In the 1980s, the idea to understand and standardize the diagnostic criteria of ALS emerged and the El Escorial criteria (EEC) were established in 1991. In April 1998, these criteria were revised at Airlie Conference center, Warrenton, Virginia, in order to increase the sensitivity. These criteria not only impacted the clinical practice but also clinical research to a much larger extent. El Escorial World Federation of Neurology (WFN) criteria recognized four regions of the body which are bulbar, cervical, thoracic, and lumbosacral.

The WFN criteria for the diagnosis of ALS are [54,55]:

- Signs of lower motor neuron (LMN) degeneration by clinical, electro-physiological, or neuropathologic examination,
- Signs of upper motor neuron (UMN) degeneration by clinical examination, and
- Progressive spread of signs within a region or other regions, together with the absence of:
  o Electrophysiological evidence of other disease processes that might explain the signs of LMN and/or UMN degenerations; and
  o Neuroimaging evidence of other disease processes that might explain the observed clinical and electrophysiological signs.

The EEC defined ALS diagnostic certainty into five categories: definitive, clinically probable, clinically probable with laboratory support, clinically possible and clinically suspected. There should be both upper and lower motor neuron signs in at least three regions to fall under the category of definitive ALS.

Another improved diagnostic criteria proposed in 2008 was Awaji Shami criteria based on the application of neurophysiological data in the diagnosis of ALS. These criteria are based on [54]:

- Evidence of lower motor neuron loss.
- Evidence of reinnervation.
- Fibrillation and sharp waves or fasciculation potentials.

These criteria reduced the diagnostic certainty from five in EEC to three categories, i.e., clinically definitive, probable, and possible.

ALS disease in its early stages is similar to wide variety of other diseases which are comparatively treatable. So, there are tests to exclude the possibility of any other disease and aid in ALS diagnosis.

Once a patient is suspected to have ALS, electromyography (EMG) is done to diagnose it critically. It is a special technique that measures the electrical activity of the muscle fibers and checks for pattern of consistency with ALS. These patterns can be seen due to acute and chronic reinnervation and denervation of the affected muscles. Another test for ALS is nerve conduction study (NCS) that measures the ability of the nerve to send signals to nerve to or muscle.

Another test which is common for any brain or spinal cord suspected disease is magnetic resonance imaging (MRI) in which the detailed image of the brain and spinal cord is produced to diagnose ALS.

Blood test using biomarkers is done routinely to identify the disease that could mimic ALS. Spinal tap or lumbar puncture may be performed in cases of pure UMN or LMN in which a needle is inserted into the back between lower two vertebrae to extract a small amount of cerebrospinal fluid for testing.

4.12 Management

ALS, being a progressive disease, requires correct diagnosis and management. With the help of medical references, several research and review papers, meta-analysis, and guidelines of The European Federation of Neurological Societies (EFNS), these certain recommendations came into picture with the consensus for the appropriate management of the suffering patients [56]:

- Early diagnosis of the disease by an experienced neurologist as soon as suggestive symptoms observed.
- Diagnosis should include clinical as well as paraclinical examinations, neurophysiological test being the highest priority.
- Inform patients about the disease and the diagnostic process by a consultant having good knowledge about the patient as well as the disease. Providing unwanted information should be avoided and upholding any diagnosis should be avoided.
- Patients as well the relatives should receive proper service care from the multidisciplinary team for the optimized diagnosis and management of the disease and the patient.
- Caregivers/relatives should be under regular support and their needs should be addressed. They should be provided with regular counseling as they are the primary resource for the patients.
- Except for patients with progressive muscular dystrophy and primary lateral sclerosis, Riluzole treatment should be administered soon after diagnosis.
- Sialorrhea, bronchial secretions, cramps, and spasticity are some of the symptoms which should be controlled to avoid complications.
- Respiratory insufficiency should be checked and gastronomy tubes should be placed before the development of the same. Percutaneous endoscopic gastronomy is used to improve nutrition.

- Autonomous stability of the patient is very important and continuous efforts should be made to maintain the same.
- A palliative and end-of-life care should be given.

4.13   Treatment

Riluzole and Edaravone are the two drugs approved by U.S. Food and Drug administration for the slowing down the progression of the disease [3]. Furthermore, antisense oligonucleotide [57,58] and virus-delivered gene therapies [58] are useful not only to slow the progression but also to improve the symptoms. Recent advances such as intrathecal injection of mesenchymal cells derived from bone marrow to see their effect on neurotrophic factors overexpression [59] and numerous other stem cell therapies are underway. The development of the disease therapy has many limiting factors, lack of biomarkers being the major [17].

4.14   Recent advances

4.14.1   Pharmacological

Besides the two drugs Riluzole (Rilutek) and Edaravone (Radicava) used for the slowing down the progression of the disease, scientists at University of Alberta have identified a new drug "telbivudine" [60] that targets SOD1 protein and reduces its toxic effect due to misfolding and dysfunctioning.

4.14.2   Neuroimaging

To date, MRI's ability to rule out potential disease causes was the neuroimaging diagnostics phase of ALS [61]. However, nowadays, multimodal neuroimaging has made a significant contribution in confirming ALS as cerebral multisystem disorder. Voxel-based morphometry is used in cerebral atrophy detection [62]. Magnetic resonance spectroscopy has the ability to detect upper motor neuron dysfunction by detecting the reduced primary motor cortex N-acetylaspartate [63,64]. In the corticospinal system of ALS patients, another tool called diffusion tensor can be used to detect reduced fractional anisotropy. The *SOD1* gene mutation brings modifications in the posterior limb of the internal capsule which can also be detected using diffusion tensor imaging [65]. A new diffusion imaging technique, Q-ball imaging, is even better than the diffusion tensor technique in detection of corticospinal tract damage which helps in distinguishing highly progressive ALS from slow progressive forms of ALS [66,67].

4.14.3   Stem cell therapy

The advent of stem cell therapy is a new potential approach to treat ALS. Induced pluripotent stem cells (iPSCs) derived from human tissues have the ability to develop into a type of support cells called astrocytes [68]. iPSC derived from ALS patients are an important source to understand the cause of ALS. Affected upper and lower motor neurons in ALS can be replaced by motor neurons created from iPSC. Identical neurons developed from iPSC can be used to find new drugs and can also be used as biomarkers [69].

4.14.4   Genetic

RNA-binding proteins contribute to gene translation and alterations in any of these proteins can alter the RNA metabolism and lead to generation of toxic protein aggregates within motor neurons which contribute to motor dysfunction and ultimately paralysis and death [70]. A research used IBM Watson-based artificial intelligence to classify five new RNA-binding proteins namely, hnRNPU, Syncrip, RBMS2, NUPL2, and CAPRIN1 that had been altered in ALS. It provides an evidence of importance of RNA metabolism in ALS [71]. ALS, in a recent study have been found to share a common unusual mutation with ALS–FTD patients in low complexity domain of the RNA-binding protein TIA1 (T-cell restricted intracellular antigen-1) gene [72].

In addition to this, our lab has also proposed to develop a method to identify novel key regulators (drivergene) of ALS genes from the network constructed from experimental gene expression data using network theory. The constructed network properties will be systematically studied within the network theoretical framework to understand how ALS network dynamics become connected to disease dynamics. Then, some of the theoretically and experimentally predicted genes will be targeted with chemical compounds of possible potential drugs, and from this data we will study the affected genes in the ALS network from this targeted gene expression data. The propagated signal of this perturbation signal induced by this drug at "local" and "global" levels will be studied systematically as the future work of this project. This analysis will give us a method to predict ALS biomarkers which will be important in ALS development. From this information, we will search for key pathways that regulate the ALS disease and mathematically model it within deterministic and stochastic frameworks. The theoretical predictions of the key pathways will provide us crucial information how the predicted ALS biomarkers control the ALS networks. Moreover, the study of the gene expression data within the framework of dynamical network theory may able to highlight the importance of the key genes with pathways during the ALS progression with the targeting compound which may help in the identification of disease and targeting compound specific biomarkers (Figure 4.3).

4.15   Conclusion

ALS is currently an implacable and incurable neurodegenerative disease leading to patient's inevitable death. Although there are numerous therapeutic strategies and approved drugs to slow down the disease progression, but a permanent cure is still to be found. At present, there is an ongoing ceaseless research to have a better understanding of the underlying mechanism and find the biomarkers of ALS. However, network theory can be a promising approach in identifying the drugtargetable biomarkers. Mathematical modeling of the biomarkers/key regulators using deterministic and stochastic framework help in understanding their role in ALS. Signal propagation can be analyzed to understand the limit of signal propagation which determines the effectiveness on the pathways.

Acknowledgments

S.K.S. and N.V. are financially supported by University Grant Commission (UGC) and Indian Council of Medical Research (ICMR), New Delhi, India.

Author contributions

S.K.S., N.V., and M.Z.M. conceptualized the work. S.S., NV., AV, S.K.S., and M. K.H. did the preparation of the figures. S.S., A.V., N.V., M.K.H, M.Z.M., and S.K. S wrote the manuscript. All the authors read and approved the manuscript.

Competing interests

The authors declare that they have no competing interests.

References

[1] Sibat, Humberto Foyaca, and Lourdes de Fátima Ibañez-Valdés, eds. Update on amyotrophic lateral sclerosis. BoD–Books on Demand, 2016.

[2] Martin, Sarah, Ahmad Al Khleifat, and Ammar Al-Chalabi. "What causes amyotrophic lateral sclerosis?" *F1000Research*, 6, p. 371, 2017.

[3] Oskarsson, Björn, Tania F. Gendron, and Nathan P. Staff. "Amyotrophic lateral sclerosis: an update for 2018." In *Mayo clinic proceedings*, vol. 93, no. 11, pp. 1617–1628. Elsevier, 2018.

[4] C. Armon, "Amyotrophic lateral sclerosis: practice essentials," *Medscape*, 2018. [Online]. Available: https://emedicine.medscape.com/article/1170097-overview. [Accessed: 11-Apr-2019].

[5] D. R. Rosen, Teepu Siddique, David Patterson, *et al.*, "Mutations in Cu/Zn superoxide dismutase gene are associated with familial amyotrophic lateral sclerosis," *Nature*, vol. 362, no. March, pp. 59–62, 1993.

[6] I. R. A. Mackenzie and R. Rademakers, "The role of transactive response DNA-binding protein-43 in amyotrophic lateral sclerosis and frontotemporal dementia," *Curr. Opin. Neurol.*, vol. 21, no. 6, pp. 693–700, 2008, doi:10.1097/WCO.0b013e3283168d1d.

[7] J. Vance, C. Rogelj, B. Hortobágyi, T. De Vos, K. J. Nishimura, A. L. Sreedharan, J., *et al.*, "Mutations in FUS, an RNA processing protein, cause familial amyotrophic lateral sclerosis type 6," *Science (80).*, vol. 323, no. 5918, pp. 1208–1211, 2009.

[8]   M. DeJesus-Hernandez, Ian R. Mackenzie, Bradley F. Boeve *et al.*, "Expanded GGGGCC hexanucleotide repeat in noncoding region of C9ORF72 causes chromosome 9p-linked FTD and ALS," *Neuron*, vol. 72, no. 2, pp. 245–256, October 2011, doi:10.1016/ j.neuron.2011.09.011.

[9]   "National Amyotrophic Lateral Sclerosis (ALS) Registry – Home." [Online]. Available: https://www.cdc.gov/als/Default.html. [Accessed: 04-Nov-2020].

[10] *"Agency for Toxic Substances and Disease Registry."* [Online]. Available: https://www.atsdr.cdc.gov/. [Accessed: 04-Nov-2020].

[11] P. Mehta, *"Prevalence of amyotrophic lateral sclerosis – United States, 2010–2011,"* vol. 63, no. 7, pp. 2010–2011, 2014.

[12] K. Horton, "Prevalence of amyotrophic lateral sclerosis – United States, 2012," vol. 65, no. December 2013, pp. 2012–2013, 2019, doi:10.15585/mmwr.ss6508a1.

[13] "Medicare.gov: The Official U.S. Government Site for Medicare | Medicare." [Online]. Available: https://www.medicare.gov/. [Accessed: 04-Nov-2020].

[14] *"Veterans Health Administration."* [Online]. Available: https://www.va.gov/ health/. [Accessed: 04-Nov-2020].

[15] *"Veterans Benefits Administration Home."* [Online]. Available: https://benefits. va.gov/benefits/. [Accessed: 04-Nov-2020].

[16] Center for Disease Control (US), and Centers for Disease Control (US). *Morbidity and mortality weekly report: MMWR.* Vol. 53. US Department of Health, Education, and Welfare, Public Health Service, Center for Disease Control, 2004.

[17] B. Oskarsson, T. F. Gendron, and N. P. Staff, "Amyotrophic lateral sclerosis: an update," *Mayo Clin. Proc.*, vol. 93, no. 11, pp. 1617–1628, 2018, doi:10. 1016/j.mayocp.2018.04.007.

[18] P. Mehta, W. Kaye, J. Raymond, R. Punjani, T. Larson, and J. Cohen, "Prevalence of amyotrophic lateral sclerosis – United States, *MMWR. Morbidity and mortality weekly report*, 2015," vol. 67, no. 46, 2018.

[19] G. Logroscino, "Amyotrophic lateral sclerosis descriptive epidemiology: the origin of geographic difference," *Neuroepidemiology*, vol. 52, no. 1–2 pp. 93–103, 2019, doi:10.1159/000493386.

[20] P. H. Gordon, B. Cheng, I. B. Katz, *et al.*, "The natural history of primary lateral sclerosis," *Neurology*, vol. 66, no. 5, pp. 647–653, 2006, doi:10.1212/ 01.wnl.0000200962.94777.71.

[21] Kiernan, M. C., Vucic, S., Cheah, *et al.*, "Amyotrophic lateral sclerosis," *Lancet*, vol. 377, no. 9769, pp. 942–955, 2011.

[22] P. J. Harwood, C. A., McDermott, C. J., and Shaw, "Physical activity as an exogenous risk factor in motor neuron disease (MND): a review of the evidence," *Amyotroph. Lateral Scler.*, vol. 10, no. 4, pp. 191–204, 2009.

[23] R. Wroe, A. Wai-Ling Butler, P. M. Andersen, J. F. Powell, and A. Al-Chalabi, "ALSOD: the amyotrophic lateral sclerosis online database," *Amyotroph. Lateral Scler.*, vol. 9, no. 4, pp. 249–250, January 2008, doi:10. 1080/17482960802146106.

[24] A. Chiò, Stefania Battistini, Andrea Calvo, *et al.*, "Genetic counselling in ALS: facts, uncertainties and clinical suggestions," *Journal of Neurology, Neurosurgery & Psychiatry*, vol. 85, no. 5, pp. 478–485, 2014, doi:10.1136/jnnp-2013-305546.

[25] P. M. Andersen and A. Al-Chalabi, "Clinical genetics of amyotrophic lateral sclerosis: What do we really know?," *Nature Reviews Neurology*. vol. 7, no. 11, pp. 603–615, 2004, 2011, doi:10.1038/nrneurol.2011.150.

[26] M. F. Hanby, Kirsten M. Scott, William Scotton, *et al.*, "The risk to relatives of patients with sporadic amyotrophic lateral sclerosis," *Brain*, vol. 134, no. 12, pp. 3451–3454, 2011, doi:10.1093/brain/awr248.

[27] E. Logroscino, G., Traynor, B. J., Hardiman, O., Chiò, A., Mitchell, D., Swingler, R. J., *et al.*, "Incidence of amyotrophic lateral sclerosis in Europe," *Neurosurg. Psychiatry*, vol. 81, no. 4, pp. 385–390, 2010.

[28] J. C. Watkins, and R. H. Evans, "Excitatory amino acid transmitters," *Annu. Rev. Pharmacol. Toxicol.*, vol. 21, no. 1, pp. 165–204, 1981.

[29] P. R. Heath and P. J. Shaw, "Update on the glutamatergic neurotransmittor system and the role of excitotoxicity in amyotrophic lateral sclerosis," *Muscle & Nerve: Official Journal of the American Association of Electrodiagnostic Medicine*, vol. 26, no. 4, pp. 438–458, 2002, doi:10.1002/mus.10186.

[30] Brian Meldrum, John Garthwaite, "Excitatory amino acid neurotoxicity and neurodegenerative disease," *Trends Pharmacol. Sci.*, vol. 11, no. 9, pp. 379–387, 1990.

[31] P. Maher and J. B. Davis, "*The role of monoamine metabolism in oxidative glutamate toxicity,*" *Journal of Neuroscience*, vol. 16, no. 20, pp. 6394–6401, 1996.

[32] K. A. S. L. Van Den Bosch, "Astrocytes in amyotrophic lateral sclerosis: direct effects on motor neuron survival," *Journal of biological physics*, pp. 337–346, 2009, doi:10.1007/s10867-009-9141-4.

[33] S. Sasaki and M. Iwata, "Mitochondrial alterations in the spinal cord of patients with sporadic amyotrophic lateral sclerosis," *Journal of Neuropathology & Experimental Neurology*, vol. 66, no. 1, pp. 10–16, 2007.

[34] M. Sasaki, and S. Iwata, "Impairment of fast axonal transport in the proximal axons of anterior horn neurons in amyotrophic lateral sclerosis," *Neurology*, vol. 47, no. 2, pp. 535–540, 1996.

[35] O. Pansarasa, M. Bordoni, L. Diamanti, D. Sproviero, S. Gagliardi, and C. Cereda, "Sod1 in amyotrophic lateral sclerosis: 'ambivalent' behavior connected to the disease," *Int. J. Mol. Sci.*, vol. 19, no. 5, pp. 1–13, 2018, doi:10.3390/ijms19051345.

[36] D. A. Bosco, "The role of SOD1 in amyotrophic lateral sclerosis," *Nat. Educ.* vol. 8, no. 3, p. 4, 2015.

[37] R. A. Saccon, R. K. A. Bunton-Stasyshyn, E. M. C. Fisher, and P. Fratta, "Is SOD1 loss of function involved in amyotrophic lateral sclerosis?," *Brain*, vol. 136, no. 8, pp. 2342–2358, 2013, doi:10.1093/brain/awt097.

[38] A. M. G. Ragagnin, S. Shadfar, M. Vidal, M. S. Jamali, and J. D. Atkin, "Motor neuron susceptibility in ALS/FTD," *Front. Neurosci.*, vol. 13, no. JUN. Frontiers Media S.A., p. 532, 27 June 2019, doi:10.3389/fnins.2019.00532.

[39]   S. H. Ou, F. Wu, D. Harrich, L. F. García-Martínez, and R. B. Gaynor, "Cloning and characterization of a novel cellular protein, TDP-43, that binds to human immunodeficiency virus type 1 TAR DNA sequence motifs," *J. Virol.*, vol. 69, no. 6, pp. 3584–96, 1995.

[40]   E. Buratti, T. Dörk, E. Zuccato, F. Pagani, M. Romano, and F. E. Baralle, "Nuclear factor TDP-43 and SR proteins promote in vitro and in vivo CFTR exon 9 skipping," *EMBO J.*, vol. 20, no. 7, pp. 1774–1784, 2001, doi:10. 1093/emboj/20.7.1774.

[41]   P. A. Mercado, Y. M. Ayala, M. Romano, E. Buratti, and F. E. Baralle, "Depletion of TDP 43 overrides the need for exonic and intronic splicing enhancers in the human apoA-II gene," *Nucleic Acids Res.*, vol. 33, no. 18, pp. 6000–6010, 2005, doi:10.1093/nar/gki897.

[42]   E. Van Deerlin, V. M., Leverenz, J. B., Bekris, L. M., Bird, T. D., Yuan, W., Elman, *et al.*, "TARDBP mutations in amyotrophic lateral sclerosis with TDP-43 neuropathology: a genetic and histopathological analysis," *Lancet Neurol.*, vol. 7, no. 5, pp. 409–416, 2008.

[43]   P. Kwiatkowski, T. J., Bosco, D. A., Leclerc, A. L., Tamrazian, E., Vanderburg, C. R., Russ, C., *et al.*, "Mutations in the FUS/TLS gene on chromosome 16 cause familial amyotrophic lateral sclerosis," *Science (80)*, vol. 323, no. 5918, pp. 1205–1208, 2009.

[44]   H. An, Lucy Skelt, Antonietta Notaro, *et al.*, "ALS-linked FUS mutations confer loss and gain of function in the nucleus by promoting excessive formation of dysfunctional paraspeckles," *Acta Neuropathol. Commun.*, vol. 7, no. 1, p. 7, 2019, doi:10.1186/s40478-019-0658-x.

[45]   NCBI, "C9orf72 C9orf72-SMCR8 complex subunit [*Homo sapiens* (human)]," *National Institutes of Health, USA.gov.* [Online]. Available: https://www.ncbi.nlm.nih.gov/gene?linkname=protein_gene&from_uid= 365906244.

[46]   M. Dejesus-Hernandez, Ian R. Mackenzie, Bradley F. Boeve, *et al.*, "Expanded GGGGCC hexanucleotide repeat in noncoding region of C9ORF72 causes chromosome 9p-linked FTD and ALS," *Neuron*, vol. 72, no. 2, pp. 245–256, 2011, doi:10.1016/j.neuron.2011.09.011.

[47]   S. Rollinson *et al.*, *"A hexanucleotide repeat expansion in C9ORF72 is the cause of chromosome 9p21-linked ALS-FTD,"* *Neuron*, pp. 257–268, 2011, doi:10.1016/j.neuron.2011.09.010.

[48]   Deng Han-Xiang, Wenjie Chen, Seong-Tshool Hong, *et al.*, "Mutations in UBQLN2 cause dominant X-linked juvenile and adult-onset ALS and ALS/ dementia," *Nature*, vol. 477, no. 7363, pp. 211–215, 2011, doi:10.1038/ nature10353.

[49]   M. J. Greenway, Peter M. Andersen, Carsten Russ, *et al.*, *"ANG mutations segregate with familial and 'sporadic' amyotrophic lateral sclerosis,"* *Nature genetics*, vol. 38, no. 4, pp. 2005–2007, 2006, doi:10.1038/ng1742.

[50]   H. Maruyama, Hiroyuki Morino, Hidefumi Ito, *et al.*, "Mutations of optineurin in amyotrophic lateral sclerosis," *Nature*, vol. 465, no. 7295, pp. 223–226, 2010, doi:10.1038/nature08971.

[51]   M. Neumann, Deepak M. Sampathu, Linda K. Kwong *et al.*, "Ubiquitinated TDP-43 in frontotemporal lobar degeneration and amyotrophic lateral sclerosis," *Science (80-.)*., vol. 314, no. 5796, pp. 130–133, October 2006, doi:10.1126/science.1134108.

[52]   A. S. Lillo Patricia, M. José Manuel, V. Daniel, V. Renato, C. José Luis, and I. Agustín, "Overlapping features of frontotemporal dementia and amyotrophic lateral sclerosis," *Rev. Med. Chil.*, vol. 142, no. 7, pp. 867–879, 2014.

[53]   V. Rajagopalan and E. P. Pioro, "2-Deoxy-2-[18F]fluoro-D-glucose positron emission tomography, cortical thickness and white matter graph network abnormalities in brains of patients with amyotrophic lateral sclerosis and frontotemporal dementia suggest early neuronopathy rather than axonopathy," *Eur. J. Neurol.*, vol. 27, no. 10, pp. 1904–1912, October 2020, doi:10.1111/ene.14332.

[54]   J. Costa, M. Swash, and M. De Carvalho, "Awaji criteria for the diagnosis of amyotrophic lateral sclerosis: a systematic review," *Arch. Neurol.*, vol. 69, no. 11, pp. 1410–1416, 2012, doi:10.1001/archneurol.2012.254.

[55]   B. R. Brooks, R. G. Miller, M. Swash, and T. L. Munsat, "El Escorial revisited: revised criteria for the diagnosis of amyotrophic lateral sclerosis," *Amyotroph. Lateral Scler. Other Mot. Neuron Disord.*, vol. 1, no. 5, pp. 293–299, January 2000, doi:10.1080/146608200300079536.

[56]   Peter M. Andersen, Sharon Abrahams, Gian D. Borasio *et al.*, *"EFNS GUIDELINES EFNS guidelines on the clinical management of amyotrophic lateral sclerosis (MALS) – revised report of an EFNS task force,"* European journal of neurology, vol. 19, no. 3, pp. 360–375, 2012, doi:10.1111/j.1468-1331.2011.03501.x.

[57]   C. F. Bennett, A. R. Krainer, and D. W. Cleveland, "Antisense oligonucleotide therapies for neurodegenerative diseases," *Annu. Rev. Neurosci.*, vol. 42, no. 1, pp. 385–406, July 2019, doi:10.1146/annurev-neuro-070918-050501.

[58]   C. V. Ly and T. M. Miller, "Emerging antisense oligonucleotide and viral therapies for amyotrophic lateral sclerosis," *Curr. Opin. Neurol.*, vol. 31, no. 5. NLM (Medline), pp. 648–654, 01 October 2018, doi:10.1097/WCO. 0000000000000594.

[59]   "Safety and Efficacy of Repeated Administrations of NurOwn® in ALS Patients - Full Text View - ClinicalTrials.gov." [Online]. Available: https:// clinicaltrials.gov/ct2/show/NCT03280056. [Accessed: 23-Aug-2019].

[60]   University of Alberta, "Biologists Identify Promising Drug for ALS Treatment," *Science Daily*, 2018. [Online]. Available: https://www.sciencedaily.com/releases/2018/12/181218123120.htm. [Accessed: 24-Jan-2020].

[61]   D. Lulé, A. C. Ludolph, and J. Kassubek, "MRI-based functional neuroimaging in ALS: an update," *Amyotrophic Lateral Sclerosis*, vol. 10, no. 5–6. Taylor & Francis, pp. 258–268, 2009, doi:10.3109/17482960802353504.

[62]   J. L. Chang, C. Lomen-Hoerth, J. Murphy, *et al.*, "A voxel-based morphometry study of patterns of brain atrophy in ALS and ALS/FTLD," *Neurology*, vol. 65, no. 1, pp. 75–80, July 2005, doi:10.1212/01.wnl.0000167602.38643.29.

[63] E. P. Pioro, J. P. Antel, N. R. Cashman, and D. L. Arnold, "Detection of cortical neuron loss in motor neuron disease by proton magnetic resonance spectroscopic imaging in vivo," *Neurology*, vol. 44, no. 10, pp. 1933–1938, October 1994, doi:10.1212/wnl.44.10.1933.

[64] O. Gredal, S. Rosenbaum, S. Topp, M. Karlsborg, P. Strange, and L. Werdelin, "Quantification of brain metabolites in amyotrophic lateral sclerosis by localized proton magnetic resonance spectroscopy," *Neurology*, vol. 48, no. 4, pp. 878–881, April 1997, doi:10.1212/WNL.48.4.878.

[65] C. A. Sage, R. R. Peeters, A. Görner, W. Robberecht, and S. Sunaert, "Quantitative diffusion tensor imaging in amyotrophic lateral sclerosis," *Neuroimage*, vol. 34, no. 2, pp. 486–499, January 2007, doi:10.1016/j.neuroimage.2006.09.025.

[66] G. Caiazzo, D. Corbo, Francesca Trojsi, *et al.*, "Distributed corpus callosum involvement in amyotrophic lateral sclerosis: a deterministic tractography study using q-ball imaging," *J. Neurol.*, vol. 261, no. 1, pp. 27–36, January 2014, doi:10.1007/s00415-013-7144-3.

[67] A. Chiò, M. Pagani, F. Agosta, A. Calvo, A. Cistaro, and M. Filippi, "Neuroimaging in amyotrophic lateral sclerosis: insights into structural and functional changes," *Lancet Neurol.*, vol. 13, no. 12. Lancet Publishing Group, pp. 1228–1240, 01-Dec-2014, doi:10.1016/S1474-4422(14)70167-X.

[68] "Stem Cell Therapy for ALS - ALS News Today." [Online]. Available: https://alsnewstoday.com/stem-cell-gene-therapy-for-als/. [Accessed: 24-Jan-2020].

[69] "The ALS Association." [Online]. Available: http://www.alsa.org/research/focus-areas/stem-cells/. [Accessed: 21-Jan-2020].

[70] M. Collins, David Riascos, Tina Kovalik, *et al.*, "The RNA-binding motif 45 (RBM45) protein accumulates in inclusion bodies in amyotrophic lateral sclerosis (ALS) and frontotemporal lobar degeneration with TDP-43 inclusions (FTLD-TDP) patients," *Acta Neuropathol.*, vol. 124, no. 5, pp. 717–732, November 2012, doi:10.1007/s00401-012-1045-x.

[71] N. Bakkar, Tina Kovalik, Ileana Lorenzini, *et al.*, "Artificial intelligence in neurodegenerative disease research: use of IBM Watson to identify additional RNA-binding proteins altered in amyotrophic lateral sclerosis," *Acta Neuropathol.*, vol. 135, no. 2, pp. 227–247, February 2018, doi:10.1007/s00401-017-1785-8.

[72] Z. Yuan, Bin Jiao, Lihua Hou, *et al.*, "Mutation analysis of the TIA1 gene in Chinese patients with amyotrophic lateral sclerosis and frontotemporal dementia," *Neurobiol. Aging*, vol. 64, pp. 160.e9–160.e12, April 2018, doi:10.1016/j.neurobiolaging.2017.12.017.

Chapter 5

Advances in health informatics for obstructive sleep apnea diagnosis and management

Anjali Thakur[1] and Gaurav Gupta[1]

Abstract

Obstructive sleep apnea (OSA), a chronic condition with a worldwide frequency of 9%–38%, affects scalability, affordability, and comprehensiveness are problems for conventional diagnostic techniques including polysomnography and home sleep apnea tests (HSATs). This chapter fills in important study gaps in diagnosing and treating OSA by combining health informatics, AI, and predictive analytics to get around the problems with old methods. Polysomnography (PSG) and other standard methods are reliable, but they are expensive and hard to get HSATs, on the other hand, don't go deep enough into the diagnosis. To overcome this gap, the chapter provides insights how to use wearable tech, AI-driven analysis, and real-time tracking to combine different types of data for a scalable, non-invasive diagnosis. It also fixes the problem of not being able to provide personalised treatment for OSA by using AI-powered decision support systems (DSS) and smart continuous positive airway pressure (CPAP) optimisation. This lets therapy be changed based on each patient's comfort and ability to stick to it. In this part, the underused potential of AI in predictive analytics is also explored. Its role in predicting treatment adherence, disease progression, and comorbidities like cardiovascular risks is shown. By combining these new ideas, the study improves the accuracy of diagnosis, the customisation of treatment, and the ease of access.

Keywords: OSA; AI; HSAT; PSG; IoT

5.1 Introduction

Obstructive sleep apnea (OSA) is a chronic sleep disease that interrupts normal airflow during sleep due to repeated episodes of partial or total obstruction of the upper airway. These blockages occur despite continued respiratory effort, leading to intermittent breathing, reduced oxygen levels in the blood, and fragmented sleep

[1]Yogananda School of AI, Computers and Data Sciences, Shoolini University, India

[1]. The illness commonly manifests as loud snoring, choking, or gasping for air, along with extreme daytime sleepiness and weariness. OSA severity is commonly quantified using the Apnea-Hypopnea Index (AHI), which evaluates the frequency of apnea (total cessation of airflow) and hypopnea (partial reduction in airflow) episodes per hour [2]. Root reasons include obesity, structural anomalies, and reduced upper airway muscle tone during sleep [3]. The incidence and implications of untreated OSA stretch far beyond sleep disruptions, severely influencing individual and society health outcomes [4]. OSA is not only a problem of interrupted sleep but a huge public health issue. Its incidence is disturbingly high, affecting 9%–38% of adults internationally, with considerably greater rates among older people, males, and those with obesity [5]. The problem has far-reaching repercussions, with untreated OSA closely related with cardiovascular diseases such as hypertension, heart attack, and stroke, as well as metabolic disorders like diabetes [6]. Beyond physical health, the cognitive and emotional toll is severe, with many patients having memory issues, sadness, and a diminished quality of life due to chronic fatigue. Additionally, the economic burden of OSA is enormous, covering increased healthcare expenses and diminished employment productivity [7]. These extensive implications show the urgency of tackling the obstacles in detecting and managing this disorder successfully.

Given the high prevalence and severe effects of OSA, prompt diagnosis and successful care are critical. However, present techniques confront substantial challenges. Polysomnography (PSG), the gold standard for diagnosis, is costly, resource-intensive, and inconvenient for patients due to its dependency on overnight stays in hospital settings [8]. While home sleep apnea tests (HSATs) offer a more accessible option, they are less complete and may overlook comorbid sleep disorders [9]. Managing OSA generally entails continuous positive airway pressure (CPAP) therapy, which, despite its efficacy, suffers from poor patient adherence due to discomfort and difficulty [10]. Alternative therapies, such as oral appliances or procedures, are not generally applicable and often have variable effects. These restrictions underscore the need for creative solutions to make OSA care more accessible, efficient, and patient-friendly.

Addressing these problems, health informatics has emerged as a transformative force in the diagnosis and treatment of OSA. By combining modern technologies such as artificial intelligence, machine learning (ML), and big data analytics, health informatics delivers scalable and tailored solutions [11]. Wearable technologies and mobile health applications now provide continuous, non-invasive monitoring of physiological parameters including oxygen saturation, heart rate, and sleep patterns, giving useful diagnostic data outside typical clinical environments. These technologies not only democratise access to diagnostic tools but also lower the demand on healthcare institutions. Furthermore, AI-powered algorithms may scan complicated datasets to deliver diagnostic insights with precision akin to PSG [12]. In management, smart CPAP devices integrate informatics to assess usage trends, dynamically modify settings, and provide meaningful feedback to both patients and therapists. Telemedicine systems bring care into patients' homes, fostering improved adherence, and allowing faster interventions. Thus, health informatics

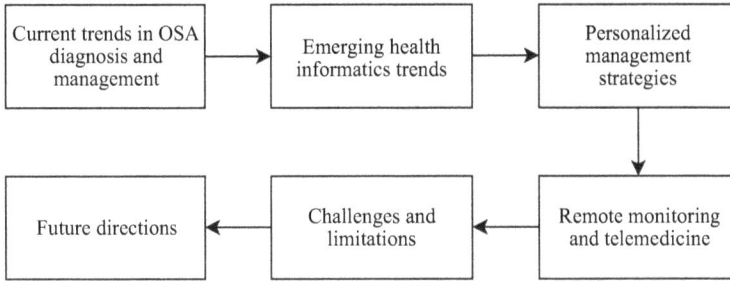

Figure 5.1   Flow of paper.

bridges the gap between present limits and the growing need for efficient, patient-centered OSA care.

Below is the flow chart of the work presented in the paper (Figure 5.1).

5.2   Current trends in OSA diagnosis and management

While pushing forward trends in health informatics that solve these challenges, this section aims to give a whole picture of the present individual techniques for OSA, critically analyses their strengths and limits. It starts by outlining accepted protocols comparable to PSG, the gold standard for OSA opinion because of its delicacy and thorough data collecting, and HSAT, a more convenient and reasonably priced choice fit for simple cases [9]. The section stresses the considerable limitations of these technologies, including the high expense, resource intensity, and vexation of PSG, as well as the reduced individual breadth and possible crimes of HSATs. These difficulties may lead to delayed judgments and undressed cases, highlighting the necessity for inventive results.

- **Polysomnography as the gold standard**

  PSG has long been considered the gold standard for diagnosing OSA due to its potential to extensively cover and dissect many physiological pointers during sleep [5]. Figure 5.2 represents how PSG is done step by step. Conducted in a professional sleep lab, PSG involves the use of colourful detectors linked to the case to gather data on brain exertion (EEG), eye movements (EOG), muscular exertion (EMG), heart rate (ECG), tailwind, respiratory problems, and blood oxygen achromatism (SpO_2). This thorough trend presents a detailed image of a case's sleep armature, letting clinicians to spot apnea and hypopnea episodes and estimate their inflexibility. PSG also aids descry other sleep ailments that may attend with OSA, similar as periodic branch movement complaint or central sleep apnea. Despite its delicacy, PSG is a resource-intensive treatment. It involves the engagement of professed technicians, technological equip, and overnight stays in therapeutic centres. Cases may also find the trend unpleasant due to the cables and electrodes, which can intrude their natural sleep habits.

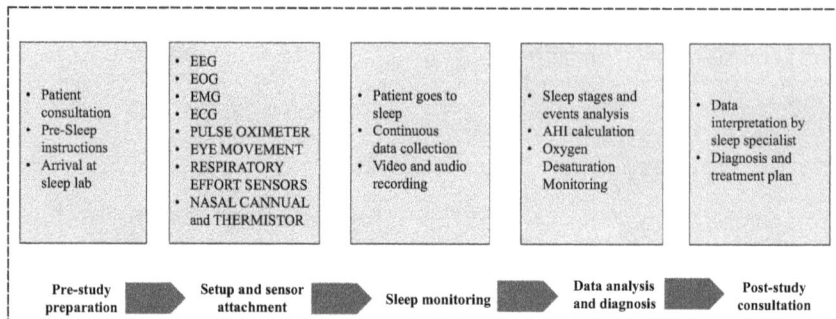

Figure 5.2   Represents the PSG process.

Moreover, the cost of PSG is high, restricting its access, especially in places with tight healthcare budgets or a shortage of sleep labs [4]. These qualities make PSG a mainly successful but not commonly practicable approach for detecting OSA.

- **Home sleep Apnea tests and their adding use**

    In response to the problems posed by PSG, HSATs have evolved as a practical and cost-effective approach for diagnosing moderate to severe OSA [13]. HSATs are movable individual bias that patients can utilise in the convenience of their houses to record critical sleep-related features, similar as respiratory problem, tailwind, and blood oxygen situations. By letting clients to conduct the test in their familiar location, HSATs decrease the anxiety and discomfort frequently linked with late stays in sleep labs. This availability has made HSATs particularly beneficial in pastoral or impoverished areas where access to PSG installations is constrained. Also, the expense of HSATs compared to PSG has made them a practicable option for healthcare businesses wanting to decrease charges while growing individual content. Still, HSATs are not without limits [14]. They yield lesser comprehensive data than PSG, as they typically do not track brain exertion, eye movements, or branch movements. This makes them less helpful in recognising complex sleep illnesses or mild anomalies. However, the rising demand of HSATs marks a tendency toward additional case-centered techniques in sleep drug, focussing on convenience and availability.

5.2.1   Limitations of traditional ways

Despite their value, both PSG and HSATs overcome substantial restrictions that hinder their scalability and usability for detecting OSA in various populations. PSG, remains precious and resource-intensive, making it unapproachable to countless instances, particularly in low-resource settings [15]. The demand for late stays in sleep labs can help cases from seeking examination, specifically if they have to travel great distances or endure scheduling detainments attributable to constrained installation vacuity. Also, the agony of sleeping with numerous detectors connected may vitiate the quality of sleep during the test, consequently altering the results [16]. On the other hand, while HSATs alleviate numerous of these limits by delivering a more provident and easier choice, their narrower compass reduces their mileage. They are best suited for initial cases with suspected moderate to severe OSA and are not advised for individuals with comorbidities or suspected central sleep apnea [17]. The dependency on patient cooperation for efficient gadget utilisation also poses challenges, since incorrect positioning of detectors could affect in missing or wrong data. These limits underscore the necessity for creative individual ways that combine the excellence of PSG with the availability of HSATs [18].

5.3 Emerging health informatics trends

This section highlights how health informatics technologies like wearables, mobile apps, AI, and cloud platforms are transforming OSA diagnosis and management by improving accessibility, accuracy, and personalised care, bridging the gap between traditional methods and modern innovations in sleep medicine as shown in Figure 5.1.

Health informatics is transubstantiating the field of sleep medication by delivering slice-cutting tools that transcend the constraints of previous individual approaches. Wearable widgets similar as smartwatches, fitness trackers, and devoted sleep observers are now able of recording sleep-related parameters non-invasively [19]. These biases descry factors including oxygen achromatism, heart rate, respiratory difficulties, and movement patterns, providing nonstop real-time data that can indicate the presence of sleep-disordered breathing. Unlike PSG or HSATs, wearables give the advantage of long-term monitoring in natural sleep surrounds, allowing croakers to record variations in sleep habits spanning multitudinous nights. Mobile health operations farther strengthen this ecosystem by permitting patients to record problems, assay their sleep data, and admit tailored advise [20]. Beyond wearables, AI and ML are transforming sleep problem opinion. Advanced algorithms evaluate data from wearables, HSATs, or other devices to descry apnea conditions and identify sleep stages with a high degree of delicacy. These AI findings reduce the need for homemade analysis of data, reducing the burden on sleep professionals and expediting the individual procedure [21]. Additionally, cloud-based platforms enable easy data integration from multitudinous sources, allowing healthcare providers to generate informed opinions grounded on comprehensive datasets. By integrating these developing technologies, health informatics islands the gap between the restrictions of existing approaches and the growing demand for effective, scalable, and case-centric solutions in sleep medication.

5.3.1 Multi-modal data integration

Integrating multiple data sources for the diagnosis and management of OSA is a game changer in sleep medicine, delivering a degree of precision and customisation that traditional methods cannot attain. It is a complicated disorder influenced by various factors, including physiological, behavioural, and environmental variables. Traditional diagnostic approaches are restricted in their capacity to capture the complicated interplay between these components. By implementing a multi-modal data integration approach, healthcare clinicians and researchers can acquire a comprehensive picture of the patient's condition, leading to more accurate diagnoses, individualised treatment strategies, and improved patient outcomes [22]. This strategy uses data from many sources, such as wearable devices, mobile health applications, and environmental sensors, and processes it using advanced analytical tools like AI and ML.

- **Physiological data: the foundation of understanding OSA**
  Physiological data has long been the cornerstone of OSA diagnosis. Key measures include heart rate, respiratory effort, oxygen saturation (SpO_2), airflow, and sometimes brain activity using electroencephalograms EEGs [23]. These parameters are frequently obtained during PSG, the gold standard for OSA diagnosis. However, improvements in wearable technology now allow for continuous and non-invasive monitoring of many of these metrics in real-world scenarios. For instance, pulse oximeters and fitness trackers can record oxygen saturation and heart rate variability throughout the night, providing insights on apnea occurrences that may occur in natural sleep environments [24]. Respiratory effort, often assessed by chest belts in traditional settings, can now be approximated using wearables equipped with accelerometers and gyroscopes [25]. This transition from one-time clinical examinations to continuous, longitudinal monitoring boosts the quality and reliability of the data, capturing variables that may be missed in a single-night study. The value of physiological data resides in its capacity to directly quantify the consequences of OSA on the organism. Oxygen desaturation, for example, is a hallmark of apnea occurrences, whereas heart rate variability may suggest the stress exerted on the cardiovascular system [26]. By integrating these measures with other data types, doctors can better understand how physiological disruptions link with behavioural and environmental factors, creating a full picture of the patient's health.

- **Behavioural data: capturing sleep patterns and habits**
  While physiological data gives a snapshot of the body's responses during sleep, behavioural data offers essential context by exposing the patient's sleep habits and patterns. This covers information such as sleep duration, bedtime consistency, sleep latency (time taken to fall asleep), and episodes of awakening during the night. Behavioural data is commonly collected using sleep diaries, mobile health apps, or wearables that measure movement and sleep cycles using accelerometers. Understanding behavioural data is crucial since irregular sleep habits can increase OSA symptoms or hide the effectiveness of therapy strategies [27]. For example, poor sleep length or unpredictable sleep cycles may aggravate exhaustion and daytime drowsiness, symptoms typically attributed to OSA but potentially influenced by lifestyle choices. Behavioural data also provides insights into how well a patient follows to prescribed

therapies, such as CPAP devices [28]. Tracking CPAP usage hours and finding patterns of non-compliance can help clinicians customise interventions to increase adherence. Integrating behavioural data with physiological measures enables the discovery of trends and triggers that could otherwise go unreported. For instance, a patient may experience higher apnea occurrences on nights when sleep duration is shorter or fragmented. This interplay between behaviour and physiology underscores the need for a multi-modal approach to completely understand and manage OSA.

- **Environmental data: the missing piece in OSA diagnosis**
  Environmental data offers a critical component to understanding OSA by accounting for environmental factors that influence sleep quality and breathing. Key parameters include bedroom temperature, humidity, air quality, noise levels, and light exposure. Studies have indicated that poor air quality, such as high levels of particulate matter or allergens, can increase airway inflammation, increasing the chance of apnea occurrences. Similarly, noise pollution from metropolitan areas or home activities can interrupt sleep continuity, increasing its symptoms [29]. Environmental sensors, frequently integrated into smart home systems or standalone devices, collect this data continuously. For example, a smart thermostat can monitor and regulate bedroom temperature and humidity, while air quality sensors may identify pollutants and allergies. Noise monitors can detect decibel levels, revealing potential interruptions that may contribute to sleep fragmentation. By linking environmental data with physiological and behavioural parameters, clinicians can spot patterns, such as an increase in apnea occurrences during periods of high humidity or poor air quality, and offer practical solutions, such as using air purifiers or soundproofing bedrooms.

5.3.1.1 The value of multi-modal data integration

The fundamental power of integrating physiological, behavioural, and environmental data comes in its capacity to provide a holistic assessment of a patient's sleep health. Each data type adds unique insights, but when combined, they show patterns and connections that would otherwise remain hidden. For instance, a patient may experience frequent apnea occurrences (physiological data) on nights when they sleep fewer than six hours (behavioural data) in a room with high levels of noise pollution (environmental data). Such findings enable clinicians to prescribe comprehensive therapies, such as increasing sleep hygiene, employing noise-cancelling equipment, and ensuring adherence to CPAP therapy [30]. Multi-modal data integration also promotes individualised care, a cornerstone of modern healthcare. By studying the interplay between multiple data sources, physicians can build treatment programmes suited to the needs and preferences of each patient. For example, a patient with OSA worsened by allergies may benefit from a combination of CPAP therapy and environmental adjustments, such as air purifiers and allergen-free bedding. Similarly, patients with variable sleep cycles may require behavioural interventions, such as cognitive-behavioural therapy for insomnia (CBT-I), in addition to physiological treatments.

5.3.1.2   Advancing technology and analytical tools

The integration of multiple data sources is made possible by developments in technology and analytical techniques. Wearable gadgets and sensors offer the hardware for data collecting, while AI and ML algorithms process the huge volumes of data created [31]. These algorithms can discover trends, forecast apnea occurrences, and offer therapies with amazing accuracy. Cloud computing platforms enable seamless data aggregation and storage, allowing clinicians to access and analyze data from multiple sources in real time [32]. For instance, an AI-powered platform might combine data from a wearable device tracking heart rate and oxygen saturation with environmental sensor data on air quality and noise levels [33]. By evaluating this data, the system might detect evenings when apnea events are more frequent and provide actionable recommendations, such as proposing the use of an air purifier or suggesting adjustments to nighttime practices. These features not only increase diagnostic accuracy but also empower patients to take an active role in managing their disease.

- **Tools: wearables, medical imaging, and mobile apps**
  The methods available today for detecting and managing OSA are more advanced and accessible than ever before, thanks to the integration of wearable devices, medical imaging, and mobile applications [34]. Wearable technology, including smartwatches, fitness trackers, and specialist sleep monitors, has become a cornerstone of modern healthcare, particularly in the field of sleep medicine. These devices allow non-invasive and continuous monitoring of vital physiological indicators such as SpO_2, heart rate, respiratory effort, and movement patterns throughout the night. Unlike traditional approaches like PSG, which are confined to clinical settings, wearables enable the flexibility to gather data in natural sleep contexts, allowing for a more realistic assessment of a patient's sleep behaviour. For example, a wristwatch equipped with pulse oximetry can detect oxygen desaturation during apnea occurrences, while a fitness tracker with an accelerometer can monitor sleep postures and movement patterns [35]. Dedicated sleep monitors go a step further by combining these capabilities with specific sensors to assess breathing effort and airflow. One of the most significant advantages of wearable technology is its capacity to do long-term monitoring. Single-night studies conducted in sleep labs sometimes fail to capture variations in sleep patterns that occur naturally over time due to factors such as stress, diet, or environmental changes [36]. Wearables overcome this restriction by enabling continuous data gathering over numerous nights or even weeks, producing a richer and more representative dataset. This longitudinal technique increases diagnostic accuracy by highlighting patterns and anomalies that could otherwise go missed. Additionally, wearables are user-friendly, inexpensive, and increasingly accessible, making them a potent tool for early detection and continuous control.
- **Medical imaging** technologies, such sophisticated MRI, CT scans, and ultrasonography, provide structural insights into the anatomy of the upper airway, which plays a crucial role [28,37]. Imaging can assist uncover physical

abnormalities, such as larger tonsils, deviated septa, or excess soft tissue, that contribute to airway obstruction during sleep. These techniques are particularly valuable for personalising surgical or non-surgical therapies depending on individual anatomical traits.

- **Mobile health applications** have altered patient engagement in the management of OSA by providing accessible platforms for tracking symptoms, examining sleep patterns, and receiving individualised insights [20,18]. These apps often sync seamlessly with wearable devices, offering a comprehensive diagnostic and management ecosystem. For example, a mobile app might integrate data from a smartwatch to display patterns in oxygen saturation, sleep duration, and heart rate, while also allowing users to register subjective complaints such as exhaustion, snoring, or morning headaches.

The interactive nature of mobile apps enables users to take an active role in their healthcare experience. By seeing their sleep data and receiving real-time feedback, patients gain a better knowledge of their condition and the factors driving their symptoms. Many apps also utilise AI-driven algorithms to give specific recommendations, such as lifestyle improvements or reminders to adhere to recommended therapy like CPAP. Furthermore, apps can include instructional content, enabling users understand about OSA and its ramifications, which promotes compliance with treatment strategies. For clinicians, mobile apps offer as essential tools for monitoring patient progress remotely. By combining data from sensors and patient inputs, these apps provide a centralised perspective of a patient's sleep health [23]. Clinicians can utilise this information to spot patterns, change treatment strategies, and connect directly with patients using app-based messaging tools. This combination of wearable data and patient participation through mobile apps not only helps the diagnostic process but also creates a more collaborative approach to managing OSA.

5.3.1.3 Benefits

- **Holistic view of patient health**
  One of the most significant benefits of employing these tools is the ability to acquire a holistic assessment of the patient's health. Traditional diagnostic methods generally focus on isolated criteria, such as the frequency of apnea events within a single night, without accounting for other relevant circumstances [38]. Wearables and smartphone apps provide a multi-dimensional perspective by collecting data on physiological indicators, sleep habits, and lifestyle aspects over time. For instance, a wearable device may indicate trends in oxygen saturation and heart rate variability that correlate with specific sleep habits or ambient factors, giving a more comprehensive picture of the patient's state. By combining data from numerous sources such as wearable devices, symptom trackers, and medical imaging clinicians can pinpoint the underlying causes of OSA with more precision. This holistic approach also permits tailored treatment strategies. For example, if imaging reveals structural anomalies, the treatment may involve surgery or positional therapy [39]. On the other hand, if wearables suggest that symptoms worsen in certain sleeping postures, practitioners can

propose positional therapy or alterations to sleep posture. By combining these findings, physicians can move beyond one-size-fits-all treatments to build targeted interventions that address each patient's individual needs.

- **Real-time data collection**

Real-time data collection is another transformational effect of these instruments. Wearables continually monitor critical physiological indicators throughout the night, delivering uninterrupted streams of data that capture small fluctuations and patterns. This functionality is especially beneficial for detecting apnea occurrences that occur sporadically or under specified settings, such as during certain sleep stages or when sleeping in a particular position. Unlike traditional approaches like PSG [35], which are confined to a single-night snapshot, wearables offer longitudinal monitoring, exposing patterns that could otherwise go undiscovered. Real-time data collecting also enables for more dynamic and responsive management of OSA. For example, wearable devices can transmit notifications to patients or physicians when oxygen levels drop below a vital threshold, encouraging rapid action. Mobile apps can evaluate this data in real-time, offering users with rapid feedback on their sleep quality and suggestions for improvement. Additionally, real-time data permits remote monitoring, enabling physicians to track patient progress and adjust treatment programmes without necessitating frequent in-person visits. This is particularly advantageous for patients in rural or underdeveloped areas who may have limited access to specialised sleep clinics.

- **Better diagnostic accuracy**

The combination of wearable technology, medical imaging, and mobile apps greatly increases diagnostic accuracy for OSA. Traditional approaches, while effective, are often confined by their environment and scope. For example, PSG is very accurate but relies on a controlled clinical environment that may not reflect the patient's natural sleep conditions. Wearables, on the other hand, collect data in real-world circumstances, obtaining a more realistic portrayal of the patient's sleep habits [40]. This improved ecological validity boosts the reliability of the diagnosis. Medical imaging offers another layer of accuracy by identifying anatomical elements that lead to airway blockage. By combining imaging data with physiological measurements from wearables, doctors can better discern between OSA subtypes (e.g., positional OSA versus anatomically driven OSA) and customise treatments accordingly. Mobile apps further boost diagnostic accuracy by merging patient-reported symptoms with objective data from wearables and imaging instruments [41]. For instance, an app may combine self-reported snoring frequency with oxygen saturation trends to detect apnea events with improved precision. The synergy between these technologies minimises the likelihood of misdiagnosis or underdiagnosis. By using data from numerous sources, practitioners may cross-validate findings and assure a more accurate and comprehensive assessment of the patient's condition. This improved diagnostic accuracy not only leads to better treatment outcomes but also fosters trust between patients and healthcare providers, as patients can see the physical relationship between their data and therapeutic decisions.

Challenges and ethical considerations

Despite its potential, multi-modal data integration creates problems. Ensuring data integrity and consistency across multiple devices and platforms is crucial, as mistakes can lead to inaccurate diagnoses or ineffective interventions. Interoperability is another key difficulty, as data from diverse devices and systems must be effortlessly incorporated into a unified analytical framework. Privacy and security are also significant concerns [33]. Collecting and analysing sensitive health data require careful compliance with legislation such as Health Insurance Probability and Accountability Act (HIPAA) and General Data Protection Regulation (GDPR) to safeguard patient confidentiality. Ethical considerations, such as gaining informed permission and providing openness in AI-driven suggestions, are vital to sustain patient trust.

5.3.2   Artificial intelligence (AI)

5.3.2.1   AI in diagnosis

AI has emerged as a transformative tool in the diagnosis of OSA, enabling innovative approaches to detect apnea occurrences with precision and speed [42]. Paper [43] has outlined the role of artificial intelligence in the diagnosis and treatment.

AI-powered algorithms examine different data sources, including audio signals, video monitoring, and wearable device outputs, to identify apnea episodes. For instance, audio signal processing entails analysing breathing sounds captured during sleep to detect unusual patterns such as pauses, gasps, or snoring, which are suggestive of apnea occurrences. These algorithms are trained on huge datasets to discern between normal and pathological breathing sounds, even in loud surroundings. Similarly, video surveillance employs AI to assess body movements and chest expansions, recording small respiratory changes that signal disturbed airflow. Wearable technologies, such as smartwatches and sleep monitors, give continuous data on oxygen saturation, heart rate variability, and movement patterns, which are processed by AI to detect apnea occurrences in real-time [16]. These approaches not only boost diagnostic accuracy but also make OSA diagnosis more accessible, especially in home settings where classical PSG might not be possible.

Deep learning, a subset of AI, further revolutionises OSA diagnosis by enabling the study of complicated and high-dimensional data using advanced models like convolutional neural networks (CNNs) [30,44]. CNNs are particularly good in analysing physiological signals, like as respiratory patterns and electroencephalogram (EEG) data, which are critical for identifying sleep problems. For respiratory signal analysis, CNNs discover irregularities in airflow, respiratory effort, and oxygen saturation trends that suggest apnea episodes. These algorithms are trained on huge datasets to distinguish patterns related with varying OSA severities, ensuring great accuracy in detecting occurrences.

For EEG signal analysis, CNNs assist classify sleep stages and identify disturbances caused by apnea [30]. Sleep architecture, which includes transitions between rapid eye movement (REM) and non-REM sleep stages, is typically disturbed in OSA patients. CNNs utilise EEG data to map these transitions and detect arousals connected to apnea occurrences. This level of study provides insights into both the frequency and impact of apnea on overall sleep quality [45]. By utilising the computing power of CNNs, deep learning models can handle complicated datasets rapidly and efficiently, making them useful in current OSA diagnosis.

5.3.2.2 Applications

- **Sleep stage categorisation:** One of the primary uses of AI and deep learning in OSA diagnosis is sleep stage categorisations [46]. Sleep stage REM and non-REM (separated into light and deep sleep) are critical for evaluating sleep quality and identifying problems. AI algorithms analyse EEG, electro-oculogram (EOG), and EMG signals to categorise these stages with high precision. Accurate classification helps clinicians understand the impact of OSA on sleep architecture, such as diminished REM sleep or frequent arousals, which are crucial for therapy planning [47].
- **Automated scoring of sleep studies:** Traditionally, scoring sleep studies entails manual evaluation by skilled sleep technologists, a time-consuming and subjective process [48]. AI-powered solutions simplify this work by evaluating raw data from PSG or wearables, recognising apnea occurrences, and producing measures like the AHI. This automation not only decreases the workload on specialists but also standardises the scoring process, lowering inter-rater variability and boosting diagnostic reliability.
- **Predictive models for OSA severity and dangers:** AI applications extend beyond diagnosis to predicting OSA severity and associated dangers. Predictive algorithms evaluate historical and real-time data to estimate the risk of severe OSA, treatment adherence issues, or other consequences such as cardiovascular events. For example, AI algorithms can detect people at higher risk of developing hypertension or atrial fibrillation due to untreated OSA, enabling earlier management [49]. These models can allow individualised care by customising treatment approaches based on projected outcomes, such as adjusting CPAP settings or offering alternative therapy for patients unlikely to adhere to CPAP.

5.3.3 Big data and predictive analytics

The application of big data in the diagnosis and management of OSA has opened new frontiers in personalised and population-level healthcare. Big data refers to the collection and analysis of massive datasets from diverse sources, such as wearable devices, mobile health applications, sleep labs, and electronic health records (EHRs) [50]. For OSA, big data enables the aggregation of information from millions of users, providing an unprecedented opportunity to identify trends, refine diagnostic algorithms, and improve treatment strategies.

Through data aggregation, patterns and correlations that are difficult to discern in smaller datasets become visible. For instance, big data can reveal how variables such as age, gender, body mass index (BMI), and comorbidities like hypertension or diabetes influence the prevalence and severity of OSA. By analysing data from a global population, researchers can uncover geographic and demographic trends, such as the higher prevalence of OSA in certain regions or among specific populations [51]. These insights inform public health strategies, such as targeted awareness campaigns or resource allocation for high-risk groups.

In addition to identifying trends, big data is crucial for refining diagnostic algorithms. The large volume of data from wearable devices, PSGs, and HSATs provides a rich training ground for AI and ML models. For example, big data can be used to train algorithms to detect apnea events with greater accuracy by exposing them to a wide variety of respiratory patterns, environmental conditions, and patient behaviours. As the dataset grows, the algorithms become more robust and generalisable, improving their performance across diverse populations. Furthermore, big data enables the continuous validation and refinement of these algorithms, ensuring that they stay up to date with evolving diagnostic standards and technologies [52]. Big data also enhances research and innovation in OSA care. By pooling de-identified data from millions of users, researchers can conduct large-scale studies on treatment efficacy, adherence patterns, and long-term outcomes. These studies help identify factors that influence the success of interventions like CPAP therapy or alternative treatments, guiding clinicians toward evidence-based decisions. Overall, big data applications empower clinicians, researchers, and policymakers to address OSA more effectively on both individual and societal levels [53].

Predictive analytics
Predictive analytics, powered by AI and ML, is revolutionising the way OSA is managed by anticipating patient outcomes, treatment adherence, and potential risks based on individual profiles. Predictive analytics involves analysing historical and real-time data to generate forecasts and actionable insights [54]. For OSA, these models process data from various sources such as wearable devices, symptom trackers, and medical imaging to predict the course of the disease and the likelihood of treatment success. One of the most impactful applications of predictive analytics in OSA is forecasting treatment adherence, particularly for CPAP therapy. CPAP is the most effective treatment for moderate to severe OSA, but adherence remains a major challenge, with many patients discontinuing use due to discomfort or inconvenience. Predictive models analyse factors such as demographic data, initial usage patterns, reported side effects, and psychological factors to identify patients who are at risk of non-adherence [55]. For example, if a model detects that a patient uses their CPAP device inconsistently during the first two weeks of therapy, it can flag this behaviour as a predictor of long-term non-compliance. This enables clin-icians to intervene early, offering alternative therapies, additional support, or adjustments to the device to improve adherence. Predictive analytics also plays a crucial role in anticipating treatment outcomes. By analysing baseline data such as the severity of OSA, coexisting medical conditions, and patient-reported symptoms models can predict how well a patient is likely to respond to a given intervention. For instance, a predictive model may suggest that a patient with mild OSA and no significant comorbidities is more likely to benefit from lifestyle changes or posi-tional therapy rather than CPAP [56]. This allows for a more tailored approach to treatment, reducing unnecessary costs and improving patient satisfaction.

In addition to adherence and outcomes, predictive analytics helps assess risks associated with untreated OSA. OSA is a known risk factor for conditions like hypertension, cardiovascular disease, and stroke. Predictive models analyse longitudinal data to estimate the probability of these complications based on the severity and duration of untreated OSA [57]. For example, a model might predict that a middle-aged patient with severe OSA and a history of hypertension has a high risk of developing atrial fibrillation within the next five years. This information enables proactive management, such as early intervention with CPAP or closer monitoring of cardiovascular health. Predictive analytics also supports population health management by identifying high-risk groups and stratifying patients based on their likelihood of developing complications [58]. This helps healthcare systems prioritise resources and design targeted prevention programmes. For instance, patients flagged as high-risk can be enrolled in closer follow-up schedules or provided with enhanced support through telemedicine.

Big data applications and predictive analytics are revolutionising the field of OSA diagnosis and management by transforming raw data into actionable insights [59]. Big data enables the aggregation of information from millions of users to identify trends, refine diagnostic algorithms, and improve treatment strategies on a global scale. Predictive analytics, on the other hand, empowers clinicians to anticipate treatment adherence, outcomes, and risks, enabling personalised and proactive care. Together, these technologies pave the way for a more data-driven, efficient, and effective approach to managing OSA, ultimately improving patient outcomes and advancing the field of sleep medicine.

Table 5.1 shows the comparison between the traditional methods and emerging trends for the OSA diagnosis and management based of different aspects.

Table 5.1   Comparison between traditional methods and emerging trends

Aspect	Traditional methods	Emerging trends
Diagnostic approach	PSG and HSATs	Wearable devices Ai-powered Diagnostics and mobile health apps
Accuracy	High (PSG), lower for HSATs	AI-driven analysis enhances accuracy comparable to PSG
Accessibility	Limited due to reliance on sleep labs and high costs	Remote monitoring and mobile health apps increase accessibility
Patient comfort	Requires overnight monitoring	Home-based monitoring with wearables
Treatment personalisation	One-size-fits-all CPAP therapy	AI-driven decision support
Cost	Very expensive	Due to its scalability and decentralised solutions, it is cost effective
Technology integration	Minimal	AI, IoT, cloud-based data processing for real-time insights
Monitoring capability	Limited to periodic assessments and clinical settings	Continuous, real-time monitoring with remote access
Data utilisation	Manual analysis	AI and big data analytics for predictive insights

5.4   Personalised management strategies

This section focuses on the revolutionary potential of informatics in generating individualised management strategies for OSA. Personalised management moves away from one-size-fits-all approaches, emphasising the need to accommodate the unique traits, needs, and preferences of each patient [60]. By integrating sophisticated health informatics tools, such as data integration systems, AI, and decision support platforms, physicians can design individualised treatment plans, optimise therapeutic effectiveness, and enhance patient adherence.

- **Individualised treatment plans**
  One of the most crucial parts of personalised management is the establishment of specific treatment programmes. Informatics facilitates the integration of diverse datasets, including demographic information (age, gender, and BMI), clinical data (severity of OSA and concomitant disorders), and behavioural patterns (sleep habits and therapy adherence). By assessing these data together, physicians can prescribe medications that are most suited to each patient's unique profile. For example, a patient with mild OSA who is overweight might benefit most from lifestyle measures, such as weight loss and exercise, along with positional therapy [61]. On the other hand, a patient with severe OSA and accompanying cardiovascular illness may require more aggressive treatment, such as CPAP therapy or surgical treatments. Behavioural data, such as irregular sleep schedules or non-compliance with past therapies, can further clarify these recommendations. Informatics systems can process this integrated data to detect patterns and connections, enabling physicians select the most effective and sustainable interventions for everyone.

  This method not only improves the likelihood of therapeutic success but also decreases unnecessary treatments, decreasing patient burden and healthcare expenses. By addressing the specific needs of each patient, personalised treatment regimens enhance both the quality and the efficiency of OSA care.

- **CPAP optimisation**
  For patients undergoing CPAP therapy, informatics plays a vital role in optimising device utilisation and promoting adherence. CPAP is considered the gold standard treatment for moderate to severe OSA, but adherence remains a substantial barrier, with many patients struggling to tolerate the equipment. Informatics technologies address this issue by monitoring patient feedback and usage patterns to make individualised adjustments to CPAP settings [62]. Modern CPAP systems are integrated with smart sensors and connectivity features that collect data on characteristics such as hours of usage, mask fit, air pressure, and the frequency of apnea occurrences during therapy. This data is transmitted to cloud-based platforms where it is examined to identify potential concerns. For instance, if a patient routinely experiences mask leaks or pain at a given pressure level, the system can flag this for the doctor, who can then alter the device settings remotely. Patient feedback, such as comments of dryness, noise, or pain, is also integrated into these optimisation efforts [63]. Informatics

tools use this feedback to offer adjustments, such as increasing humidification, attempting different mask designs, or altering pressure levels during specific sleep periods. These tailored changes not only enhance the comfort and effectiveness of CPAP therapy but also improve patient adherence, increasing the likelihood of long-term therapeutic success.

• **Decision support systems**
  Decision support systems (DSS) are a crucial component of individualised OSA management, providing doctors with evidence-based advice tailored to individual patients [64]. These systems employ informatics and AI to process complex datasets, including patient history, diagnostic results, and real-time therapy data, to support clinical decision making. For example, a DSS might examine a patient's diagnostic data, such as the AHI, oxygen desaturation levels, and comorbidities, and prescribe a specific therapy plan. It can propose CPAP therapy for patients with severe OSA or positional therapy for those with primarily positional apnea. Additionally, DSS tools incorporate the newest clinical standards and research findings, ensuring that suggestions are both individualised and consistent with current best practices. In the context of CPAP therapy, DSS can uncover patterns of non-adherence and recommend targeted interventions, such as switching to bilevel positive airway pressure (BiPAP) for patients who struggle with fixed-pressure devices [58]. For patients with persisting complaints despite therapy, the system might offer further diagnostic tests or alternative therapies, such as mandibular advancement devices or surgical examination.

  DSS technologies help promote communication between clinicians and patients by offering clear, data-driven explanations for treatment decisions. This transparency creates confidence and encourages patients to take an active role in their care. By simplifying complex decision-making procedures and lowering the likelihood of errors, DSS guarantees that doctors can deliver high-quality, tailored OSA care efficiently and effectively.

5.5   Remote monitoring and telemedicine

This section highlights how improvements in technology have transformed remote care for OSA, enabling more efficient and accessible management of the disorder. By integrating Internet of Things (IoT) devices, telemedicine platforms, and real-time monitoring tools, remote care solutions improve patient engagement, adherence, and outcomes while lowering the dependency on in-person consultations and hospital visits. Below is an in-depth description of the specific tools and their usefulness.

5.5.1   IoT devices

IoT devices have become a cornerstone of remote OSA treatment, allowing real-time monitoring and better connectivity between patients and healthcare providers [65]. Two significant uses of IoT in OSA control are smart CPAP devices and wearable monitors.

- **Smart CPAP machines:** Modern CPAP machines are integrated with IoT-enabled sensors that measure numerous parameters, such as usage hours, pressure settings, mask fit, and the frequency of apnea occurrences during therapy [66]. This data is automatically transferred to cloud-based platforms, so clinicians can access it remotely. Smart CPAP machines allow healthcare providers to monitor adherence and therapeutic effectiveness in real time, recognising issues like inconsistent usage or poor pressure settings. When problems are recognised, physicians can intervene swiftly by offering remote coaching or modifying settings without requiring an in-person visit. This proactive strategy not only improves therapy adherence but also guarantees that patients receive the full benefits of CPAP treatment. Figure 5.3 [67] presents the block diagram of CPAP system for the better understanding.
- **Wearable monitors:** Wearable devices, such as smartwatches, fitness trackers, and dedicated sleep monitors, continuously track vital signs, including SpO_2, heart rate, and breathing effort [19,68]. These devices provide a non-invasive and simple way to monitor patients' physiological responses to treatment over time. For instance, a wearable can detect nightly oxygen desaturation, urging early intervention if a patient's condition worsens. The capacity to monitor these indicators outside of a clinical setting provides for a more accurate assessment of therapy efficacy in real-world situations. Additionally, wearable gadgets generally link with mobile apps, enabling patients to analyse their own data and stay engaged with their therapy [69].

5.5.2 *Telemedicine platforms*

Telemedicine stages have revolutionised the way OSA care is delivered, advertising patients and doctors the ease of virtual interviews and real-time observing tools. These phases make OSA administration more accessible to underprivileged individuals by addressing barriers to care, like geographic removal or limited availability to specialised rest clinics.

- **Online meetings:** Telemedicine lets patients refer to rest masters or respiratory specialists from the comfort of their homes. Clinicians can audit data gathered from IoT devices, discuss indications, and offer individualised advice within virtual conversations [70]. This eliminates the need for patients to travel

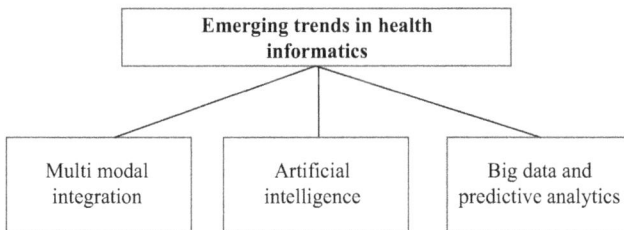

Figure 5.3   Emerging trends in health informatics.

great distances for follow-up appointments, therefore saving time and lowering related expenses. Virtual meetings also promotes better understanding and participation, as they enable for more visit intelligent and opportune reactions to silent issues [71]. Moreover, these phases routinely include safe informing systems, thus enabling ongoing communication between suppliers and patients.

- **Real-time checking tools:** Numerous telemedicine stages synchronised real-time checking apparatuses that alarm doctors to fundamental occasions, such as noteworthy reductions in oxygen immersion or unexpected variations in heart rate [72]. These cautions empower fast mediation, anticipating difficulties, and assuring continuous security. For illustration, in case a sharp CPAP machine identifies a delayed period of non-use or visit apnea situations despite therapy, the framework can alert the practitioner, who can at that point reach out to the calm to resolve the issue [73]. Real-time observation devices empower physicians to track patterns and designs over time, promoting data-driven modifications to treatment approaches. Benefits of Inaccessible Care Innovation Patients and healthcare providers gain clearly from the combination of IoT devices with telemedicine platforms.

- **Expanded persistent adherence:** Real-time monitoring and criticism components excite patients to remain reliable with their treatment [74]. When patients can observe their advance through wearable information or acquire convenient mediations for problems like veil inconvenience, they are more inclined to follow their treatment regimens. Improved adherence leads to improved wellness results and a lower probability of problems connected to untreated OSA.

- **Decreased healing center visits and related costs:** Farther care essentially minimises the need for in-person interviews, follow-ups, and hospitalisations. By addressing difficulties proactively through virtual interviews and real-time watching, physicians might foresee complications that would require crisis treatment or healing centre confirmations. Not only does this spares patients time and cash, but also reduces the pressure on healthcare systems, making care more effective and cost-effective.

5.6   Challenges and limitations

Information privacy, security, and moral contemplations are key issues in harnessing wellbeing informatics for controlling OSA. The sensitive nature of wellbeing information, such as rest designs, oxygen levels, and restorative histories, makes it impotent to unauthorised access, possibly driving to breaches of private privacy. Guaranteeing compliance with information assurance directions like HIPAA (Wellbeing Protections Compactness and Responsibility Act) within the United States and GDPR (Common Information Security Direction) in Europe is vital to safeguarding understanding data. These directives order stringent steps for information encryption, secure capacity, and controlled get to, making organisations

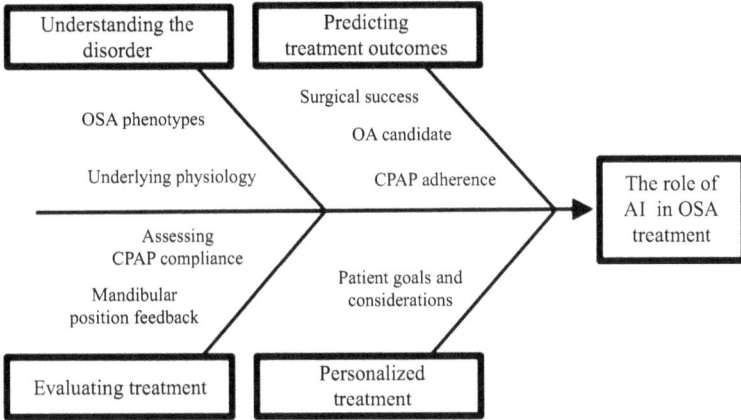

Figure 5.4   Role of AI in diagnosis of OSA [43]

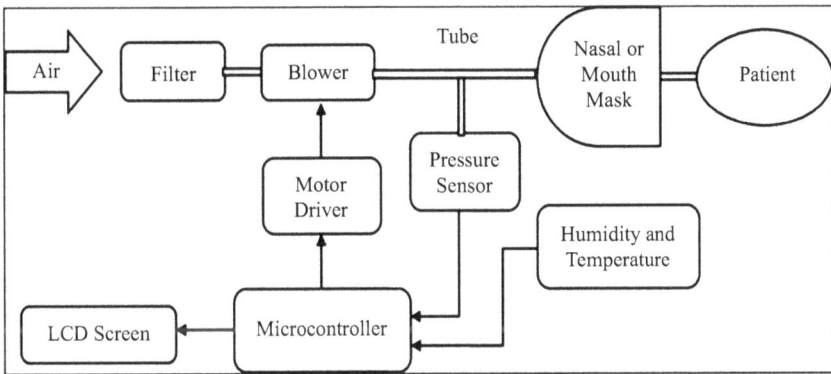

Figure 5.5   Block diagram of CPAP system [67]

responsible for information abuse. Past security, the issue of interoperability developed when combining information from different gadgets and frameworks, such as wearable displays, flexible apps, and electronic wellbeing records (EHRs). Standardising information designs and communication conventions is crucial to guarantee consistent trade and examination; however, the need of widespread rules regularly leads to fracturing, limiting compelling care coordination. Also, moral challenges appear, notably with the use of AI in wellbeing informatics. AI models may demonstrate propensity due to uneven prepared knowledge, possibly coming about in erroneous expectations for underrepresented groups. Tending to these inclinations demands ongoing refinement of computations to assure decency and inclusion. Besides, the moral standards of silent consent and independence must be maintained in data-driven healthcare. Patients ought to be entirely aware almost how their information is utilised, have the capacity to select out, and hold management over their data. Adjusting advancement with these protection, security, interoperability, and moral contemplations is necessary to build trust and ensure the reliable conveying of wellness informatics innovations in OSA administration.

5.7   Future directions

The future of controlling OSA lies in harnessing expanded advances that empower exact, adaptable, and individualised care. Wearable and sensor improvements are set to transform long-term observing, promoting non-invasive gadgets that supply continuous following of physiological measures such as oxygen immersion, heart rate, and respiratory activity. Progresses in tiny, high-fidelity sensors will make these gadgets comfier and more exact, recording nuanced information for increased intervals in distinctive rest situations. These advancements will give more profound bits of knowledge about rest designs and empower early finding of OSA or its worsening. Additionally, the combination of genetic and epigenomic information will play a transformational impact within the tailoring of OSA treatment. By recognising inherited predisposition to OSA and comprehending gene-environment intuitive, physicians will be able to develop custom-made mediations. For illustration, patients with hereditary indicators associated to upper aviation route collapsibility may advantage from focused treatments, whilst those influenced by natural triggers like allergens can obtain combined natural and pharmaceutical intercessions. Half breed cloud arrangements will satisfy the increasing demand for versatile information capacity and handling skills, permitting real-time analytics of huge wellness datasets collected from wearables, portable apps, and genomic thinks about. These phases will promote regular worldwide collaboration, letting analysts and doctors to communicate discoveries and develop demonstration and rehabilitative ways. AI-driven experiences will promote improve care conveyance, with next-generation computations capable of delivering profoundly individualised proposals based on a patient's coordinates information profile. Independent frameworks will monitor and adjust therapies in real time, such as forcefully altering CPAP settings depending on live understanding input and physiological reactions. These headways will make a more proactive and versatile care show, decreasing the load of OSA and entirely progressing lasting results. Together, these advancements hint to a worldview move in OSA administration, offering a future of exactness medicine, improved openness, and transforming treatment.

References

[1]   J. Lim, Khan, S. S., Pandya, A., *et al.*, "Diagnosis of obstructive sleep apnea during wakefulness using upper airway negative pressure and machine learning," *Proc. Annu. Int. Conf. IEEE Eng. Med. Biol. Soc. EMBS*, pp. 1605–1608, 2019, doi: 10.1109/EMBC.2019.8856754.

[2]   A. T. Thornton, P. Singh, W. R. Ruehland, and P. D. Rochford, "AASM criteria for scoring respiratory events: interaction between apnea sensor and hypopnea definition," *Sleep*, vol. 35, no. 3, pp. 425–432, 2012, doi: 10.5665/sleep.1710.

[3]   L. J. Palmer, S. G. Buxbaum, E. Larkin, *et al.*, "A whole-genome scan for obstructive sleep apnea and obesity," *Am. J. Hum. Genet.*, vol. 72, no. 2, pp. 340–350, 2003, doi: 10.1086/346064.

[4]   K. N. V. P. S. Rajesh, R. Dhuli, and T. S. Kumar, "Obstructive sleep apnea detection using discrete wavelet transform-based statistical features," *Comput. Biol. Med.*, vol. 130, no. December 2020, p. 104199, 2021, doi: 10.1016/j.compbiomed.2020.104199.

[5]   A. Kulkas, B. Duce, T. Leppänen, C. Hukins, and J. Töyräs, "Gender differences in severity of desaturation events following hypopnea and obstructive apnea events in adults during sleep," *Physiol. Meas.*, vol. 38, no. 8, pp. 1490–1502, 2017, doi: 10.1088/1361-6579/aa7b6f.

[6]   C. Arnaud, T. Bochaton, J. L. Pépin, and E. Belaidi, "Obstructive sleep apnoea and cardiovascular consequences: pathophysiological mechanisms," *Arch. Cardiovasc. Dis.*, vol. 113, no. 5, pp. 350–358, 2020, doi: 10.1016/j.acvd.2020.01.003.

[7]   P. E. Peppard, T. Young, J. H. Barnet, M. Palta, E. W. Hagen, and K. M. Hla, "Increased prevalence of sleep-disordered breathing in adults," *Am. J. Epidemiol.*, vol. 177, no. 9, pp. 1006–1014, 2013, doi: 10.1093/aje/kws342.

[8]   H. Ting, Y. T. Mai, H. C. Hsu, H. C. Wu, and M. H. Tseng, "Decision tree based diagnostic system for moderate to severe obstructive sleep apnea," *J. Med. Syst.*, vol. 38, no. 9, pp. 1–10, 2014, doi: 10.1007/s10916-014-0094-1.

[9]   I. M. Rosen, Douglas B. Kirsch, Kelly A. Carden, *et al.*, "Clinical use of a home sleep apnea test: an updated American academy of sleep medicine position statement," *J. Clin. Sleep Med.*, vol. 14, no. 12, pp. 2075–2077, 2018, doi: 10.5664/jcsm.7540.

[10] R. K. Kakkar and R. B. Berry, "Positive airway pressure treatment for obstructive sleep apnea," *Chest*, vol. 132, no. 3, pp. 1057–1072, 2007, doi: 10.1378/chest.06-2432.

[11] C. Jimenez-Mesa, J. E. Arco, F. J. Martinez-Murcia, J. Suckling, J. Ramirez, and J. M. Gorriz, "Applications of machine learning and deep learning in SPECT and PET imaging: general overview, challenges and future prospects," *Pharmacol. Res.*, vol. 197, no. October, p. 106984, 2023, doi: 10.1016/j.phrs.2023.106984.

[12] J. Zhu, A. Zhou, Q. Gong, Y. Zhou, J. Huang, and Z. Chen, "Detection of sleep apnea from electrocardiogram and pulse oximetry signals using random forest," *Appl. Sci.*, vol. 12, no. 9, 2022, doi: 10.3390/app12094218.

[13] C. R. Laratta, N. T. Ayas, M. Povitz, and S. R. Pendharkar, "Diagnosis and treatment of obstructive sleep apnea in adults," *Cmaj*, vol. 189, no. 48, pp. E1481–E1488, 2017, doi: 10.1503/cmaj.170296.

[14] D. A. Pevernagie, B. Gnidovec-Strazisar, L. Grote, *et al.*, "On the rise and fall of the apnea−hypopnea index: a historical review and critical appraisal," *J. Sleep Res.*, vol. 29, no. 4, pp. 1–20, 2020, doi: 10.1111/jsr.13066.

[15]   N. Freedman, "Doing it better for less: incorporating OSA management into alternative payment models," *Chest*, vol. 155, no. 1, pp. 227–233, 2019, doi: 10.1016/J.CHEST.2018.06.033.

[16]   A. Green, N. Nagel, L. Kemer, and Y. Dagan, "Comparing in-lab full poly-somnography for diagnosing sleep apnea in children to home sleep apnea tests (HSAT) with an online video attending technician," *Sleep Biol. Rhythms*, vol. 20, no. 3, pp. 397–401, 2022, doi: 10.1007/S41105-022-00384-7/TABLES/3.

[17]   C. Hawco, A. Bonthu, T. Pasek, K. Sarna, L. Smolley, and A. Hadeh, "The correlation of computerized scoring in home sleep apnea tests with technician visual scoring for assessing the severity of obstructive sleep apnea," *J. Clin. Med.*, vol. 13, no. 14, 2024, doi: 10.3390/JCM13144204.

[18]   J. T. Jagielski, N. Bibi, P. C. Gay, *et al.*, "Evaluating an under-mattress sleep monitor compared to a peripheral arterial tonometry home sleep apnea test device in the diagnosis of obstructive sleep apnea," *Sleep Breath.*, vol. 27, no. 4, pp. 1433–1441, 2023, doi: 10.1007/S11325-022-02751-7.

[19]   M. De Zambotti, N. Cellini, A. Goldstone, I. M. Colrain, and F. C. Baker, "Wearable sleep technology in clinical and research settings," *Med. Sci. Sports Exerc.*, vol. 51, no. 7, pp. 1538–1557, 2019, doi: 10.1249/MSS.0000000000001947.

[20]   C. S. Wood, M. R. Thomas, J. Budd, *et al.*, "Taking connected mobile-health diagnostics of infectious diseases to the field," *Nature*, vol. 566, no. 7745, pp. 467–474, 2019, doi: 10.1038/S41586-019-0956-2.

[21]   G. W. Pien, L. Ye, B. T. Keenan, *et al.*, "Changing faces of obstructive sleep apnea: treatment effects by cluster designation in the Icelandic sleep apnea cohort," *Sleep*, vol. 41, no. 3, 2018, doi: 10.1093/SLEEP/ZSX201.

[22]   Y. Yang, H. Jiang, H. Yang, *et al.*, "Multimodal data integration enhance longitudinal prediction of new-onset systemic arterial hypertension patients with suspected obstructive sleep apnea," *Rev. Cardiovasc. Med.*, vol. 25, no. 7, p. 258, 2024, doi: 10.31083/J.RCM2507258.

[23]   F. Mendonça, S. S. Mostafa, A. G. Ravelo-García, F. Morgado-Dias, and T. Penzel, "Devices for home detection of obstructive sleep apnea: A review," *Sleep Med. Rev.*, vol. 41, pp. 149–160, 2018, doi: 10.1016/J.SMRV.2018.02.004.

[24]   P. Sharma, A. Jalali, M. Majmudar, K. S. Rajput, and N. Selvaraj, "Deep-learning based sleep apnea detection using SpO2 and pulse rate," *Proc. Annu. Int. Conf. IEEE Eng. Med. Biol. Soc. EMBS*, vol. 2022, July, pp. 2611–2614, 2022, doi: 10.1109/EMBC48229.2022.9871295.

[25]   D. J. Gottlieb, C. W. Whitney, W. H. Bonekat, *et al.*, "Relation of sleepiness to respiratory disturbance index: the sleep heart health study," *Am. J. Respir. Crit. Care Med.*, vol. 159, no. 2, pp. 502–507, 1999, doi: 10.1164/AJRCCM.159.2.9804051.

[26]   A. Undrajavarapu, A. Prasath, M. Varghese, P. George, and K. H. Kisku, "Correlation between oxygen desaturation index measured by overnight oximetry and apnea-hypopnea index measured by polysomnography

in patients diagnosed with obstructive sleep apnea," *Cureus*, vol. 16, no. 10, 2024, doi: 10.7759/CUREUS.71895.

[27] T. da Silva Gusmão Cardoso, S. Pompéia, and M. C. Miranda, "Cognitive and behavioral effects of obstructive sleep apnea syndrome in children: a systematic literature review," *Sleep Med.*, vol. 46, pp. 46–55, 2018, doi: 10.1016/J.SLEEP.2017.12.020.

[28] N. J. Douglas, "Systematic review of the efficacy of nasal CPAP," *Thorax*, vol. 53, no. 5, pp. 414–415, 1998, doi: 10.1136/THX.53.5.414.

[29] N. S. Marshall and C. T. Cowie, "Completely scoobied: the confusing world of temperature and pollution effects on sleep apnoea," *Eur. Respir. J.*, vol. 46, no. 5, pp. 1251–1254, 2015, doi: 10.1183/13993003.01155-2015.

[30] A. Rajabrundha, A. Lakshmisangeetha, and A. Balajiganesh, "Analysis of sleep apnea considering electrocardiogram data using deep learning algorithms," *J. Phys. Conf. Ser.*, vol. 2318, no. 1, 2022, doi: 10.1088/1742-6596/2318/1/012009.

[31] C. F. Kuo, C. Y. Tsai, W. H. Cheng *et al.*, "Machine learning approaches for predicting sleep arousal response based on heart rate variability, oxygen saturation, and body profiles," *Digit. Heal.*, vol. 9, 2023, doi: 10.1177/20552076231205744.

[32] Z. Xu, G. C. Gutiérrez-Tobal, Y. Wu, *et al.*, "Cloud algorithm-driven oximetry-based diagnosis of obstructive sleep apnoea in symptomatic habitually snoring children," *Eur. Respir. J.*, vol. 53, no. 2, 2019, doi: 10.1183/13993003.01788-2018.

[33] E. J. Topol, "High-performance medicine: the convergence of human and artificial intelligence," *Nat. Med.*, vol. 25, no. 1, pp. 44–56, 2019, doi: 10.1038/S41591-018-0300-7.

[34] T. Giallorenzi and M. Lesky, "OSA imaging and applied optics congress support," 2017, Technical Report by Optical Society of America Washington United States Accession Number: AD1028558.

[35] M. Marino, Y. Li, M. N. Rueschman, *et al.*, "Measuring sleep: accuracy, sensitivity, and specificity of wrist actigraphy compared to polysomnography," *Sleep*, vol. 36, no. 11, pp. 1747–1755, 2013, doi: 10.5665/sleep.3142.

[36] C. L. Fraser, "Update on obstructive sleep apnea for neuro-ophthalmology," *Curr. Opin. Neurol.*, vol. 32, no. 1, pp. 124–130, 2019, doi: 10.1097/WCO.0000000000000630.

[37] N. B. Kribbs, A. I. Pack, L. R. Kline, *et al.*, "Effects of one night without nasal CPAP treatment on sleep and sleepiness in patients with obstructive sleep apnea," *Am. Rev. Respir. Dis.*, vol. 147, no. 5, pp. 1162–1168, 1993, doi: 10.1164/AJRCCM/147.5.1162.

[38] L. La Fisca, C. Jennebauffe, M. Bruyneel, *et al.*, "Enhancing OSA assessment with explainable AI," *Proc. Annu. Int. Conf. IEEE Eng. Med. Biol. Soc. EMBS*, 2023, doi: 10.1109/EMBC40787.2023.10341035.

[39] A. Gunes, D. Sigirli, I. Ercan, S. Turan Ozdemir, Y. Durmus, and T. Yildiz, "Evaluation of the corpus callosum shape in patients with obstructive sleep

apnea," *Sleep Breath.*, vol. 26, no. 3, pp. 1201–1207, 2022, doi: 10.1007/S11325-021-02502-0.

[40]   Z. F. Udwadia, A. V. Doshi, S. G. Lonkar, and C. I. Singh, "Prevalence of sleep-disordered breathing and sleep apnea in middle-aged urban Indian men," *Am. J. Respir. Crit. Care Med.*, vol. 169, no. 2, pp. 168–173, 2004, doi: 10.1164/RCCM.200302-265OC.

[41]   S. G. Hershman, B. M. Bot, A. Shcherbina, *et al.*, "Physical activity, sleep and cardiovascular health data for 50,000 individuals from the MyHeart Counts Study," *Sci. Data*, vol. 6, no. 1, 2019, doi: 10.1038/S41597-019-0016-7.

[42]   H. Liang, B. Y Tsui, H. Ni, *et al.*, "Evaluation and accurate diagnoses of pediatric diseases using artificial intelligence," *Nat. Med.*, vol. 25, no. 3, pp. 433–438, 2019, doi: 10.1038/S41591-018-0335-9.

[43]   H. L. Brennan and S. D. Kirby, "The role of artificial intelligence in the treatment of obstructive sleep apnea," *J. Otolaryngol. – Head Neck Surg.*, vol. 52, no. 1, pp. 1–6, 2023, doi:10.1186/S40463-023-00621-0/FIGURES/1.

[44]   Z. Strumpf, W. Gu, C. W. Tsai, *et al.*, "Belun Ring (Belun Sleep System BLS-100): deep learning-facilitated wearable enables obstructive sleep apnea detection, apnea severity categorization, and sleep stage classification in patients suspected of obstructive sleep apnea," *Sleep Heal.*, vol. 9, no. 4, pp. 430–440, 2023, doi: 10.1016/J.SLEH.2023.05.001.

[45]   Y. K. Loke, J. W. L. Brown, C. S. Kwok, A. Niruban, and P. K. Myint, "Association of obstructive sleep apnea with risk of serious cardiovascular events: a systematic review and meta-analysis," *Circ. Cardiovasc. Qual. Outcomes*, vol. 5, no. 5, pp. 720–728, 2012, doi: 10.1161/CIRCOUTCOMES.111.964783.

[46]   A. N. Olesen, P. Jennum, P. Peppard, E. Mignot, and H. B. D. Sorensen, "Deep residual networks for automatic sleep stage classification of raw polysomnographic waveforms," *Proc. Annu. Int. Conf. IEEE Eng. Med. Biol. Soc. EMBS*, vol. 2018, July, pp. 3713–3716, 2018, doi: 10.1109/EMBC.2018.8513080.

[47]   S. Güneş, K. Polat, and Ş. Yosunkaya, "Efficient sleep stage recognition system based on EEG signal using k-means clustering based feature weighting," *Expert Syst. Appl.*, vol. 37, no. 12, pp. 7922–7928, 2010, doi: 10.1016/J.ESWA.2010.04.043.

[48]   A. M. Koupparis, V. Kokkinos, and G. K. Kostopoulos, "Semi-automatic sleep EEG scoring based on the hypnospectrogram," *J. Neurosci. Methods*, vol. 221, pp. 189–195, 2014, doi: 10.1016/j.jneumeth.2013.10.010.

[49]   E. S. Muxfeldt, V. S. Margallo, G. M. Guimarães, and G. F. Salles, "Prevalence and associated factors of obstructive sleep apnea in patients with resistant hypertension," *Am. J. Hypertens.*, vol. 27, no. 8, pp. 1069–1078, 2014, doi: 10.1093/AJH/HPU023.

[50]   "Big Data in sleep apnoea: Opportunities and challenges – Pépin – 2020 – Respirology – Wiley Online Library." https://onlinelibrary.wiley.com/doi/full/10.1111/resp.13669 (accessed Jan. 31, 2025).

[51]   S. Ancoli-Israel, D. F. Kripke, M. R. Klauber, W. J. Mason, R. Fell, and O. Kaplan, "Sleep-disordered breathing in community-dwelling elderly," *Sleep*, vol. 14, no. 6, pp. 486–495, 1991, doi: 10.1093/SLEEP/14.6.486.

[52]   P. Olivera, S. Danese, N. Jay, G. Natoli, and L. Peyrin-Biroulet, "Big data in IBD: a look into the future," *Nat. Rev. Gastroenterol. Hepatol.*, vol. 16, no. 5, pp. 312–321, 2019, doi: 10.1038/S41575-019-0102-5.

[53]   N. Norori, Q. Hu, F. M. Aellen, F. D. Faraci, and A. Tzovara, "Addressing bias in big data and AI for health care: a call for open science," *Patterns*, vol. 2, no. 10, 2021, doi: 10.1016/j.patter.2021.100347.

[54]   A. I. Pack, "Application of personalized, predictive, preventative, and participatory (P4) medicine to obstructive sleep apnea a roadmap for improving care?," *Ann. Am. Thorac. Soc.*, vol. 13, no. 9, pp. 1456–1467, 2016, doi: 10.1513/ANNALSATS.201604-235PS.

[55]   K. Liu, S. Geng, P. Shen, L. Zhao, P. Zhou, and W. Liu, "Development and application of a machine learning-based predictive model for obstructive sleep apnea screening," *Front. Big Data*, vol. 7, 2024, doi: 10.3389/FDATA.2024.1353469/FULL.

[56]   J. P. Bakker, R. Wang, J. Weng, *et al.*, "*Motivational enhancement for increasing adherence to CPAP: a randomized controlled trial,*" Chest, vol. 150, no. 2, pp. 337–345, 2016, doi: 10.1016/J.CHEST.2016.03.019.

[57]   C. A. Kushida, B. Efron, and C. Guilleminault, "A predictive morphometric model for the obstructive sleep apnea syndrome," *Ann. Intern. Med.*, vol. 127, no. 8 I, pp. 581–587, 1997, doi: 10.7326/0003-4819-127-8_PART_1-199710150-00001.

[58]   R. Dutta, G. Delaney, B. Toson, *et al.*, "A novel model to estimate key obstructive sleep apnea endotypes from standard polysomnography and clinical data and their contribution to obstructive sleep apnea severity," *Ann. Am. Thorac. Soc.*, vol. 18, no. 4, pp. 656–667, 2021, doi: 10.1513/ANNALSATS.202001-064OC.

[59]   M. Naughton, P. A. Cistulli, P. De Chazal, J.-L. Pépin, S. Bailly, and R. Tamisier, "Big data in sleep apnoea: opportunities and challenges," *Respirology*, vol. 25, no. 5, pp. 486–494, 2020, doi: 10.1111/RESP.13669.

[60]   D. Geng, Z. Qin, J. Wang, Z. Gao, and N. Zhao, "Personalized recognition of wake/sleep state based on the combined shapelets and K-means algorithm," *Biomed. Signal Process. Control*, vol. 71, 2022, doi: 10.1016/j.bspc.2021.103132.

[61]   R. C. Heinzer, C. Pellaton, V. Rey, *et al.*, "Positional therapy for obstructive sleep apnea: an objective measurement of patients' usage and efficacy at home," *Sleep Med.*, vol. 13, no. 4, pp. 425–428, 2012, doi: 10.1016/J.SLEEP.2011.11.004.

[62]   B. W. Rotenberg, D. Murariu, and K. P. Pang, "Trends in CPAP adherence over twenty years of data collection: a flattened curve," *J. Otolaryngol. – Head Neck Surg.*, vol. 45, no. 1, 2016, doi: 10.1186/S40463-016-0156-0.

[63]   P. Anderer, G. Gruber, S. Parapatics, *et al.*, "An E-health solution for automatic sleep classification according to Rechtschaffen and Kales: validation study of the Somnolyzer 24 × 7 utilizing the siesta database," *Neuropsychobiology*, vol. 51, no. 3, pp. 115–133, 2005, doi: 10.1159/000085205.

[64]   K. Kawamoto, C. A. Houlihan, E. A. Balas, and D. F. Lobach, "Improving clinical practice using clinical decision support systems: a systematic review of trials to identify features critical to success," *Br. Med. J.*, vol. 330, no. 7494, pp. 765–768, 2005, doi: 10.1136/BMJ.38398.500764.8F.

[65]   R. A. Raisa *et al.*, "Deep and shallow learning model-based sleep apnea diagnosis systems: a comprehensive study," *IEEE Access*, vol. 12, pp. 122959–122987, 2024, doi: 10.1109/ACCESS.2024.3426928.

[66]   M. R. Bonsignore, M. C. Suarez Giron, O. Marrone, A. Castrogiovanni, and J. M. Montserrat, "Personalised medicine in sleep respiratory disorders: focus on obstructive sleep apnoea diagnosis and treatment," *Eur. Respir. Rev.*, vol. 26, no. 146, 2017, doi: 10.1183/16000617.0069-2017.

[67]   "(PDF) CPAP Hardware/Simulation and Control Design for Respiratory Disorders: A Review." https://www.researchgate.net/publication/357478815_ CPAP_HardwareSimulation_and_Control_Design_for_Respiratory_Disorders_ A_Review (accessed Jan. 31, 2025).

[68]   J. Zhang and Y. Wu, "Complex-valued unsupervised convolutional neural networks for sleep stage classification," *Comput. Methods Programs Biomed.*, vol. 164, pp. 181–191, 2018, doi: 10.1016/j.cmpb.2018.07.015.

[69]   A. Malhotra, I. Ayappa, N. Ayas, *et al.*, "Metrics of sleep apnea severity: beyond the apnea-hypopnea index," *Sleep*, vol. 44, no. 7, 2021, doi: 10.1093/ SLEEP/ZSAB030.

[70]   K. Li, J. Rollins, and E. Yan, "Web of Science use in published research and review papers 1997–2017: a selective, dynamic, cross-domain, content-based analysis," *Scientometrics*, vol. 115, no. 1, pp. 1–20, 2018, doi: 10.1007/ S11192-017-2622-5.

[71]   C. Davies, J. Y. Lee, J. Walter, *et al.*, "A single-arm, open-label, multi-center, and comparative study of the ANNE sleep system vs polysomnography to diagnose obstructive sleep apnea," *J. Clin. Sleep Med.*, vol. 18, no. 12, pp. 2703–2712, 2022, doi: 10.5664/JCSM.10194.

[72]   J. M. Montserrat, M. Suárez-Girón, C. Egea, *et al.*, "Spanish society of pulmonology and thoracic surgery positioning on the use of telemedicine in sleep-disordered breathing and mechanical ventilation," *Arch. Bronconeumol. (English Ed.)*, vol. 57, no. 4, pp. 281–290, 2021, doi: 10.1016/J.ARBR. 2021.02.001.

[73]   L. Pinilla, I. D. Benitez, F. MArtos, *et al.*, "Plasma profiling reveals a blood-based metabolic fingerprint of obstructive sleep apnea," *Biomed. Pharmacother.*, vol. 145, 2022, doi: 10.1016/J.BIOPHA.2021.112425.

[74]   O. Tsinalis, P. M. Matthews, and Y. Guo, "Automatic sleep stage scoring using time-frequency analysis and stacked sparse autoencoders," *Ann. Biomed. Eng.*, vol. 44, no. 5, pp. 1587–1597, 2016, doi: 10.1007/S10439-015-1444-Y.

Chapter 6

Optimized audio signal reconstruction for AI-driven diagnosis of chronic respiratory conditions

S. Mayakannan[1], K. Prabhu[2], T. Kanagasabapathy[2], G. Meena Devi[3], Y. Zh. Akimbayev[4], C. Selvaraj[5] and Hadeel Alsolai[6]

Abstract

The characteristic feature of chronic obstructive pulmonary disease (COPD) is the liberal failure in lungs functioning over a long duration. In terms of worldwide public health concern, these have surpassed cancer in the past 20 years. There are still a lot of obstacles to overcome in the field of COPD, particularly in terms of predicting how often specific patients' conditions would worsen, keeping tabs on their health, and identifying signs of lung function degradation early on. We urgently require scalable AI-powered data-driven ways to tackle this critical challenge in the treatment of COPD in the modern day. This study lays the experimental groundwork for intelligent diagnosis and monitoring of COPD for individual patients by collecting and generating data from biological observations, using optimal behavior signal processing, and applying machine learning (ML). We also conducted machine classification on lung audio signals in two separate studies and looked into multi-resolution analysis and compression. First, those involving the "Healthy or "COPD" classes, and second, those involving the "Healthy", "COPD", or "Pneumonia" classes individually. The original audio recordings were also tested and signal reconstruction was done using the retrieved features for ML. They outperformed the chosen ML-based classifiers across a range of metrics. The classifications of Healthy and COPD as well as Healthy, COPD, and Pneumonia

[1]Department of Mechanical Engineering, Rathinam Technical Campus, India
[2]Department of Computer Science and Engineering, Dr. Mahalingam College of Engineering and Technology, India
[3]Department of Mathematics, St. Joseph's College of Engineering, India
[4]Department of Defense Research, Center for Military-Strategic Research, Republic of Kazakhstan
[5]Jain Online, Jain Deemed to be University, India
[6]Department of Information Systems, College of Computer and Information Sciences, Princess Nourah Bint Abdulrahman University, Saudi Arabia

have shown encouraging results in this research. The discussion now turns to the findings, their practical implementation, an examination of classification methods, and concludes with future work recommendations and a brief summary of the results.

Keywords: Compressed sensing; signals reconstruction; machine learning; dictionary learning; COPD; artificial intelligence

6.1   Introduction

Globally, chronic obstructive pulmonary disease (COPD) ranked fifth in terms of cause of mortality in the year 2000, as reported by the World Health Organization (WHO) [1]. However, COPD surpassed all other killers in 2018 and is expected to overtake all others by 2030. A degenerative inflammatory illness that persistently restricts airflow and causes various lung problems characterizes COPD, a complicated respiratory ailment [2].

However, patients with COPD are more likely to require readmission to the hospital following their first release due to acute exacerbations, which can necessitate emergency hospitalization. As the prevalence of COPD continues to rise, the already high cost of healthcare for this condition is projected to continue rising sharply [3]. COPD was a £1.9 billion annual expense for the NHS in the United Kingdom. Therefore, it is crucial for healthcare providers to prioritize the inhibition, premature recognition, and organization of COPD situations [4]. As a result, novel decision support systems are required to aid doctors in tracking, recognizing, and comprehending COPD symptoms, which in turn allows for the early prevention of exacerbation episodes. Timely medication delivery to the homes of COPD patients is only one example of how these tools will help the rest of the healthcare community with their own unique care operations [5,6]. It is not yet practical to send enough medical professionals to keep tabs on COPD patients in their homes and hospitals. Consequently, it is essential to choose alternative strategies to address the current and future therapy requirements of COPD patients. Knowledge extraction and big data analysis have become more easier in the last 20 years due to the proliferation of wearable sensors, advancements in information and communication technology, and critical health issues and processes. With the use of real-time data obtained via sensors and measurements, there is an opportunity to assess and diagnose patients' symptoms and subconditions related to chronic respiratory illnesses. We can learn more about the risk of severe exacerbation episodes with lung function failures in patients with COPD and other respiratory illnesses if we confirm these disorders using machine learning (ML) and classification technologies [7].

Adventitious lung sounds developed on top of normal lung sounds that can be caused by airway or lung injury or obstruction. Typically, there are two basic ways to classify them based on observations and measurements: continuous for around 250 ms and discontinuous for approximately 25 ms [8]. Wheezing and other continuous lung sounds are common in COPD, while crackles and other discontinuous sounds are common in pneumonia [9]. An inadequate signal-to-noise ratio is also caused by the presence of internal and external noise, as well as the heart's and digestive system's background noise, within the respiratory auscultations. In addition, the noises overlap in both the frequency and time domains, and the signals are inherently non-stationary due to the breathing rhythm, with statistics that change over time.

It is difficult to reconstruct and classify pulmonary auscultation signals when those signals are not steady, when they are fleeting and comprise discontinuous crackling sounds, or when the noises overlap in both time and frequency. But the noises come from all across the three-dimensional lung organ, and auscultation records only come from one spot. Disentangling these blended noises becomes much more difficult as a result of this [10].

A low band or pass filter will be used frequently to differentiate lung sounds from heart sounds [11]. On the other hand, aliasing and other undesired effects are possible with low pass filters. Not removing the heart sounds also had little impact on the outcomes. So, it's possible that isolating the heart sounds is not necessary in this case. For the purpose of time-frequency analysis of lung illnesses and adventitious sounds, researchers have utilized a variety of transform approaches, including empirical mode decomposition (EMD), short-time Fourier transforms (STFT), and wavelet transforms (WTs) [12]. It has been shown that STFT and EMD struggle to extract features from overlapping, non-stationary signals, and that Fourier-based algorithms are unable to detect transitory signals [13]. While EMD is effective on signals that do not overlap, it is unable to differentiate between crackles in lung sounds that overlap. Furthermore, compared to STFT alone, the method that combined STFT with continuous WT yielded better results. More nuanced signal features can be captured by WT through multi-resolution analysis [14]. Mechanical equipment structural health monitoring can also benefit from multi-resolution analysis's ability to detect impulsive and transient signals in noise [15]. The WT multi-resolution form's abundance of features facilitates the use of EMD and STFT for feature extraction from acoustic auscultation of the lungs.

Although STFT, EMD, and WT offer inverse transformations about signal reconstruction of respiratory auscultations utilizing significant representative qualities. Recent developments in compressed sensing and signal reconstruction have made it feasible to transmit pulmonary auscultation sounds directly from a sensor to a mobile device. Signal reconstruction relies on selecting characteristics that can match input and output features to extract the most important information from original audio signals. Consequently, this study relies heavily on signal reconstruction before applying ML to classify respiratory disorders based on their most salient properties. The following is the intended structure and presentation of this chapter: research data, data cleaning, data transformation, feature reduction approaches, and reconstruction outcomes. We next move on to a discussion of the findings, their implementation, and a study of classification methods, before concluding with recommendations for future work and providing a brief summary of the findings.

6.2   Case study

This study relied on the International Conference on Biomedical and Health Informatics (ICBHI) Respiratory challenge database for its data [16]. There are 126 patients' audio recordings (830 total) in the dataset. The audio samples range from 30 to 90 s in length, have a sampling rate between 4000 and 44,100 Hz, and include either one or two channels. Each patient's demographics and diagnosis are included in the accompanying material. We employed the Healthy, COPD, and Pneumonia auscultation diagnostic classes for this study. You can see the demographic breakdown of the classes used in Table 6.1.

The audio samples were not uniform in length, which was a problem for the modeling process. To account for this, we randomly chose a seven-second segment that would capture a breathing cycle, which can have any number of revolutions per minute between twelve and eighteen [17]. Due to imbalances between the classes, two of the five data augmentation approaches were utilized to ensure that the Healthy and Pneumonia audio parts were unique from each other. As far as augmentation goes, there are a few options: time-stretching (it entails adjusting the tempo of the sound), pitch-shifting (which contains moving the audio frequency up or down), additional noise (which involves adding extra noise), time-shifting (which involves rolling forward or backward the audio), and no augmentation at all [18,19]. In order to supplement each sample in a unique way, two of the five choices provided permutations of up to 20 distinct options. Both the Healthy and Pneumonia classes saw increase in audio sample size; the former went up from 35 to 735 samples, while the latter went up from 39 to 740.

6.2.1   Audio cleaning and normalization

Noise is reduced and all samples are placed into a normalized format during the pre-processing cleaning stage. The following steps are included in the process:

• Normalization
• Signal smoothening
• Normalizing audio loudness
• Detrending
• Thresholding

Down sampling the audio samples to 4000 Hz before loading them guarantees that they are all at the same sample rate. Thresholding was used to decrease the audio

Table 6.1   Breakdown of chosen classes in the ICBHI 2017 challenge database

Criteria	No. of recordings	Age range (years)		Biological sex (count)	
		Min.	Max.	Female	Male
Pneumonia	37	4	81	7	30
Healthy	35	0.25	16	20	15
COPD	793	45	93	266	512

amplitude outliers that were anticipated due to the stethoscope contact movement. The signal amplitude needs to be reduced to the mean after being thresholded above four standard deviations in order for crackles to be visible within four standard deviations (SDs). Audio noise can be reduced by using a smoothing filter in conjunction with downsampling and outlier removal. To reduce background noise in lung sounds, the Savol filter was chosen [20,21]. This moving filter has a polynomial function. Since the audio samples can show trends and are not stationary, detrending makes them less so [22]. There are two parts to the respiratory audio: the turbulence in the air and the structural noises of the lungs. These parts clash when heard from various places. Consequently, we employ the EUB R128 standardization. The last step is to normalize the numbers so that they all fall within the similar range.

6.2.2   Wavelet transform

The formula shown in (6.1) is used to partition multi-resolution audio streams into separate frequency bands. Ψ^* is the parent wavelet that has a distribution that resembles the transitory crackling with an abrupt peak, hence it is selected as the Morlet wavelet [23].

$$\omega_n(s) = \sum_{n'=0}^{N-1} x_{n'} \psi^* \left[\frac{(n'-n)\delta t}{s} \right] \tag{6.1}$$

This study will employ the complicated Morlet wavelet to analyze the real and imaginary components. Due to WT's inverse transform, noise resilience, and ability to locate audio features, this analysis lends credence to the goals [23,24]. It is possible to convert the output of a multi-resolution analysis into an audio signal by using the inverse transform.

6.2.3   Compressed sensing

A key component of sparse encoder dictionary learning is compressed sensing. Here are the key concepts of compressed sensing:

- Incoherence
- Sparsity

With no geographical or temporal relationship between the samples, the incoherence property makes time-frequency localization even more of a challenge, or the uncertainty problem. This is because there is a greater dispersion and sparsity of samples over the domain [25]. In contrast, there is no correlation between the row and column values in compressed sensing matrices [26]. The distribution of samples is called sparsity, and it allows for the elimination of very small values that are close to zero. Because of this, the data can have a few non-zero elements. Restricting sparsity in compressed sensing allows for a relaxation of the transition from an over-complete solution to the discovery of a unique one [27]. Matrix forms in compressed sensing automatically preserve the restricted isometric property (RIP), which translates linearly to sparsity constraints. Meeting the goal of signal reconstruction is made possible by sub-sampling from subspace, which facilitates feature reduction with a sample rate lower than the Nyquist.

6.2.4 *Dictionary learning*

Dictionary learning (DL) uses a combination of compressed sensing and sparsity factors to relax linear constraints [28]. It does this by using an error-bound element and an incoherence factor over all columns of the dictionary. Algorithms such as orthogonal matching pursuit (OMP) and gradient descent are utilized by DL to assist in finding a sparse representation and repair procedure [29,30]. These techniques pick atom samples from the dictionary that are highly linked. Equation (6.2) is used to calculate dictionary learning.

$$[u, v] = argmin \frac{1}{2} ||X - uV||_{L2} + \alpha \times ||u||_{L1}(u, V) \ \ with \ \ ||V_k||_2$$
$$\leq 1 \ for \ all \ 0 \leq k < n\_components \tag{6.2}$$

By using DL, the matrix for analysis at several resolutions can be decomposed into fewer components. By multiplying these components and applying the transform, an approximation of the original matrix can be recreated.

6.2.5 *Singular value decomposition*

Singular value decomposition (SVD) is a technique that can simplify matrices by reducing them to a set of three, regardless of their complexity. Mostly used in signal processing, it takes complex signals and transforms them into feature matrices that are more reflective of their original state, making their handling much easier. In particular, the approach reveals numerous intriguing and crucial representational properties of signals derived from the initial matrix. Here is an example of SVD in action, using the particular case of real matrices:

$$A = u \sum V^T \tag{6.3}$$

Here, A is the result of multiplying prime numbers with natural numbers. The left-singular vectors are the columns in an orthogonal matrix U, which has dimensions of $(n \times n)$. The diagonal of $\sum$, which is same size as A $(n \times p)$, contains the so-called singular values. To describe the rows of the matrix $(p \times p)$, V^T denotes the transposal matrix of V, the terms "right singular vectors" are employed. Further, for SVD calculations to be performed, the eigenvalues and vectors of AAT and ATA need to be retrieved. Both V and U's columns are based on the eigenvectors. The diagonal entries in the $\sum$ matrix stand for the singular values. The standard order is from highest to lowest. Similar to AAT or ATA, the square roots of the eigenvalues are applicable here. We also mention that SVD can help with noise reduction signals by decomposing matrices based on their characteristics; this yields the most interesting aspects for the signal and guarantees that the original matrix can be recovered using SVD matrices operations.

6.2.6   *Metrics of signal reconstruction*

By contrasting the preprocessed and rebuilt signals, one may see how well the signal reconstruction worked. Consequently, statistics like the correlation coefficient and the mean square error (MSE) can be employed to measure the similarity of signals.

Equation (6.4) [31] calculates the variation between the original signal (A) and the regenerated signal (B), and the standard deviation of this variation is measured by the MSE. The MSE displays the typical variation in the separation of two signals.

$$Mean\ Square\ Error = \frac{\sum_n (A[n] - B[n])^2}{m} \qquad (6.4)$$

Equation (6.5) calculates the coefficient of correlation among signals A and B, which is another measure of signal similarity [32].

$$Corr\ Coef = \frac{\sum (A_i - \overline{A}) \sum (B_i - \overline{B})}{\sqrt{\sum (A_i - \overline{A})2(B_i - \overline{B})2}} \qquad (6.5)$$

when A denotes the mean of the real signal and B denotes the mean of the retrieved signal. There is a linear relationship between the signals, as shown by the correlation coefficient.

6.2.6.1   Overview of extracted features

There were 153 features in U, 90 in V^T, and 9 in S, according to the features recovered by the framework. The signal's real and imaginary parts had the same feature count.

6.2.6.2   Outcomes of signal reconstruction

Table 6.2 displays the findings of the signal reconstruction.

6.2.6.3   Signal reconstruction overview

The precision of the reconstruction is almost identical to that of the preprocessed and reconstructed signals, as indicated by the MSE results, which range from an average of 3×10^{-3} to a maximum of 5×10^{-4}. Same goes for the correlation

Table 6.2   Analyzed results for signal reconstruction using features

Stats	Coefficient of correlation	MSE
Std	0.150377	0.012137
Mean	0.576079	0.030668
Count	2268	2268
Minimum	0.014053	0.005188
0.3	0.488954	0.022598
0.6	0.582712	0.029151
0.9	0.682031	0.036262
Maximum	0.924803	0.142799

coefficients; the average is 0.57 and the maximum is 0.92. Findings show that the rebuilt audio signal is very close to the original, unprocessed signal.

6.2.7   Classification

Both "Healthy"/"COPD" and "Healthy", "COPD", and "Pneumonia" were considered in the research. The adventitious noises in pneumonia are primarily crackles, which enables for distinction between the two classes, while COPD is characterized by wheezing. The complex Morlet wavelet classifies each signal component by revealing its real and imaginary parts. As far as classification models go, there are four options: RFC, Support Vector Machine (SVM), decision tree classifier, and Gaussian mixture model (GMM).

One possible explanation for the similarity in the frequencies of lung sounds is the GMM, a classification system that permits clusters with overlapping borders and a Gaussian distribution. To achieve objectivity in its classifications, Decision Tree Classifier (DTC) employs a divide-and-conquer technique, which promotes openness. The SVM employs a borderline parting or, in the case of high-dimensional data, a hyper-plane separation to sort data into linear, polynomial, quadratic, or higher-order categories. One efficient data mining technique is the RFC ensemble approach, which uses a number of trees to generate a bias-variance decomposition-like result. The use of random bagging with replacement from training data and features bootstrapping both provide credence to its performance-enhancing claims [33]. Random forests can also inform how essential certain features are, making them perfect for classification. In order to discover the optimal model parameters, grid search iteratively cycles through them. The model's efficiency is enhanced as a result. The RFC's grid search settings allow for an estimation count between 100 and 600, with 50-unit increments, and a depth between 10 and 100, with 10-unit increments.

6.2.7.1   Metrics classification

By comparing the models' false positives (F.P.), true negatives (T.N.), false negatives (F.N.), and true positives (T.P.), we may analyze their performance. We utilized area under the curve (AUC), accuracy, and F1 scores for this study. By comparing one class to the other two, the one-versus-all categorization will be used to generate the ROC curves for the Healthy, COPD, and Pneumonia categories. In order to thoroughly examine the model's performance, we employ five-fold cross-validation and utilize the five-fold averages and the cross-validated SD as per outputs [34,35]. Statistical confidence ranges of 95% are provided with reports on the model's performance level.

6.3   Results

6.3.1   Classification results of health and COPD

The study presents an outcome graphically, including ROC, baseline data, and AUC plots, and optimization outputs for the model's parameters. The preliminary results for both the healthy and COPD groups are indicated in Table 6.3.

Table 6.3 Baseline results for Healthy versus COPD categorization

Specifications	Model classification	Accurateness	F1-score
Singular value decomposition V^T, Imaginary	Depth (d) = 500, RFC, No. of estimators (e) = 280	72	71
	DTC	59	59
	GMM, components = 2	48	47.5
	SVC	54	53.5
Singular value decomposition V^T, real	Depth (d) = 500, RFC, No. of estimators (e) = 280	72	71
	DTC	59	59
	GMM, components = 2	47	35
	C = 3000, SVC	54	53.5
Singular value decomposition S, Imaginary	Depth (d) = 500, RFC, No. of estimators (e) = 280	72	71
	DTC	64	63
	GMM, components = 2	54	35
	C = 3000, SVC,	54	35
Singular value decomposition S, real	Depth (d) = 500, RFC, No. of estimators (e) = 280	71	71
	DTC	60	60
	GMM, components = 2	38	35.5
	C = 3000, SVC,	54	35
Singular value decomposition U, Imaginary	Depth (d) = 500, RFC, No. of estimators (e) = 280	79	78.5
	DTC	69	69
	Components = 2, GMM	53	37
	C = 3000, SVC	70	70
Singular value decomposition U, real	Depth (d) = 500, RFC, No. of estimators (e) = 280	80	78.5
	DTC	70	69.5
	GMM, components = 2	44	33.5
	C = 3000, SVC	69	68.5

Table 6.4 displays the outcomes of further parameter tuning of the SVD and random forest. Confidence intervals and cross-validation scores are provided [36].

A classification model's discriminative power can be seen in a ROC curve. Figure 6.1 compares the various models. In addition, the Random Forest classifier is used to show an analysis of the actual and imaginary components. Figure 6.2 displays the results of the ROC curve.

Figures 6.3 and 6.4 illustrate the findings of the ROC curvature for the pneumonia, COPD, and Health classes.

Table 6.4   Classification of parameter tuning outcomes for Healthy versus COPD

Classification	Model classification	Accurateness	Macro F1-score	CV score	CI 95%	CV std.
Singular value decomposition U, image	Depth = 30, RFC, No. of estimators = 390	80	79.5	76	74–79	5
	C = 80, SVC, 190.1	70	70			
Singular value decomposition U, real	Depth = 25, RFC, No. of estimators = 390	79	78.5	76	73–78	5
	C = 2265.8, SVC	69	68.5			
Singular value decomposition V^{T}, image	Depth = 20, RFC, No. of estimators = 400	80	79.5	76	74–79	5
	C = 58, SVC, 523.6	70	70			
Singular value decomposition V^{T}, real	Depth = 20, RFC, No. of estimators = 400	73	72.5	68	65–70	5
	C = 17,911.6, SVC,	54	53.5			
Singular value decomposition S, image	Depth = 30, RFC, No. of estimators = 400	72	71	73	70–74	5
	C = 2764.8, SVC	54	35			
Singular value decomposition S, Real	Depth = 25, RFC, No. of estimators = 390	72	72	73	70–75	6
	C = 1251.9, SVC	54	35			

6.3.2   COPD, pneumonia, and healthy classification results

Table 6.5 displays the initial outcome for the diagnosis of healthy, COPD, and pneumonia.

Based on the results provided in Table 6.6, the best performing classifiers to move forward with parameter tuning are the random forest and SVC.

6.3.3   Overview of classification results

The most effective models were random forest models for the good health against COPD against good Health and good Health against COPD against Pneumonia categories. The random forest model utilized several features for classifying healthy individuals versus those with COPD, with the SVD VT and U features pertaining to the imaginary element of examination audio proving most significant, achieving 80% accuracy, 0.87 area under the receiver operating characteristic (ROC) curves, and 0.87 discrimination. The random forest classifier performed a fantastic job comparing the healthy, COPD, and pneumonia groups, as seen in Figure 6.3. S (Singular) SVD values yielded ideal features that improved auscultation recording accuracy to 70% for the actual component and 68% for the imaginary component. Results for the actual component (between 0.83 and 0.89) and the imaginary component (0.82–0.84), as shown in Figure 6.3(c) and (f), respectively, for class discrimination on SVD S elements using the random forest model were quite close.

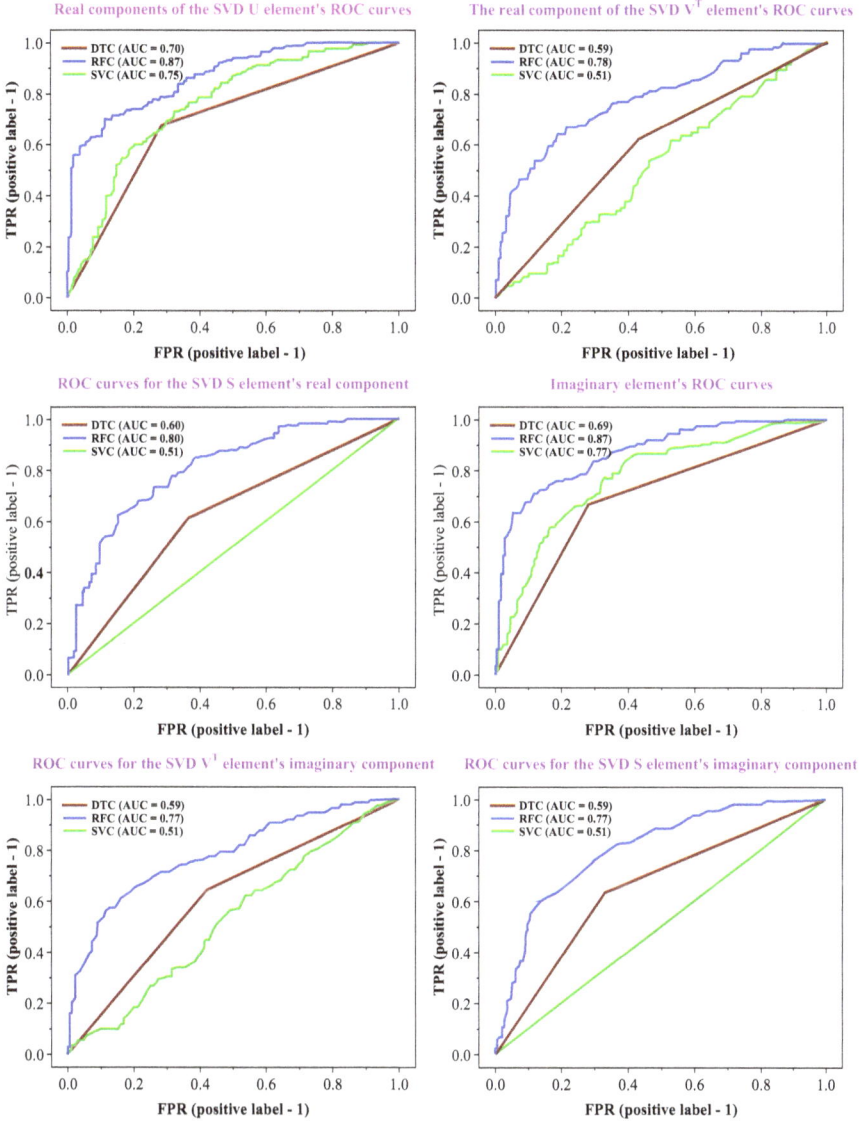

Figure 6.1   ROC curves for SVD component-specific classification models

6.4   Discussion

Compared to both the healthy and COPD patient groups, the orthogonal SVD elements and imaginary components of the signal perform better. COPD wheezes may have a role in this. Seeing the classification results is a positive thing. To achieve a reasonable 80% accuracy in the Healthy versus COPD category at 95% confidence

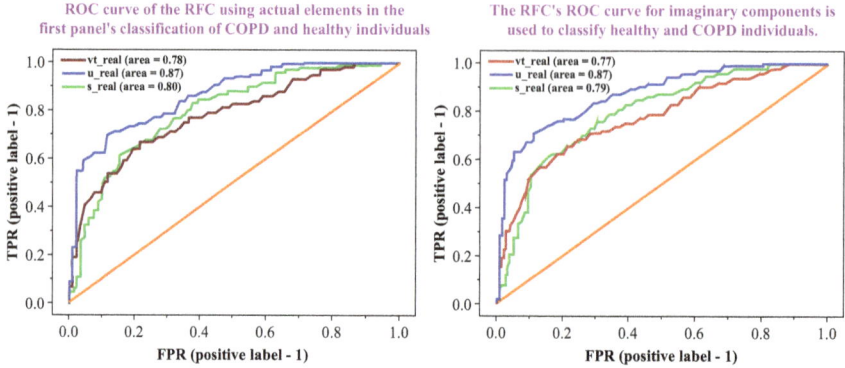

Figure 6.2   ROC graph for the arrangement of healthy individuals and those with COPD, pertaining to each signal component and singular value decomposition elements

levels of 76–79%, the singular value decompositions U and V^T essentials were trained using the imaginary components of the audio signals. There was a 70% success rate when it came to health issues, COPD, and pneumonia. The condition discrimination was good, and the 95% confidence range for the real audio signal components on the SVD's S (single values) was 66–70%. An ideal result for signal reconstruction would be an average score of 3.0×10^{-2}, a correlation value of 0.92, and an MSE of 5.2×10^{-3}. The signal recovery seems to be top-notch. We saw that the real signal component with the singular value decompositions element—which pertains to the strength of the signal —performed better when the confusion matrix had higher classification numbers for COPD and pneumonia, respectively, when we looked closely at the results in the good Health, COPD, and Pneumonia groups.

When it came to recognizing typical lung sounds, wheezes, and crackles, WT received scores between 39.97 and 40.96% on the ICBHI database. This study's diagnostic accuracy was higher for benign, COPD, and pneumonia than for other possible causes of auditory hallucinations. Furthermore, they endeavored to categorize normal, wheezes, crackles, and hybrids thereof with the aim of exceeding the 50% accuracy threshold established for the ICBHI 2017 challenge database. There was a patient's audio from a respiratory illness that was detected in the dataset, but there was probably an issue with the annotations that failed to detect any adventitious sound. However, a lack of adventitious sounds does not necessarily indicate health. F1 score of 82.75% was similar to the 83% achieved by the best models for Healthy versus COPD; we used deep learning and discrete WTs to classify the ICBHI 2017 challenge database as healthy or unhealthy. Unhealthy referred to a wider variety of disorders than just COPD, which was the primary focus of this investigation. A 17-layered two-dimensional CNN achieved an impressive accuracy of 92.30% in classifying auscultation recordings from the ICHBI dataset into their corresponding illnesses by utilizing MFCC and spectrogram features.

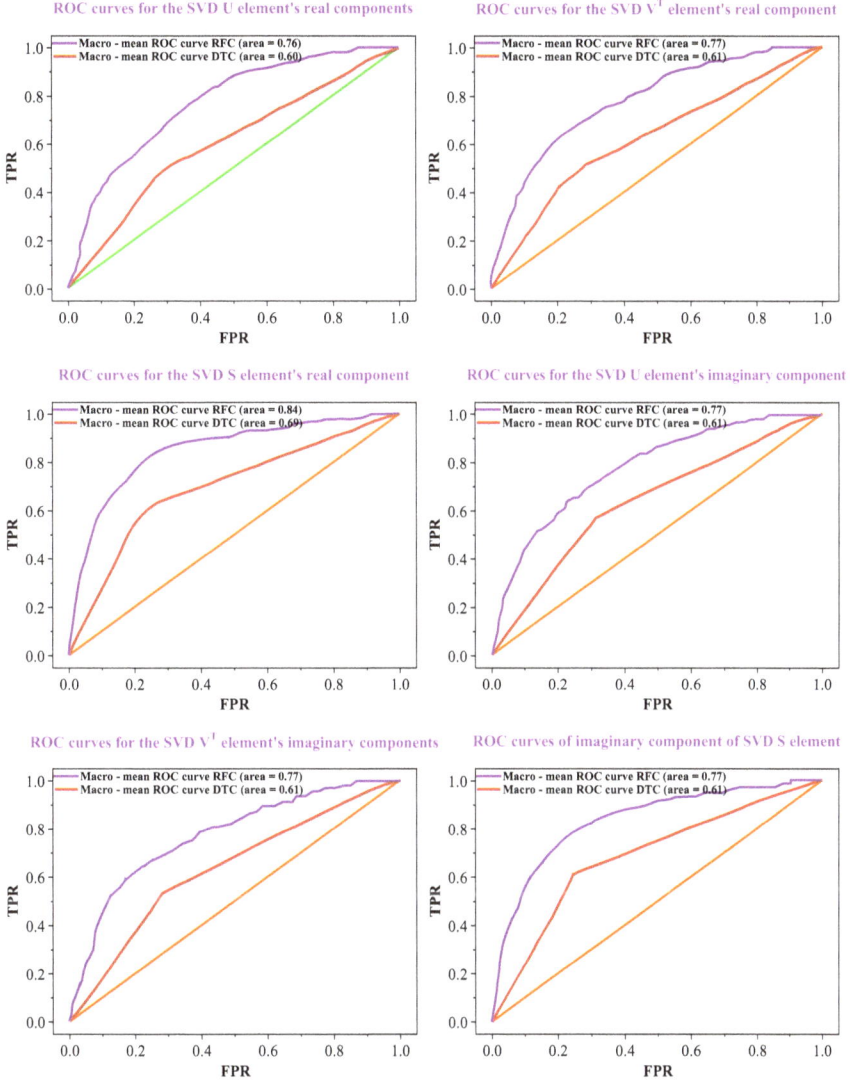

Figure 6.3 Evaluating the real and imaginary components of the signals for each SVD element and classifying them as healthy, COPD, or pneumonia using DTC and RFC

One major advantage of the proposed method is its ability to effectively recover and reconstruct signals, enabling highly accurate approximations of the original signal. Achieving such a high level of precision in signal recovery through ML facilitates a deeper understanding of human respiratory issues, while also delivering exceptional accuracy in health condition classification. In the context of respiratory auscultation classifications, the method proves to be highly practical

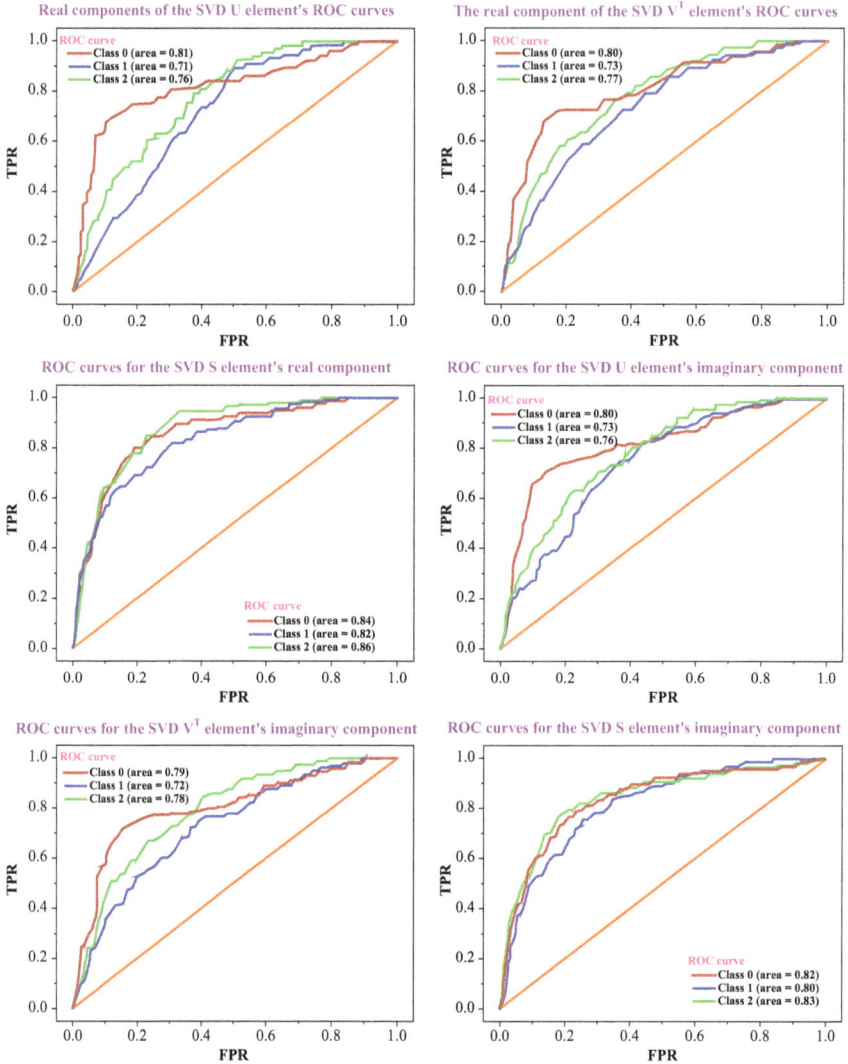

Figure 6.4   RFC categorization of Healthy, COPD, and Pneumonia based on one-versus-rest ROC graph for each singular value decomposition element, taking into account both real and imaginary signals

and supports reliable assumptions about health conditions. When it comes to respiratory auscultation classifications, this method is very practical and lends credence to the health condition assumptions.

Furthermore, prior research has primarily concentrated on methods based on statistics and neural networks; nevertheless, this research findings unveil an innovative strategy for classifying auscultation by compressed sensing. However, in

Table 6.5   Initial classification outcomes for Healthy, COPD, and Pneumonia

Specification	Model classification	Accuracy	F1-score
SVD S, real	Depth (d) = 500, RFC, No. of estimators (e) = 280	70	69.7
	Components = 2 GMM	40	22
	DTC	56	55.3
	C = 3000, SVC	39	19
SVD U, real	d = 500, RFC, e = 280	51	59.7
	Components = 2 GMM	37	30.3
	DTC	60	50.7
	C = 3000, SVC	46	45
SVD V^T, real	d = 500, RFC, e = 280	60	59.3
	Components = 2 GMM	32	31
	DTC	46	46
	C = 3000, SVC	45	44.7

Note: All baseline results were acquired utilizing the subsequent parameter setups.

Table 6.6   Classification of parameter tuning results for Healthy, COPD, and Pneumonia

Types	Model classification	F1-score	Accurateness	CV std	CV score	95% CI
SVD U, image	d = 30, RFC, e = 500	61.4	62	5.0	59	57–62
	C = 1536.9, SVC	47	48			
SVD U, real	d = 20, RFC, e = 300	59.8	60	4	59	57–60
	C = 1143.9, SVC	44.8	46			
SVD V^T, image	d = 20, RFC, e = 400	63.4	64	3.9	60	58–62
	C = 1536.9, SVC	50	51			
SVD V^T, real	d = 40, RFC, e = 500	61.4	62	5	60	58–62
	C = 1839.8, SVC	47.8	48			
SVD S, image	d = 20, RFC, e = 300	68.8	69	4.3	69	66–71
	C = 1536.9, SVC	20.8	40			
SVD S, real	d = 20, RFC, e = 400	71	71	4.6	69	67–71
	C = 1536.9, SVC	22				

order to improve the precision of machine classifications and signal recovery, more modification of the extraction process must be implemented in tandem with massive experimental datasets. Improving COPD diagnosis and prognosis should be the primary goal of future research utilizing multi-modal data and lexicon learning.

6.5   Conclusions

In addition to producing high-performing machine classification of respiratory lung sounds, the established benchmark work in this study offers very accurate signal reconstruction. We talk about this in light of the chronic health issues that often accompany them. When it comes to recognizing "Healthy" and "COPD" examples, the random forest classifier achieves the best performance in machine classification, with accuracies of about 80%. Classifying instances, such as "Healthy", "COPD", and "Pneumonia", it achieves accuracies of around 70%. The stability of the models was demonstrated by the confidence intervals that were calculated for all of these. Although there are limits, the ROC curves demonstrate the classifiers' discriminatory ability. This research findings may also be useful for a variety of other respiratory illness categories and possibly beyond. Nevertheless, additional effort is required, since we must initially validate the classifiers on significantly bigger and more varied datasets in order to enhance their performance to higher levels. In order to improve the accuracy and efficacy of this method at scale, we plan to conduct future research on chronic respiratory disorders affecting the obstructive pulmonary system utilizing bigger datasets. We will also try to figure out which lung noises are associated with different types of COPD subtypes, with a focus on the ones that are likely to cause patients' exacerbations. Efforts will soon be made to autonomously predict, using appropriate parameters, the likelihood of such potentially dangerous events happening ahead of time, so that medical treatments to patients in critical breathing situations can be expedited.

References

[1] K. G. Isangula and R. J. Haule, "Leveraging AI and machine learning to develop and evaluate a contextualized user-friendly cough audio classifier for detecting respiratory diseases: protocol for a diagnostic study in rural Tanzania," *JMIR Res Protoc*, vol. 13, no. 1, 2024, doi:10.2196/54388.

[2] F. A. Mostafa, L. A. Elrefaei, M. M. Fouda, and A. Hossam, "Diagnosis of lung diseases from chest X-ray images using different fusion techniques," in *2023 11th International Conference on Information and Communication Technology, ICoICT 2023*, 2023, pp. 429–435. doi:10.1109/ICoICT58202.2023.10262761.

[3] E. Melekoglu, U. Kocabicak, M. K. Uçar, C. Bilgin, M. R. Bozkurt, and M. Cunkas, "A new diagnostic method for chronic obstructive pulmonary disease using the photoplethysmography signal and hybrid artificial intelligence," *PeerJ Comput Sci*, vol. 8, 2022, doi:10.7717/PEERJ-CS.1188.

[4] S. Baheti, S. Daware, A. Warokar, J. Arora, M. Nayak, and R. Bawane, "A review on diagnosis of chronic obstructive pulmonary disease," *Journal of Medical Pharmaceutical and Allied Sciences*, vol. 12, no. 2, pp. 5693–5695, 2023, doi:10.55522/jmpas.V12I2.4561.

[5] A. U. Haq, Jian Ping Li, Inayat Khan, Bless Lord Y. Agbley, Sultan Ahmad and M. Irfan Uddin, "DEBCM: deep learning-based enhanced breast invasive ductal carcinoma classification model in IoMT healthcare systems," *IEEE J Biomed Health Inform*, vol. 28, no. 3, pp. 1207–1217, 2022.

[6] S. Saif, P. Das, S. Biswas, S. Khan, M. A. Haq, and V. Kovtun, "A secure data transmission framework for IoT enabled healthcare," *Heliyon*, vol. 10, no. 16, 2024.

[7] M. Azrour, J. Mabrouki, A. Guezzaz, S. Ahmad, S. Khan, and S. Benkirane, *IoT, Machine Learning and Data Analytics for Smart Healthcare.* CRC Press, 2024.

[8] R. Girimurugan, Pon. Maheskumar, G. Sahoo, A. Sivalingam, and S. Mayakannan, "Effect of nano alumina powder and water hyacinth stem powder addition on tensile properties of polypropylene matrix hybrid composites – An experimental study," *in Materials Today: Proceedings*, 2022, pp. 2099–2104. doi: 10.1016/j.matpr.2022.01.477.

[9] M. J. Antony, B. P. Sankaralingam, S. Khan, A. Almjally, N. A. Almujally, and R. K. Mahendran, "Brain–computer interface: the HOL–SSA decomposition and two-phase classification on the HGD EEG data," *Diagnostics*, vol. 13, no. 17, p. 2852, 2023.

[10] B. Ramesh, S. Sathish Kumar, A. H. Elsheikh, S. Mayakannan, K. Sivakumar, and S. Duraithilagar, "Optimization and experimental analysis of drilling process parameters in radial drilling machine for glass fiber/nano granite particle reinforced epoxy composites," *in Materials Today: Proceedings*, 2022, pp. 835–840. doi:10.1016/j.matpr.2022.04.042.

[11] C. Sharma, S. Khan, H. S. Alsagri, A. Almjally, B. I. Alabduallah, A. A. Ansari, "Lightweight security for IoT," *Journal of Intelligent & Fuzzy Systems*, no. Preprint, pp. 1–17, 2023.

[12] G. Sathiaraj, R. Mani, M. Muthuraj, and S. Mayakannan, "The mechanical behavior of Nano sized Al2O3-reinforced Al-Si7-Mg alloy fabricated by powder metallurgy and forging," *ARPN Journal of Engineering and Applied Sciences*, vol. 11, no. 9, pp. 6056–6061, 2016, [Online]. Available: https://www.scopus.com/inward/record.uri?eid=2-s2.0-84968883775&partnerID=40&md5=bc716c47341a5effe624f46afb05bfd9

[13] L. AlSuwaidan, S. Khan, R. Almakki, A. R. Baig, P. Sarkar, and A. E. S. Ahmed, "Swarm intelligence algorithms for optimal scheduling for cloud-based fuzzy systems," *Mathematical Problems in Engineering*, vol. 2022, no. 1, p. 4255835, 2022.

[14] R. Yousef, S. Khan, G. Gupta, B. M. Albahlal, S. A. Alajlan, and A. Ali, "Bridged-U-Net-ASPP-EVO and deep learning optimization for brain tumor segmentation," *Diagnostics*, vol. 13, no. 16, p. 2633, 2023.

[15] R. Girimurugan, P. Selvaraju, P. Jeevanandam, M. Vadivukarassi, S. Subhashini and N. Selvam, "Application of deep learning to the prediction of solar irradiance through missing data," *International Journal of Photoenergy*, vol. 2023, no. 1, p. 4717110, 2023.

[16] D. D. Solomon, S Khan, S Garg *et al.*, "Hybrid majority voting: prediction and classification model for obesity," *Diagnostics*, vol. 13, no. 15, p. 2610, 2023.

[17] R. M. Saleem *et al.*, "Internet of Things based weekly crop pest prediction by using deep neural network," *IEEE Access*, vol. 11, pp. 85900–85913, 2023.

[18] R. Girimurugan, S. Mayakannan, V. M. Madhavan, and C. Shilaja, "Static structural analysis of roof ventilator turbine blades using ANSYS," in *AIP Conference Proceedings*, AIP Publishing, 2023.

[19] B. Gaddala, S. R. Kandavalli, G. Raghavendran, A. Sivaprakash and R. Rallabandi *et al.*, "Exploring the impact of hybridization on green composites: pineapple leaf and sisal fiber reinforcement using poly (furfuryl alcohol) bioresin," *Zeitschrift für Physikalische Chemie*, no. 0, 2024.

[20] P. Zhang, A. Swaminathan, and A. A. Uddin, "Pulmonary disease detection and classification in patient respiratory audio files using long short-term memory neural networks," *Front Med (Lausanne)*, vol. 10, 2023, doi:10. 3389/fmed.2023.1269784.

[21] L. Brunese, F. Mercaldo, A. Reginelli, and A. Santone, "A neural network-based method for respiratory sound analysis and lung disease detection," *Applied Sciences (Switzerland)*, vol. 12, no. 8, 2022, doi:10.3390/ app12083877.

[22] M. S. Rao, S. Modi, R. Singh, K. L. Prasanna, S. Khan, and C. Ushapriya, "Integration of cloud computing, IoT, and Big Data for the development of a novel smart agriculture model," in *2023 3rd International Conference on Advance Computing and Innovative Technologies in Engineering (ICA-CITE)*, IEEE, 2023, pp. 2779–2783.

[23] A. A. Shah, S. K. Devana, C. Lee, R. Kianian, M. van der Schaar, and N. F. SooHoo, "Development of a novel, potentially universal machine learning algorithm for prediction of complications after total hip arthroplasty," *Journal of Arthroplasty*, vol. 36, no. 5, pp. 1655–1662.e1, 2021, doi:10.1016/j.arth.2020.12.040.

[24] S. Kaur, E. Larsen, J. Harper *et al.*, "Development and validation of a respiratory-responsive vocal biomarker–based tool for generalizable detection of respiratory impairment: independent case-control studies in multiple respiratory conditions including asthma, chronic obstructive pulmonary disease, and COVID-19," *Journal of Medical Internet Research*, vol. 25, 2023, doi:10.2196/44410.

[25] J. Sanjana, P. P. Naik, M. A. Padukudru, S. G. Koolagudi, and J. Rajan, "Attention-Based CRNN Models for Identification of Respiratory Diseases from Lung Sounds," in *2023 14th International Conference on Computing Communication and Networking Technologies, ICCCNT 2023*, 2023. doi:10. 1109/ICCCNT56998.2023.10306490.

[26] M. Mosuily, L. Welch, and J. Chauhan, "MMLung: Moving Closer to Practical Lung Health Estimation using Smartphones," in *Proceedings of the Annual Conference of the International Speech Communication Association, INTERSPEECH*, 2023, pp. 2333–2337. doi:10.21437/Interspeech.2023-721.

[27] G. S. Marepalli, P. K. Kollu, and M. D. Inavolu, "Early Detection of Chronic Obstructive Pulmonary Disease in Respiratory Audio Signals Using CNN and LSTM Models," in *Proceedings of InC4 2024 - 2024 IEEE International Conference on Contemporary Computing and Communications*, 2024. doi: 10.1109/InC460750.2024.10648991.

[28] S. Khan and M. Alshara, "Development of Arabic evaluations in information retrieval," *International Journal of Advanced Applied Sciences*, vol. 6, no. 12, pp. 92–98, 2019.

[29] P. Yadav, V. Rastogi, A. Yadav, and P. Parashar, "Artificial Intelligence: a promising tool in diagnosis of respiratory diseases," *Intelligent Pharmacy*, vol. 2, no. 6, pp. 784–791, 2024, doi:10.1016/j.ipha.2024.05.002.

[30] J. Batlle Garcia and I. Benítez *et al.*, "GATEKEEPER's strategy for the multinational large-scale piloting of an eHealth platform: tutorial on how to identify relevant settings and use cases," *Journal of Medical Internet Research*, vol. 25, 2023, doi:10.2196/42187.

[31] M. Mashika and D. van der Haar, "Mel Frequency Cepstral Coefficients and Support Vector Machines for Cough Detection," in *Lecture Notes in Computer Science (including subseries Lecture Notes in Artificial Intelligence and Lecture Notes in Bioinformatics)*, 2023, pp. 250–259. doi:10.1007/978-3-031-35748-0_18.

[32] T. Yamane, Y. Yamasaki, W. Nakashima, and M. Morita, "Tri-axial accelerometer-based recognition of daily activities causing shortness of breath in COPD patients," *Physical Activity and Health*, vol. 7, no. 1, pp. 64–75, 2023, doi:10.5334/PAAH.224.

[33] C. S. Kumari and K. Seethalakshmi, "Synthesizing radiological insights: enhancing lung disease classification through multimodal imaging," *International Journal of Pharmaceutical Quality Assurance*, vol. 14, no. 4, pp. 1126–1135, 2023, doi:10.25258/ijpqa.14.4.47.

[34] Q. Abbas, M. E. A. Ibrahim, S. Khan and A. R. Baig, "Hypo-driver: a multiview driver fatigue and distraction level detection system," *CMC-computers Mater Contin,* vol. 71, no. 1, pp.1999–2017, 2022.

[35] K. Sattar, T. Ahmad, H. M. Abdulghani, S. Khan, J. John and S. A. Meo, "Social networking in medical schools: medical student's viewpoint," *Biomedical Research,* vol. 27, no. 4, pp.1378–1384,2016.

[36] S. Khan, and S. Alqahtani, "Hybrid machine learning models to detect signs of depression," *Multimedia Tools and Applications,* vol. 83, no. 13, pp. 38819–38837, 2024.

Chapter 7

Emotion recognition using speech

Renu Dalal[1], Manju Khari[2],
Jyoti[1], Samanvay Jatana[3] and Vijay Joshi[3]

Abstract

Detecting emotions in person-to-person interactions is often straightforward, as they can be inferred from face expressions, body language, and voice patterns. However, in human–machine conversations, deciphering human feelings is the major challenge. To enhance this conversation, the concept of "speech emotion recognition" has introduced, aiming to identify emotions based solely on vocal inflections. In this work, based on machine learning approach "deep learning" system for speech emotion recognition has been introduced, incorporating efficient data augmentation techniques. The system's performance is evaluated using two distinct datasets: TESS and RAVDESS. The methodology involves employing various approach such as Mel frequency cepstral coefficients (MFCC), zero crossing rate (ZCR), Mel spectrograms, root mean square value (RMS), and chroma. The foundation of speech emotion identification system rests on the utilization of a convolutional neural network (CNN). Thus, the proposed work attains an impressive accuracy rate of 88.11%.

Keywords: Speech emotion identification; human–computer conversation; deep learning; CNN

7.1 Introduction

Speech emotion recognition (SER) used to recognize the emotional nuances within speech, disregarding the semantic information. Although humans effortlessly excel at this skill in everyday speech interactions, the capacity for automated execution using programmable devices remains an active area of investigation. Infusing machines with emotions is acknowledged as a pivotal aspect of endowing them

[1]University School of Automation and Robotics (USAR), Guru Gobind Singh Indraprastha University, East Delhi Campus, India
[2]Department of Computer and Systems Science, Jawaharlal Nehru University, India
[3]Department of Information Technology, Maharaja Surajmal Institute of Technology, Guru Gobind Singh Indraprastha University (GGSIPU), India

with human-like attributes and behaviours. Enabling robots to comprehend emotions holds the potential for them to offer fitting emotional responses and showcase emotive personalities. In specific scenarios, human roles could potentially be supplanted by computer-generated characters adept at engaging in highly natural and persuasive conversations by tapping into human emotions. It is imperative for machines to decipher the emotional undercurrents carried by speech. Only by attaining this ability can a truly meaningful dialogue, characterized by mutual trust and comprehension between humans and machines, be attained. SER finds applications across a diverse spectrum of fields. For instance, the detection of anger can enhance the efficacy of voice portals and call centres, allowing for tailored service provision aligned with clients' emotional states. This capability extends to domains like civil aviation, where monitoring the stress levels of aircraft pilots contributes to reducing the risk of potential accidents. Furthermore, in the domain of mental health care, an innovative approach involves introducing a psychiatric counselling service facilitated by a chat-bot. This service involves analysing input text data, voice data, and visual cues to discern the individual's psychiatric condition and subsequently offer insights into diagnosis and treatment options. Thus, SER has evolved into not just a challenging subject of study, but also a pivotal research area with far-reaching implications. This work also works as an applications of various fields like opportunistic network, and wireless networks [1–5].

This study aims to create an automatic speech recognition system capable of receiving human speech signals imbued with expressions as input, and subsequently identifying and categorizing them into different emotional states. The research delineates specific goals as follows:

1. To collect and pre-process a dataset of audio recordings containing a diverse range of emotional states.
2. To develop the deep learning model for classification of the emotional states from audio data.
3. To measure the performance factors of the developed model on the collected dataset.

7.2   Related work

SER has garnered significant attention within the academic realm, prompting an array of investigations. This segment delves into a selection of these scholarly articles, shedding light on their accomplishments as shown in Table 7.1.

7.3   Methodology

The SER methodology put forth encompasses the subsequent stages: gathering data, preparing data, enhancing data through augmentation, extracting features, creating models utilizing deep learning methodologies, training and assessing the models thus generated, culminating in the emotion classification process. In this proposed SER approach, a CNN model has been adopted. Additionally, distinct

Table 7.1   Comparison between existing approaches

Title of study	Year	Methodology	Findings
"A CNN-Assisted Enhanced Audio Signal Processing for Speech Emotion Recognition" [6]	2019	It introduced deep stride CNN architecture. The test is conducted on the two datasets: IEMOCAP and RAVDESS datasets.	It was stated that using spectrograms on the enhanced speech signals increases the accuracy and minimize the calculative complexity. This model attains the accuracy of 81.75% with IEMOCAP dataset, and RAVDESS dataset achieved 79.5% accuracy. To add using Depthwise Separable Convolutional Neural Network (DS CNN), the dataset-size is also minimized.
"Convolutional Neural Networks for Speech Emotion Recognition" [7]	2020	Dataset: SAVEE Feature extracted: MFCC (Mel Frequency Cepstral Coefficients) Machine Learning Technique: CNN.	The accuracy is 84.31% with MFCC and SAVEE better than previous work which showed accuracy between 55% and 75%.
"Speech Emotion Recognition with Deep Convolutional Neural Networks" [8]	2020	Dataset: RAVDESS, EMO-DB, IEMOCAP Feature extracted: MFCC, chromagram, Mel-scale spectrogram, Tonnetz representation and spectral contrast features Machine Learning Technique: 1D CNN.	The accuracy was found to be 71.61% for RAVDESS, 86.1% for EMO-DB, and 64.3% for IEMOCAP. Thus, making this model outperform previous works.
"Speech Emotion Recognition using Emotion Perception Spectral Feature" [9]	2019	Dataset: CASIA, EMO-DB, FAU AIBO Feature extracted: Custom feature extraction (perception spectral feature) Machine Learning Technique: SVM classifier.	The accuracy was outstanding compared to MFCC feature techniques. 78.6%, 81.5%, 54.6% accuracy were achieved while using CASIA, EMO-DB, and FAU AIBO datasets, respectively.
"Continuous Speech Emotion Recognition with Convolutional Neural Networks" [10]	2020	In this work, the authors used AESDD (Acted Emotional Speech Dynamic Database) to train with the CNN architecture. The CNN input vectors information can be 1D signal as pulse code modulation (PCM) and 2D signal by using spectro-temporal conversion and Mel-scale coordinate.	The proposed CNN model was 69.2% more accurate. It works better than the SVM technique. It doesn't affect the performance with data enlargement but it showed improved robustness.

(Continues)

Table 7.1　(Continued)

Title of study	Year	Methodology	Findings
"Generalization and Robustness Investigation for Facial and Speech Emotion Recognition using Bio-inspired Spiking Neural Network" [11]	2021	Publicly available datasets, CNN spiking neural network, unsupervised learning, cross dataset evaluation.	Facial emotion recognition (FER) has an accuracy of 89%, while speech emotion recognition (SER) has 70%.
"Speech Emotion Recognition in Neurological Disorders using CNN" [12]	2020	The CNN model is introduced. This methodology used particular tonal traits with RAVDESS and MFCCs dataset used for train the model.	This technique attains more accuracy as compared to the existing conventional models. It also can work with manifests the emotions of the person disabled neurologically.
"Speech Emotion Recognition based on SVM and ANN" [13]	2018	Acoustic and statistical two features were classified and evaluated by SVM and ANN. To minimize the dimension principal component analysis (PCA) method were utilized.	With and without PCA; SVM and ANN were compared. SVM attains more accuracy; 46.67% with PCA and 76.67% without PCA as compared to ANN. It performs well by using feature dimension minimization.
"Emotion Recognition of EEG Signals based on the Ensemble Learning method: AdaBoost" [14]	2021	The technique used for emotion recognition is EEG signals. It is the ensemble learning approach called Ada-Boost is introduced. The diverse areas were considered and non-linear features used for emotion were selected from pre-processing EEG signals. After this the eigenvector matrix were utilized the fused features.	The method proposed in this chapter is tested on the DEAP dataset. This approach works effectively to recognize the emotions with 88% accuracy.

feature selection algorithms—MFCC, ZCR, Mel Spectrogram, RMS, and Chroma—are incorporated as inputs to enhance the effectiveness of the formulated methods. The diagram below provides an overview of the structural framework underpinning the envisioned speech emotion recognition system.

(A)　**Dataset used**
- The Toronto emotional speech set

  The Toronto emotional speech set (TESS) encompasses the collection of 200 designated words, uttered within the framework of the phrase "Say the word _" by two actresses (aged 26 and 64). Recordings of this set were executed to encapsulate a spectrum of seven emotions: anger, disgust, fear, happiness, surprise, sadness, and neutrality. The entirety of the dataset comprises 2800 individual data points, each represented by an audio file. The dataset's arrangement involves the allocation of each of the two female performers, along with their respective emotional expressions, within dedicated folders. The audio files are formatted in WAV format.

- The Ryerson Audio-Visual Database of Emotional Speech and Song
  The Ryerson audio-visual database of emotional speech and song (RAVDESS) is a comprehensive compilation of speech and song, encompassing both audio and video components (amounting to 24.8 GB in size). Within this context, a specific subset of RAVDESS is utilized herein, housing a collection of 1440 files. This number is derived from a calculation involving 60 trials per actor, with a total of 24 actors, leading to the total of 1440 files. RAVDESS comprises a roster of 24 proficient actors (12 female and 12 male), articulating two linguistically matched statements in a neutral North American accent. The speech feelings/ emotions encapsulated with the following expressions: calm, happy, sad, angry, fearful, surprised, disgusted, and neutral.

(B) **Data preparation**

In this important step, the following operations are performed:

- **Data loading:** The datasets comprise data samples stored in the (.wav) format. However, this format is unsuitable for serving as input to the envisioned machine learning models. Thus, the initial step involves uploading the sample-data and converting it from their audio file format to a representation based on time-series.

- **Data labelling:** Alternatively referred to as data annotation, plays a pivotal role in the supervised approach to SER. This process is essential in enhancing the accuracy rate and effectiveness of the machine learning models. This involves assigning specific labels to each sample within the datasets. For instance, let's assign a numerical value (e.g., 1) to characterize emotions like "happy" and extend this approach to other emotions accordingly.

- **Data augmentation:** Two data augmentation strategies have been implemented—specifically, noise incorporation and spectrogram adjustment—to enrich the dataset content employed in this study.

- **Data splitting:** During this final phase of data preparation, the datasets are divided, allocating 75% for training and validation purposes, while reserving 25% for testing.

(C) **Data augmentation**

- **White noise addition:** The rationale underlying this strategy is to enhance the model's adaptability to real-world noise challenges. This involves introducing white noise, generated by infusing random values sourced with the normal distribution, negligible standard deviation and a mean of zero, into each sample within the audio signal. This integration of controlled noise remarkably bolsters the generalization capabilities of the recommended deep learning model.

- **Spectrogram shift:** The concept of transposing a spectrogram involves analysing the changing frequency components of a spoken signal throughout its duration. A spectrogram functions as the visual depiction of how the signal frequency spectrum variation over time. Through transposing the spectrogram along the temporal axis, these nuances can be replicated, generating fresh training instances that exhibit slight deviations from the original dataset.

(D) **Feature extraction process**

The objective of this process is to minimize the dimensionality of input data, when retaining maximum possible information.

- **Mel frequency cepstral coefficients**

  MFCC stand out as the prevailing choice among speech features, primarily due to their exceptional precision in estimating speech parameters and their streamlined computational framework. Renowned for their efficacy, the MFCC method captures speech signals by transforming its short-term power spectrum into a linear cosine conversion of the logarithmic power spectrum, all mapped onto the nonlinear Mel frequency scale. As a widely adopted approach, MFCC plays a pivotal role in extracting pertinent attributes from audio signals (Figure 7.1).

- **Mel spectrogram**

  A Mel spectrogram, often referred to as a mel-frequency spectrogram, visually depicts the frequency composition of an audio signal. It finds extensive application in dissecting the acoustic attributes of sound within

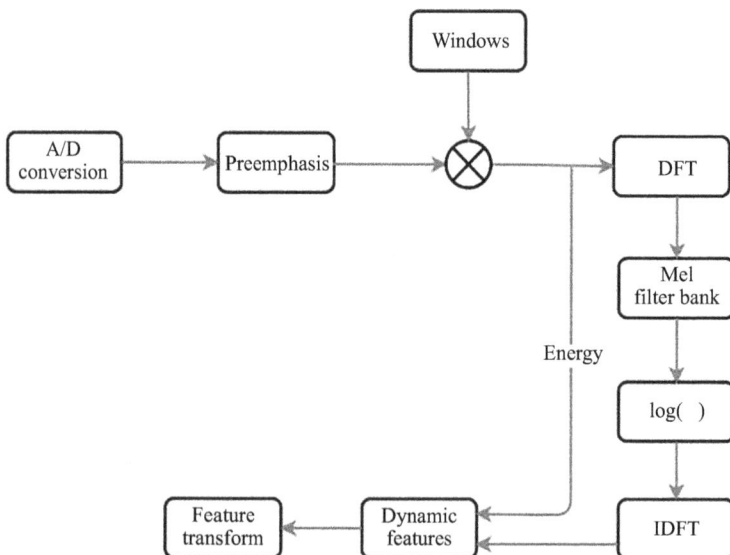

Figure 7.1   MFCC workflow

speech and music processing. Unlike a traditional spectrogram, the Mel spectrogram portrays the temporal evolution of an audio signal's frequency components, utilizing a non-linear frequency scale that closely aligns with human auditory perception.

- **Zero crossing rate**

  The zero crossing rate (ZCR) constitutes the basic characteristic of the signal, denoting the count of instances where the signal transitions between positive values, negative values, and zero values. This property proves valuable for detecting brief, intense sounds within a signal and discerning minute alterations in its amplitude. Through scrutinizing the temporal distribution of ZCR values, one can assess the spectral attributes of the speech signal, shedding light on the overall frequency composition and energy dispersion. Hence, ZCR stands as a crucial instrument for dissecting speech and diverse audio signal categories across varied domains, encompassing speech analysis and audio manipulation.

- **Chroma**

  Chroma features, alternatively known as "pitch class profile features," offer a captivating and potent means of sound representation, transforming spectra into a configuration of 12 distinct compartments, each corresponding to a halftone within the musical octave. This mechanism of feature extraction serves to create a chromagram, formed by the arrangement of 12 pitch classes. The dynamic fluctuations of these pitch classes throughout the audio duration yield an extra set of 12 features, attributed to the diverse pitch classes (Figure 7.2).

```python
def extract_features(data):
    # ZCR
    result = np.array([])
    zcr = np.mean(librosa.feature.zero_crossing_rate(y=data).T, axis=0)
    result=np.hstack((result, zcr))

    # Chroma_stft
    stft = np.abs(librosa.stft(data))
    chroma_stft = np.mean(librosa.feature.chroma_stft(S=stft, sr=sample_rate).T, axis=0)
    result = np.hstack((result, chroma_stft))

    # MFCC
    mfcc = np.mean(librosa.feature.mfcc(y=data, sr=sample_rate).T, axis=0)
    result = np.hstack((result, mfcc))

    # Root Mean Square Value
    rms = np.mean(librosa.feature.rms(y=data).T, axis=0)
    result = np.hstack((result, rms))

    # MelSpectogram
    mel = np.mean(librosa.feature.melspectrogram(y=data, sr=sample_rate).T, axis=0)
    result = np.hstack((result, mel))

    return result
```

Figure 7.2   Function for extracting features from audio

(E)   **Proposed CNN model**

For researchers delving into the realm of speech emotions, a central obstacle arises: the extraction of features that effectively capture emotional nuances. Deep neural networks, exemplified by CNNs, provide a streamlined avenue for extracting features capable of attaining remarkable levels of performance.

The CNN architecture is the composition of many layers, with convolutional layers, pooling layers, complete-connected layers, and Soft-Max classification layer. At the core of CNN are the convolution layers, it employs filters to execute convolution process on input data. Subsequently, the output of these convolution-layers is directed to pooling-layers, tasked with reducing output resolution and computational demand. The output is then proceeds to the complete-connected layer, then it undergoes flattening and classification through employment of the SoftMax unit. The proposed work is structured around four local feature learning blocks (LFLBs). Each block encompasses a convolutional layer succeeded by an activation layer (ReLU). In the initial LFLB, let's tailor the convolutional layer with 16 filters, doubling this count with each iteration. Following data flattening from the max-pooling layer, dense layer is employed, utilizing the Soft-Max activation function to ultimately classification of the processed features.

7.4   Results

This study aims to create a comprehensive system for emotion recognition through speech analysis. The accuracy and loss progression of the suggested CNN model during both training and validation stages is illustrated in Figure 7.3. The horizontal axis presents the count of epochs, and the vertical axis depicts the accuracy rate and loss metrics of the proposed model with respect to training and validation. A baseline of 50 epochs for model training has been established, with the accuracy

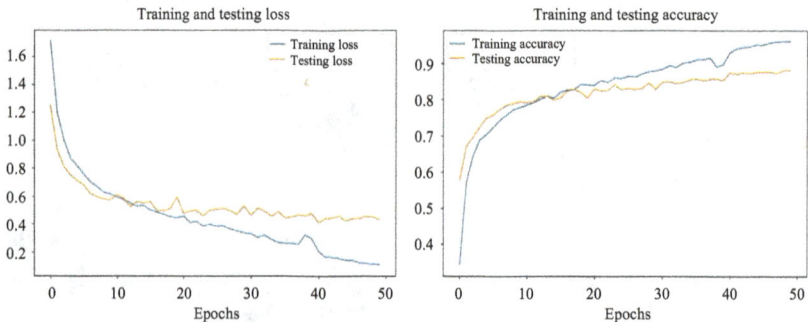

Figure 7.3   *Training and validation of the proposed model: (a) loss and (b) accuracy*

ratio depicted on the y-axis. Following each epoch, both training and validation accuracies exhibit a progressive increment, eventually stabilizing after a number of epochs. Across the 50-epoch span, the accuracy reached 88.11%, signifying the alignment of the proposed model with data fluctuations.

Utilizing the Adam optimizer with a learning rate of 0.0000001 and incorporating a decay value, the confusion matrix has been constructed for the CNN model illustrated in Figure 7.4. This matrix provides a numerical depiction of the outcomes achieved. Within the matrix, the y-axis labels correspond to the true emotions, while the x-axis labels represent the predicted emotions.

Performance metrics includes precision value, recall value, and f1-scores value, were also calculated for the model and are shown in Table 7.2.

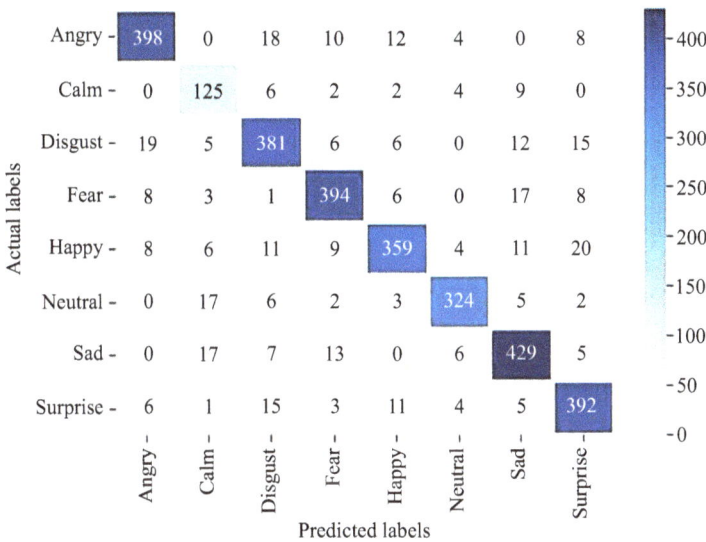

Figure 7.4   Confusion matrix of the introduced model

Table 7.2   Evaluation metrics of the model

Emotions	Precision	Recall	F1-score
Angry	0.91	0.88	0.90
Disgust	0.86	0.86	0.86
Fear	0.90	0.90	0.90
Happy	0.90	0.84	0.87
Sad	0.88	0.90	0.89
Surprise	0.87	0.90	0.88
Neutral	0.94	0.90	0.92
Calm	0.72	0.84	0.78

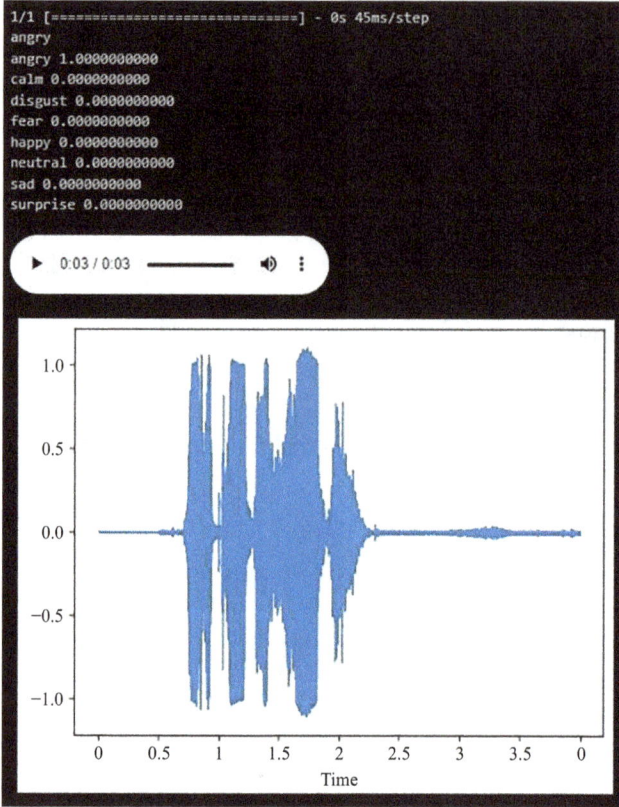

Figure 7.5   Model output for angry audio file

The outputs of the proposed SER model are shown in Figure 7.5 on the input files. The input audio files used here were of the emotion: angry.

7.5   Conclusion and future scope

In this chapter, the comprehensive exploration of a speech emotion/feeling recognition (SER) system has been presented, employing a neural network model called CNN alongside a multitude of acoustic features. The architectural framework of this system is meticulously crafted to harness the potency of several acoustic attributes, encompassing MFCCs, ZCR, Chroma, Mel spectrogram, and RMS, all contributing to precise speech emotion classification. Moreover, two data augmentation methods are incorporated, specifically noise addition and spectrogram shifting, which enhance dataset quality and diversity. The experimentation unfolds across two datasets, TESS and RAVDESS, revealing the prowess of the proposed SER system. The CNN model showcased in this study exhibits commendable performance, achieving an average accuracy of 88.11%. Evidently, this work underscores the capacity of the devised SER system, proficiently utilizing diverse acoustic features and neural network models, to aptly discern and categorize speech emotions.

While the acoustic attributes investigated in this research demonstrate effectiveness in speech emotion recognition, the inclusion of additional features could potentially enhance the accuracy of the SER system further. Notably, while the datasets utilized in this study boast diversity and representation, their size remains relatively limited. To ascertain the broader applicability of the suggested SER system, subsequent inquiries could assess its performance across more expansive datasets, encompassing a wider array of speakers, speech styles, and emotional expressions. Although this study primarily centres on the recognition of speech emotions in English, the prospects for extending the proposed SER system to other languages are worth exploring. Future investigations might delve into the feasibility of adapting the system to alternative languages and subsequently gauge its efficacy in cross-lingual contexts. To recapitulate, through the investigation of supplementary acoustic attributes, assessment on more expansive datasets, and expansion into diverse languages, forthcoming studies possess the potential to enhance both the accuracy and versatility of the suggested SER system.

References

[1] R. Dalal, M. Khari, and M. Hernandez, "Persuasive simulation of optimized protocol for OppNet" 2021, *Dynamic Systems and Applications*, *30*(5), 865–900.

[2] R. Dalal, M. Khari, J.P. Anzola, and V. García-Díaz, "Proliferation of opportunistic routing: a systematic review" 2021, *IEEE Access*, *10*, 5855–5883.

[3] R. Dalal, M. Khari, A. Kumar, and D.K. Sharma, "Evaluation of association rule-based routing protocol for Oppnet" 2022, *Mechatron Syst Control*, 50, 74–80.

[4] R. Dalal, and M. Khari, "February. Peculiar Effectual Approach: Q-Routing in Opportunistic Network" 2022, In *Proceedings of International Conference on Industrial Instrumentation and Control: ICI2C 2021* (pp. 609–615). Singapore: Springer Nature Singapore.

[5] R. Dalal, A. Sangwan and M. Khari, "The bibliometrics assessment of opportunistic network protocols and simulation tools" 2023, *Telematics and Informatics Reports*, *11*, 100082, ISSN 2772-5030.

[6] N. Mustaqeem and S. Kwon, "A CNN-assisted enhanced audio signal processing for speech emotion recognition" 2019, *Sensors*, 20(1), 183.

[7] S. Garg and G. Kumar, "Convolution neural network for speech emotion recognition" 2020, *International Journal of Creative Research Thoughts*, 8 (6), 2786–2792

[8] D. Issa, M. F. Demirci and A. Yazici, "Speech emotion recognition with deep convolutional neural networks" 2020, *Biomedical Signal Processing and Control*, 59, 101894.

[9] L. Jiang, P. Tan, J. Yang, X. Liu and Chao Wang, "Speech emotion recognition using emotion perception spectral feature" 2019, *Concurrency and Computation: Practice and Experience*, 33(11), e5427.

[10] N. Vryzas, L. Vrysis, M. Matsiola, R. Kotsakis, C. Dimoulas and G. Kalliris, "Continuous speech emotion recognition with convolutional neural networks" 2020, *Journal of Audio Engineering Society*, 68(1/2), 14–24

[11] E. Mansuori-Benssassi and J. Ye, "Generalisation and robustness investigation for facial and speech emotion recognition using bio-inspired spiking neural network" 2021, *Soft Computing*, *1*, 1432–7643

[12] S. N. Zisad, M. S. Hossain and K. Andersson "Speech emotion recognition in neurological disorders using convolutional neural networks" 2020, In *International conference on brain informatics*, 287–296.

[13] X. Ke, Y. Zhu, L. Wen, and W. Zhang, "Speech emotion recognition based on SVM and ANN" 2018, *International Journal of Machine Learning and Computing*, 8(3), 198–202.

[14] Y. Chen, R. Chang, J. Guo, "Emotion recognition of EEG signals based on the ensemble learning method: AdaBoost" 2021, *Mathematical Problems in Engineering*, *2021*(1), 8896062.

Chapter 8

Enhanced security and privacy in IoMT: a hierarchical federated learning approach using Dew-Cloud with HLSTM for hostile attack mitigation

K. Vijayalakshmi[1], N. Prasath[2], P.M. Sithar Selvam[3], L.R. Sujithra[4], S.A. Arunmozhi[5], Chetna Vaid Kwatra[6], Gaurav Gupta[7], Shakir Khan[8,9] and S. Mayakannan[10]

Abstract

Due to the coronavirus pandemic, doctors have had to treat patients remotely while medical facilities are overwhelmed. Also, as a result of COVID-19, people are far more concerned about their health, which has increased demand for internet-connected medical devices. Because of its incredible growth in value, the internet of medical things (IoMT) has caught the attention of cybercriminals. Many people's private health data and other very sensitive documents are safe on the dark web. Regardless, the trespassers were able to take advantage of the patient's health information because it was not adequately protected. The system administrator cannot tighten security since resource-constrained network devices do not have enough space or processing power. The primary objective is to study the expanding hostile attacks before they jeopardize the health system's security, while there are several supervised and unsupervised machine learning techniques that can detect outliers. This study's methodology utilizes Dew-Cloud to provide hierarchical

[1]Department of Electrical and Communication Engineering, College of Engineering, National University of Science and Technology, Muscat
[2]Department of Networking and Communications, School of Computing, Faculty of Engineering and Technology, SRM Institute of Science and Technology, India
[3]Department of Mathematics, KCG College of Technology, India
[4]Department of Artificial Intelligence and Data Science, Sri Eshwar College of Engineering, India
[5]Department of Electronics and Communication Engineering, Saranathan College of Engineering, India
[6]Department of Computer Science Engineering, Lovely Professional University, India
[7]Yogananda School of AI, Computers and Data Sciences, Shoolini University, India
[8]College of Computer and Information Sciences, Imam Mohammad Ibn Saud Islamic University (IMSIU), Saudi Arabia
[9]University Centre for Research and Development, Chandigarh University, India
[10]Department of Mechanical Engineering, Rathinam Technical Campus, India

federated learning (HFL). More availability of critical IoMT application(s) and enhanced data privacy are two advantages of the proposed Dew-Cloud idea. The hierarchical LSTM (long short-term memory) concept is used by distributed Dew servers that employ cloud computing for their backend implementation. Training the proposed model with the data pre-processing feature results in a low loss of 0.034 and a high accuracy of 99.31%. The suggested HFL-HLSTM model surpasses other methods in a number of performance parameters, such as f-score, recall, accuracy, and precision.

Keywords: Federated learning (FL); disease; intrusion detection system (IDS); IoMT; Dew computing

8.1   Introduction

The internet of medical things (IoMT), a promising new technology pushed by the Internet of Things (IoT) is quickly becoming standard in the healthcare sector [1]. The IoT has gained widespread recognition due to its widespread use in smart grids, smart cities, and smart industries. On the other hand, the IoMT's ability to assist medical personnel in preventing fatalities has recently garnered a lot of attention [2]. With the use of the IoMT, remote monitoring of patients' vital signs is now possible. Among the 5.8 billion gadgets that make up the IoT, 40% are located in healthcare facilities, according to a survey [3].

The IoT allows for real-time monitoring of vital signs in a smart healthcare setting, including insulin levels, temperature, heart rate, etc., and storage of this data in the cloud for analysis, feature sampling, and disease diagnosis [4]. Cloud servers provide the massive amounts of health data needed for accurate patient health records scanning. As a result of these servers' lack of proper security measures, the confidentiality and integrity of healthcare institutions are at risk. To add insult to injury, research has shown that almost 50% of clinical devices that use IoMT are susceptible to malware and attacks [5]. Life-threatening cyberattacks on IoMT systems are possible. A cybercriminal could harm patients health by gaining illegal access to their IoT medical devices and then overdosing them [6]. Not to mention that cybercriminals pose a threat to the availability and integrity of IoMT networks by injecting fake information and causing congestion in the network [7].

According to many studies, cyberattacks on IoMT could put patients' lives in jeopardy. When it comes to privacy, availability, authorization, and security, for instance, cloning, MITM, phishing, and denial of service can all play a role [8]. As compared to other IoT applications, IoMT ones are more delicate due to the fact that human lives are at stake. Not to mention the serious consequences that could ensue for patients and the healthcare organization in the event of a data breach involving their medical records. Furthermore, clinical data fetches 50 times the price of financial data on the dark web and black market [9]. Protecting patients' private health information from prying eyes is an important part of intelligent healthcare systems that aim to provide objective data analysis.

8.1.1   Research trends in intrusion detection for IoMT

Recent literature highlights significant challenges in securing IoMT environments. Among the most critical limitations in existing intrusion detection methods are issues related to scalability, reliability, and resource efficiency. Many conventional techniques rely heavily on centralized architectures and resource-intensive computations, making them less suitable for heterogeneous and constrained IoMT infrastructures.

To address these limitations, researchers have begun exploring architectures that integrate Dew computing with cloud-based federated learning frameworks. These hybrid models are capable of identifying intrusion patterns in real-time while preserving data privacy and minimizing computational burdens on edge devices. One such approach involves using a Dew-Cloud infrastructure to develop intrusion detection systems (IDS) that require minimal training and support distributed processing. This design not only facilitates timely detection of network anomalies but also enhances resilience against cyber threats.

Furthermore, hierarchical federated learning (HFL) has been adopted as a mechanism to improve model training by distributing computation across multiple layers—ranging from edge nodes to centralized servers. When combined with long short-term memory (LSTM) networks, this approach enables the detection of complex temporal patterns in network traffic, which are characteristic of many IoMT-related attacks.

Empirical studies utilizing benchmark datasets such as NSL-KDD and TON_IoT have demonstrated that HFL–LSTM frameworks can outperform traditional IDS models. Evaluation metrics including training loss, recall, precision, and F1-score consistently indicate that hierarchical models offer enhanced detection accuracy with reduced latency and system overhead. These developments underscore the growing viability of federated and hierarchical models as scalable and privacy-preserving solutions for securing IoMT infrastructures.

Security for the cloud-based IoMT architecture was the goal of the authors [10], who devised a method to detect fraudulent nodes. Using as a starting point, privacy-preserving support vector machine (SVM)-based approach was created that guarantees easy training and efficient execution of crypto algorithms [11]. Similarly, a federated learning system was developed based on layers to limit communication costs while maintaining anonymity. Applying their method to three industry-standard datasets, the authors examined its performance. In contrast, a clustering-based approach to cyber threat protection for IoMT networks that guarantees the privacy of medical data was presented by the researchers [12]. Healthcare systems should benefit from a secure data transfer paradigm that is both efficient and time-efficient, rather than just concentrating on privacy. A fuzzy-based approach has been used by researchers to stop Sybil attacks, the authors [13] introduced Deep EDN, a new method of encrypting data, to protect medical records in IoMT apps from hackers. Because wireless IoT networks can be accessed remotely, masqueraders have been lured to undermine the security and privacy of important applications. This has led to a thorough review of the many security options for IoMT applications.

A method for securing medical networks using ciphers and signatures was suggested. In addition, lightweight security mechanisms were proposed as a solution to the resource constraints faced by IoMT nodes [14]. There have been instances where unauthorized individuals have gained access to medical records and made changes that could compromise the patient's treatment or integrity. Consequently, the authors came up with a method that anonymously aggregates user-trained models. Using common clinical signs, a Explainable Artificial Intelligence (XAI)-based model is presented for the early detection of abnormal behaviors [15], [16]. A security model based on SVMs is also created to ascertain how often U2R and DoS attacks occur on IoMT networks. By using an ensemble learning–inspired sequential ELM strategy based on a fog framework, the authors [17] were able to distinguish between typical and out-of-the-ordinary events. The purpose of developing an anomaly detection system is to identify data outliers using stacked auto-encoders and soft-max classifiers, which are based on a two-stage deep learning model [18].

The authors used a combination of Improved Conditional Variational Autoencoder (ICVAE) and deep neural networks (DNNs) to detect shellcodes, U2R, and R2L cyber assaults, drawing inspiration from their intrusion detection efforts [19]. In contrast, a federated deep learning method that is both computationally and communicationally efficient to identify breaches in healthcare networks. The authors [20] looked at the current intrusion detection methods for medical applications that used DNNs rather than coming up with a new machine learning-based solution. A number of DNN–IoMT methods were found to make improper use of feature data and perform incorrect classifications. In order to improve upon existing DNN-based surveillance systems, the authors devised an enhanced data processing method. The Paillier cryptosystem is used to encrypt the model parameters before they are shared, guaranteeing confidentiality and anonymity [21]. Contrarily, a convolutional neural network (CNN) and the soft max approach were used to train the network in the aforementioned work using 49,4021 data. Then, using 31,029 samples from the KDD99 dataset, they put the network through its paces.

These methods are designed to detect and prevent intrusions in IoMT networks, but (a) some nodes in the ecosystem have limited resources like processing power, storage capacity, and battery life, and (b) other nodes are linked through inefficient message exchange protocols, so they cannot afford the excessive resource demands of complex and standard security measures [22,23]. Not only are state-of-the-art schemes resource intensive, but they also have poor reliability, a low accuracy rate, and restricted security features. Because of their high computational, power, and time costs, old approaches also negatively affect the energy and service accessibility of the IoMT nodes.

8.1.2   Challenges in secure IoMT systems

The global healthcare infrastructure has experienced considerable strain due to surges in medical cases, prompting an increased reliance on digital platforms such as the IoMT. While these technologies have facilitated remote diagnosis and continuous monitoring of less critical conditions, they have also introduced new challenges. Specifically, the expanded connectivity has contributed to increased communication overhead and network congestion across healthcare systems.

Despite their advantages, many existing IoMT frameworks remain dependent on centralized servers for data storage and transmission. This centralization, coupled with insufficient implementation of cryptographic safeguards, renders such systems susceptible to privacy breaches, denial-of-service (DoS) attacks, and other forms of cyber threats. These vulnerabilities underscore the urgency of developing robust, decentralized, and privacy-aware security mechanisms for healthcare environments.

To advance secure IoMT architectures, researchers must address several critical theoretical and practical questions:

1. Data privacy and infrastructure efficiency: How can patient information be protected from unauthorized access without overburdening already constrained cyber-physical infrastructures?
2. Accuracy and resilience of IDS: What methods can be used to design IDS that are both precise and reliable, yet lightweight enough to operate on heterogeneous IoMT devices?
3. Efficient model training: How can IDS models be trained in a time-efficient manner that supports real-time or near-real-time threat detection?
4. Defense against unknown threats: What strategies can be employed to anticipate and mitigate previously unseen or evolving cyberattacks in the IoMT landscape?

Addressing these open challenges is essential not only for ensuring the safety and privacy of patient data but also for reinforcing public trust in healthcare technologies. Building secure and scalable IDS frameworks is a foundational step toward enabling resilient and intelligent healthcare delivery systems in the future [24,25].

8.2   Improved IoMT structure

Using cloud computing as an IoMT edge layer, we present the Dew-Cloud IoMT framework, a novel approach to IoMT. The proposed system aims to securely monitor patients' vital signs. Hackers can talk to the bio-sensors that are already in place by gaining access to the IoMT ecosystem through local servers. The data that is transmitted by the gateway also reaches the servers at the edge, in addition to the hacker server. By imitating the actions of the authorized edge servers, the hacker server is able to redirect communication—specifically, medical data—towards itself. Dew is considering introducing an intelligent service (DIS) to prevent such negative events. Figure 8.1 depicts a flow process of internet of medical things interface.

Hierarchical architecture is the result of combining decentralized and centralized design principles. Problems with complexity, scalability, and system

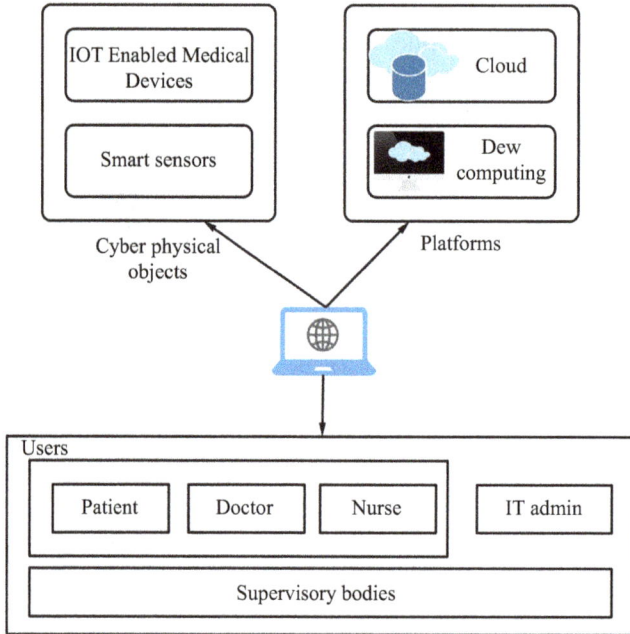

Figure. 8.1   A process of internet of medical things interface

breakdowns are all within its capabilities. A form of federated learning known as "hierarchical federated learning" is used in hierarchical architecture. The suggested Dew-Cloud IoMT framework's heterogeneity is well-suited to HFL. Frameworks based on fog and edge computing already make use of HFL, which offers a novel privacy protection feature [26]. Acquiring data, analyzing it, and using HLSTM to categorize the acquired health information are all parts of the DIS process.

A suggested DIS is used to gather and preprocess data before training the HLSTM model. The Dew-Cloud architecture makes use of DIS to detect abnormalities in network traffic and keep tabs on it. Together, the data reader and the planned DIS services actively seek out possible outliers. The suggested framework makes use of the following functional components.

Through the use of implanted mobile devices, such as sensors and actuators, gathers information from IoMT. These devices take the patient's vital signs and record when they take their medication at set intervals. To allow for wireless communication between devices at a short range, technologies like Zigbee or Bluetooth are utilized. This data is gathered and aggregated by the devices located at the edge. The information is further encrypted before being transmitted to the hospital's Dew server. The suggested intrusion detection method uses federated learning in conjunction with big data handled by a server in the cloud. The hospital's Wi-Fi connects all the equipment, and the high-speed internet allows Dew and cloud servers to communicate with each other. An application programming interface (API) allows for the updating of caregivers' and patients' mobile applications, while a cloud server is in charge of the electronic medical record (EMR). Reviewing reports for hardware component failures and software updates is the administrator's responsibility. Websites and mobile apps are part of the proposed system's service offering, which includes backup, report production, medical analytics, and contact with other parties like accounts and insurance. It also offers robust administration. The administrator checks the whole procedure by hand and can fix any problems they find. The administrator directs the vendors and maintenance crews to do what is needed to keep the system running smoothly.

8.3   Hierarchical frameworks for IDS in IoMT

In the context of the IoMT, the ecosystem typically includes users, gateways, mobile applications, and cloud infrastructure, all interacting in a dynamic and distributed environment. This complexity gives rise to potential anomalies triggered by various sources such as compromised devices, unauthorized access, and unpredictable user behavior. These risks are further compounded by the critical nature of healthcare data and the real-time requirements of medical services.

To address these challenges, recent developments have focused on employing HFL combined with LSTM networks for effective intrusion detection. HFL enables decentralized model training across distributed nodes, preserving data privacy and reducing communication overhead. When integrated with Dew computing—an edge-centric model that operates closer to the data source—this framework supports lightweight, scalable, and robust anomaly detection in resource-constrained IoMT environments. The combined use of federated learning principles and temporal sequence modeling through LSTM allows for early and accurate identification of intrusion patterns without centralized data aggregation.

8.3.1   Preprocessing data

To improve the HFL–HLSTM model, we feed it raw network traffic. Afterwards, the data is cleaned and normalized as part of the data preprocessing that the proposal model does. Quantitative and categorical data are generated by IoMT. Steps including feature reduction, scaling, and labeling unlabeled fields are part of data preprocessing. The HFL–HLSTM model saves memory and computing costs by removing features from the dataset that are not statistically significant.

Network traffic cleaning: One of data mining's primary functions is to filter out irrelevant data packets before training a model. The NSL–KDD benchmark dataset has been utilized. Each entry has the value 0 for the "num_outbound_cmds" parameter. Hence, removing this parameter from the dataset is crucial. The NSL–KDD dataset yielded 40 features following cleaning.

Clean data: Following the cleaning process, each record in the network security level-KDD dataset contains 40 parameters. There are a total of 37 parameters, including 17 label parameters and 33 integers. Also, the label factors are used to construct numerical fields. Just by changing the values of the "is_guest_login," "is_hot_login," and "logged_in" parameters from "yes" to "no"—1 and 0, respectively. Thus, "flag" contains 11 names, "service" seventy, and "protocol_type" three. To transform these factors into binary integers, the one-hot training method is employed. Using the same procedure, every record label uses one-hot encoding.

Normalize the numerical range: It is important to give the numerical range to eliminate the effect of fractional fluctuation so that the proposed model may be trained efficiently. We can find the minimum and maximum values using y_{min} and y_{max}, respectively, when y is used as the descriptor for every value in the collection. To normalize the $y^{\vee}$ according to (8.1), we use the [0,1] range [27].

$$\hat{y} = \frac{(y - y_{min})}{(y_{max} - y_{min})} \tag{8.1}$$

Dimensional reduction: Principle component analysis (PCA) simplifies data fields.

$$x_i = A_{i1} \cdot V_1 + A_{i2} \cdot V_2 + \cdots + A_{in} \cdot V_n \tag{8.2}$$

Principal component analysis is represented by x_i, and the eigenvectors are A and the input value, V, respectively.

8.3.2 LSTM hierarchy

The traditional LSTM model uses a vector of input values ($V = v_1, v_2, \ldots, v_n$) to generate a vector of predicted values ($U = u_1, u_2, \ldots, u_n$). For the purpose of making sequential token predictions, we employ the softmax function in (8.3) and (8.4) [28].

$$S(U|V) = \Pi_{(t \in [1, n_u])} s(u_t | v_1, v_2, \ldots, v_t, u_1, u_2, \ldots, u_{t-1}) \tag{8.3}$$

$$S(U|V) = \Pi_{(t \in [1, n_u])} \frac{\exp(f(p_{t-1}, e_{u_t}))}{\exp\left(\sum_{\hat{u}} f(p_{t-1}, e_{\hat{u}})\right)} \tag{8.4}$$

The activating function is represented by $f(p_t - 1, e_{\hat{u}})$, whereas the output of LSTM is represented by $p_t - 1$.

Hierarchical LSTM may learn the complex sequence on multiple layers. In the initial layer of the recurrent networking, the sentence vector is built according to (8.5).

$$p_t^s = \text{LSTM}(p_{t-1}^s, e_t^s) \tag{8.5}$$

In order to simplify things, LSTM operations are defined via functions. The hidden vector is denoted as, p_{t-1}^s, while the word level embedding is denoted as, e_t^s. The second layer of the recurrent network is then used to create the document vector according to the instructions given in (8.6).

$$p_t^d = \text{LSTM}(p_{t-1}^d, e_t^d) \tag{8.6}$$

The sentence-level embedding and hidden vector in the LSTM model are represented by, p_{t-1}^d. The document vector maintains the structure at the level of sentences [29].

The suggested Dew-Cloud based HFL framework uses HLSTM to improve intrusion detection accuracy and computational performance. Counting the records allows one to create a grayscale picture. The Telemetry, Operational data, and Network data (TON) internet of things dataset contains 43 features, while the NSL−KDD dataset contains 40 features. It is possible to create a 7×7 picture for every record. The initial layer of an HLSTM model is encoded as a shape column vector of length 64 bytes [30]. Additionally, all seven column vectors can be encoded to form a complete image using a structure of (7, 64). Finally, with all of the layers connected, we can make an accurate prediction. The scalability of the HLSTM model is based on the fact that LSTM is most effective when applied to hierarchies that allow it to learn across several layers. We have an IoMT setup where each layer does its own thing before going on to the next. The outcome is a pipeline-like structure where the gradient is passed from one layer of LSTM to another after each layer calculates its own local data. Federated learning is helpful for developing a global, top-down model. Cloud computing demands can be mitigated by utilizing Dew computing infrastructure at every layer of LSTM.

8.3.3   A model for detecting intrusions

Figure 8.2 depicts the suggested Dew-Cloud intrusion detection system. Hospitals, patients, physicians, healthcare employees, etc. are some of the K objects included in the suggested framework. In addition, there are M number of IOT devices linked to these items. These IOT devices have local storage LSi, where i ranges from 1 to $n - 1$, n. The federated learning system's local model is trained with the data provided by these devices and stored in the Dew servers of the linked smart hospital. In addition to shielding the system from potential threats, this also guarantees the security of user data.

1. The suggested model takes into account that smart hospitals are legally recognized model training institutions with a signed contract.
2. As data is aggregated, various medical institutions send their local model gradients to a cloud server.
3. Dew-Cloud medical institutions, like hospitals, receive the global weight from the cloud server.

The HFL-long term memory model is one way to detect intrusions in the IoMT ecosystem. A number of different sources feed data into the system. These include ambulances, hospital wards, faraway patients, and caretaker. The linked healthcare provider's Dew server receives the TrainingData (δ) from the participants (δNa, δNb). The number of layers and units of the NN are evaluated by the size of the health institution's infrastructure in relation to the Dew server. But this solution is scalable since it uses cloud servers for computational offloading. The suggested method takes into account the amount of computing resources available in

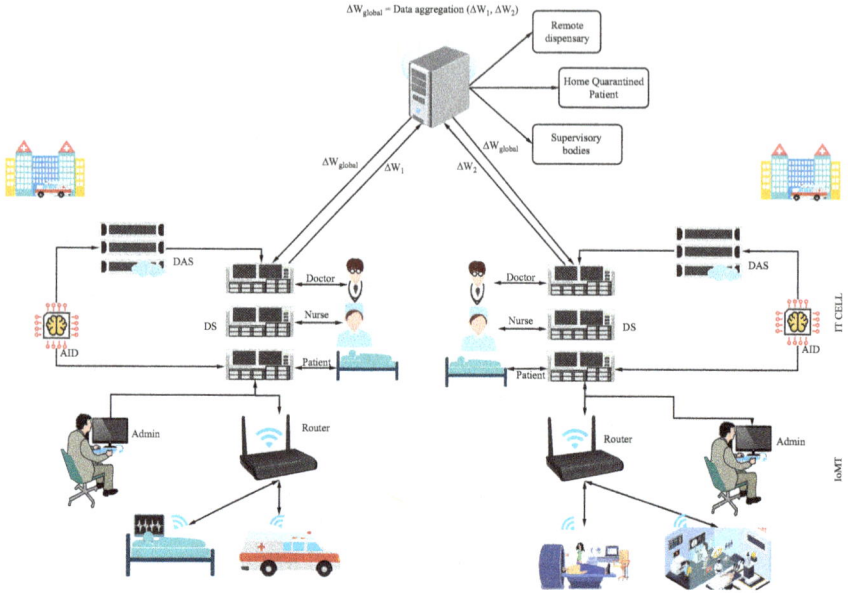

Figure 8.2   The Dew-Cloud internet of medical things is a proposed framework.

healthcare facilities to dynamically modify the batch size of incoming traffic. This guarantees improved performance with reduced processing unit requirements and great throughput.

A number of healthcare facilities have Dew servers set up, and they get the global model that was trained on the cloud. A non-essential framework, including healthcare facilities, patients, physicians, and other medical professionals. As an added bonus, these items are linked to M number of internet of things devices. Internet of things devices have resident storage LSi, where i ranges from 1 to $n - 1$. If the linked smart hospital's Dew servers store the data collected by these devices, the local model can be trained within the federated learning system. Not only does this prevent unauthorized access to user data, but it also protects the system from various security risks.

Algorithm 1 A Dew-Cloud Federated Learning IoMT System using the Pseudo-Code of an HLSTM.

Input:
  IoMT devices generate data $D = \{D_1, D_2, \ldots, D_n\}$
  Dew servers $S = \{S_1, S_2, \ldots, S_m\}$
  Cloud server C
  Parameters: Hidden layers HL, Rounds R, Local minibatch size η, Learning rate φ

Output:

Global HLSTM model parameters W_global

Phase 1: Initialization

1. Initialize HLSTM model with parameters W_global on Cloud server C.
2. Distribute IoMT device data across Dew servers.
3. Preprocess datasets: normalize, numeralize, and reduce dimensionality using PCA.

Phase 2: Hierarchical Federated Learning

For each round r = 1 to R do:

Step 2.1: Distribute Global Model

1. Broadcast W_global to all Dew servers S_i.

Step 2.2: Local Training at Dew Servers

For each Dew server S_i in parallel do:

1. Retrieve local dataset D_i.
2. Perform local training on HLSTM:
   (a) Split data into minibatches.
   (b) For each epoch σ:
       (i) Feed minibatch data to HLSTM layers.
       (ii) Compute gradients and update local model W_i.
3. Return updated local model W_i to Cloud server.

Step 2.3: Aggregation at Cloud

1. Aggregate local models to update global model:

$$\text{W\_global} \leftarrow \sum (|D_i|/|D|) * W_i$$

2. Evaluate updated W_global using performance metrics (accuracy, recall, F1-score).

Phase 3: Deployment

1. Deploy final W_global to Dew servers for intrusion detection in IoMT systems.

Key HLSTM Processes:

Forward Pass:

For each time step t:

1. Forget Gate: $f\_t = \sigma(W\_f * [h\_t\text{-}1, x\_t] + b\_f)$
2. Input Gate: $i\_t = \sigma(W\_i * [h\_t\text{-}1, x\_t] + b\_i)$
3. Candidate Memory: $g\_t = \tanh(W\_c * [h\_t\text{-}1, x\_t] + b\_c)$
4. Update Memory: $c\_t = f\_t \odot c\_t\text{-}1 + i\_t \odot g\_t$
5. Output Gate: $o\_t = \sigma(W\_o * [h\_t\text{-}1, x\_t] + b\_o)$
6. Hidden State: $h\_t = o\_t \odot \tanh(c\_t)$

Backward Pass:

Use gradient descent to update weights:

$W \leftarrow W - \varphi * \nabla L(W)$, where L is the loss function.

Termination

1. Repeat the federated learning process until performance metrics are optimized or R rounds are completed.
2. Deploy trained model for real-time intrusion detection in IoMT environments.

Considerations such as these are made in the suggested model:

1. A signed deal recognizing smart hospitals as official model training institutions;
2. For data aggregation, several medical institutions transmit their local model gradients to a server in the cloud;
3. The medical institutions Dew-Cloud, such as hospitals, receive the global weight from the cloud server.

The HLSTM−HFL method is one way to detect intrusions in the IoMT ecosystem. A number of different sources feed data into the system. These include ambulances, hospital wards, faraway patients, and caretakers. The linked healthcare provider's Dew server receives the Training Data (δ) from the participants (δNa, δNb). The neural network's layer and unit count are dictated by the health institution's infrastructure size in respect to the Dew server. But this solution is scalable since it uses cloud servers for computational offloading. The suggested method takes into account the amount of computing resources available in healthcare facilities to dynamically modify the batch size of incoming traffic. This guarantees improved performance with reduced processing unit requirements and great throughput.

More medical centers can access the globally trained model using Dew servers. Along with the epoch value, a random value is given a range of weights upon starting. The HLSTM pseudo-code is identified as Algorithm 1 in the Dew-Cloud based HFL model. While waiting for the optimal number of inputs, the LSTM model gathers data in preparation for training. The Dew computing hierarchical LSTM algorithm is invoked with input data samples once enough information is obtained [31]. A number of local variables, including the hidden layer, rounds, batch size, and global model parameters, must be defined before moving forward. Iterative execution is defined by the number of rounds R. Each record's global model is based on the local gradients obtained from the Dew servers, and the dataset is chosen at random. The newly built global model is transmitted to the medical facilities' Dew servers. Iterating through the steps yields either a very accurate model or a specific number of iterations.

The following steps are taken by our proposed model after data collection from IoT devices:

1. Determining chunk size using the Dew-layer resources;
2. Selecting data at random;
3. Training the model using the hierarchical LSTM approach.

Algorithm 2 Dew Server Local Update (Iterative) up to date.

Input:

Local dataset D_i on Dew server S_i

Global model parameters W_global received from Cloud

Learning rate φ, no of epochs σ, minibatch size η

Bias term β

Output:

Updated local model parameters W_local$_i$

Procedure LocalUpdate (S_i, W_global)

1. Initialize W_local$_i$ ← W_global
2. Preprocess local dataset D_i:
   (a) Normalize and numeralize data.
   (b) Apply dimensionality reduction (e.g., PCA).

3. Split dataset D_i into minibatches of size η.
4. For each epoch e = 1 to σ do:
   (a) For each minibatch B ∈ D_i do:
   (i) Perform forward pass in HLSTM:
   Compute hidden states h_t and cell states c_t for each time step t:

   f_t = σ(W_f * [h_t-1, x_t] + b_f) (Forget Gate)

   i_t = σ(W_i * [h_t-1, x_t] + b_i) (Input Gate)

   g_t = tanh(W_c * [h_t-1, x_t] + b_c) (Candidate Memory)

   c_t = f_t ⊙ c_t-1 + i_t ⊙ g_t (Memory Update)

   o_t = σ(W_o * [h_t-1, x_t] + b_o) (Output Gate)

   h_t = o_t ⊙ tanh(c_t) (Hidden State)

   (ii) Compute loss L(W_local$_i$) on minibatch B.
   (iii) Perform backward pass to calculate gradients:

   $$\nabla L(\text{W\_local}_i).$$

   (iv) Update local model parameters:

   $$\text{W\_local}_i \leftarrow \text{W\_local}_i - \varphi * \nabla L(\text{W\_local}_i).$$

5. End For (Minibatches)
6. End For (Epochs)
7. Return updated W_local$_i$ to Cloud server.

The training continues in the Dew servers of all the health institutions in accordance with Algorithm 2 until the minimum number of σ epochs is reached. At the cloud layer, Algorithm 1 takes the weights from all the different Dew-servers and feeds them back into itself. At the level of the health institution, the model is aggregated from different branches using HFL. At the level of each health institution, it is aggregated all the way up to the level of the cloud. The global model Wglobal notifies the appropriate Dew servers of the updated model after all of the local models have been compiled. Training the intrusion detection model at many levels—local, institutional, and global—with the help of public and private health organizations ensures its correctness. Once the Dew layers' local model receives the global parameters Wglobal, training is halted. The process of gradient aggregation is iterative. The local training ends when either the local minibatch η is filled with epochs or the local model detects traffic incursions as well as it can. The network receives its examples from the minibatch. A back-propagation is executed after each and every example. By averaging these gradients, the weight is revised. Raising the minibatch size to lower the learning rate range is necessary to attain substantial test performance and convergence. For mission-critical settings like IoMT, we saw substantial performance gains of up to 32 mini-batches. The suggested HFL−HLSTM model for intrusion detection uses data generated continually in health institutions, making it the ideal fit for the IoMT ecosystem.

8.4   Comparative performance analysis

Based on the Extended Simple Recurrent Unit (XSRU)−IoMT model's experimental evaluation, the suggested HFL−HLSTM model is tested with the TON_IoT dataset [32]. Table 8.1 details the experimental setup.

8.4.1   Dataset

With the help of the real-world data source TON_IoT, all of the models—both current and future—are tested. With its three-tiered structure, the framework makes it easy for LinuxOS and IoT devices to communicate with one another. These levels are IoMT, Dew, and Cloud. The IoT ecosystem has been the target of countless different types of attacks. Injection, man-in-the-middle (MITM), backdoor, password, distributed denial of service (DDoS), ransomware, doS, scanning, and more are all included in the attacked category of the dataset, which also includes regular categories. Model training takes place on the benchmarked KDDTrainC dataset, whereas default parameter testing takes place on the KDDTest-21 and KDDTest+ datasets.

For each sort in the TON_IoT dataset, the total number of records is listed in Table 8.2. There are 43 things that are present in every data sample. The latest

Table 8.1   Experimental setup

Configuration	Feat
CPU	Intel Core i3 3.30 GHz 12100F 12th Gen
Operating system (OS)	Ubuntu 20.04.3 LTS
Librari	Scikit-learn, Keras, Numpy TensorFlow,
GPU	NVIDIA RTX 3050 8GB
Programming	Python
RAM	64 GB

Table 8.2   Total count of records for each TON IOT
           dataset category

Type	No. of records
DDoS	26,000
Normal	246,000
XSS	6200
Password	36,000
Ransomware	16,200
Backdoor	36,000
Scanning	4000
Injection	36,000

dataset includes attack vectors based on IoMT_IoT. Newest dataset TON_IoT includes many real-world threat vectors. A more precise picture of the IoMT ecology can be provided by this dataset.

8.4.2   Results and discussion

A number of tests on the network security level-KDD and TON_IoT benchmarked datasets confirm that the suggested HFL−HLSTM model is valid.

(1) Validation and testing with the TON_IoT dataset: The suggested HFL−HLSTM model is tested and validated using the ablation experiment as its basis. Increasing the number of layers in a deep learning or machine learning model causes the computation time to grow [33]. So, to get better results, we run the tests with few hidden layers and minibatches.

In terms of training loss and accurateness, HLSTM's performance is contrasted with that of LSTM. In addition, Figure 8.3 shows the evaluation of HLSTM's computational cost in comparison to LSTM and gated recurrent unit (GRU) [34].

Training loss and accuracy of trajectory models are displayed in Figures 8.4 and 8.5, respectively. We also compare the models' accuracy to that of numerous cutting-edge intrusion detection methods. A neural network's calculation expenses are directly proportional to its layer count. By drastically cutting the number of layers at the Dew layer in the IoMT framework, HFL help to improve the calculation cost.

The suggested HFL−HLSTM models are shown in Table 8.3. Compared to another state-of-the-art model, HFL−HLSTM obviously performs better with a 99.72% accuracy, 99.27% precision, 98.66% recall, and 98.93% F-Score. While presenting different IDSs, researchers have paid more attention to the effectiveness of attack routes than to the usefulness of supplementary algorithms. The majority of these are focused on making things more accurate, but there are a few that are more concerned with making things more detectable while reducing computation costs and false alarms. In this comparative study, we use the TON_IoT to evaluate the improved HFL−HLSTM model against various cutting-edge intrusion detection techniques for a range of attacks. In the context of the IoMT ecosystem, the improved

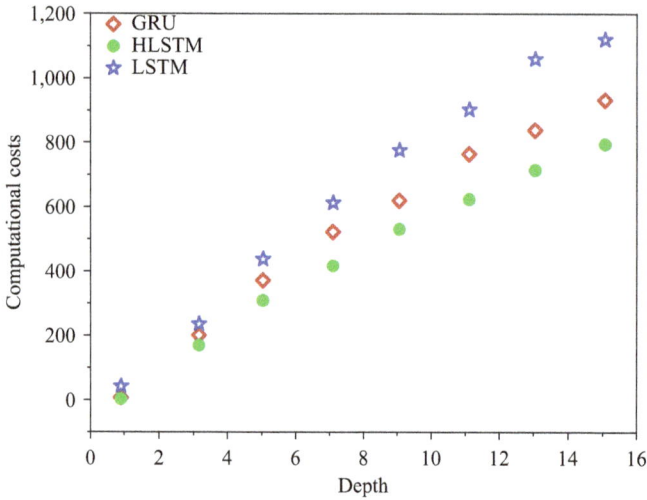

Figure 8.3   Evaluation of computing expenses

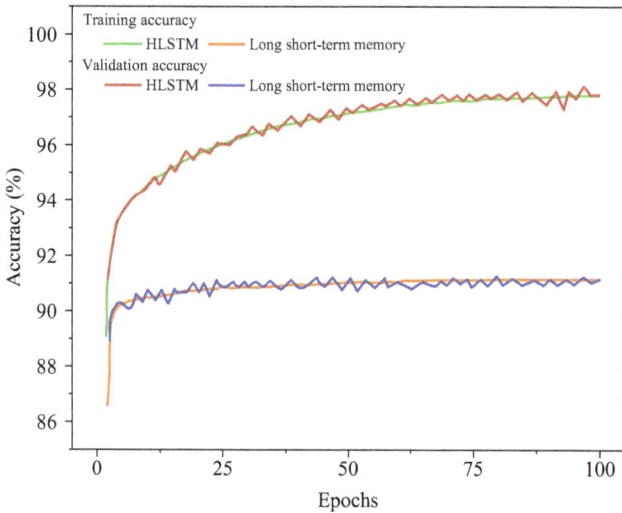

Figure 8.4   Evaluation of the accuracy of training

solution is beneficial. Ongoing service availability and threat detection are the lynchpins of IDSs. By combining Dew and Cloud Computing, our system offers very accurate and highly available services. With the computing power of Dew servers on-premises, our solution can continue to deliver the same level of service in the event that the connection to the cloud is disrupted. The proposed HFL−LSTM model

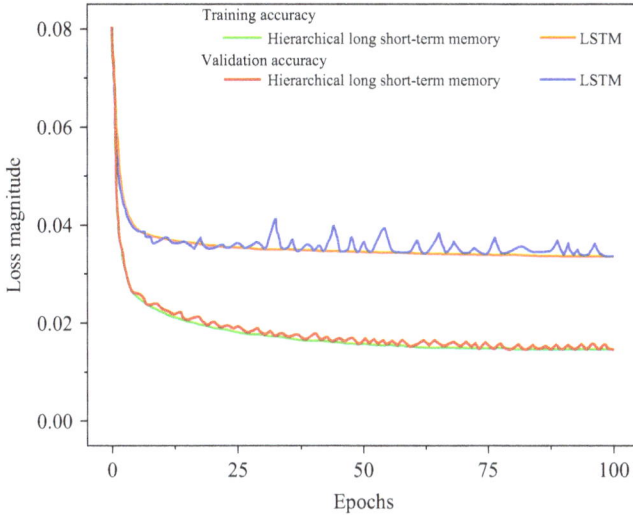

Figure 8.5   Training loss analysis

Table 8.3   Evaluation of current and prospective models

Measurement	Our model
Recall	98.66
Accuracy	99.72
Precision	99.27
F-Score	98.93

outperforms the status quo in terms of service quality and performance. Both the suggested model and the admin's ability to respond on threat prediction—thinking at a human level—help the autonomic decisions get better with time. Taken together, the findings demonstrate that the proposed approach outperforms all others when it comes to discovering IoMT ecosystem breaches [35,36]. This is why the recommended HFL-LSTM intrusion detection model is the way to go for effectively handling the enormous data sets generated by an IoMT environment.

(2) Validation and testing with the network security level-KDD database: The classification of binary in the network security level-KDD dataset is tested using learning rates of 0.001, 0.1, and 0.5. Improved model training at the Dew level is achieved by varying the number of hidden nodes (20, 60, 80, 120). In addition, up to 100 epochs can be defined. Data preprocessing is initiated using the 41-dimensional attributes as input. The learning rate and amount of hidden nodes significantly impact the model's accuracy. Figure 8.6 displays the results of binary classification using various models, including the proposed HFL−HLSTM, SVM, Random Forest, and MLP. Our suggested model achieves an accuracy of 89.91%

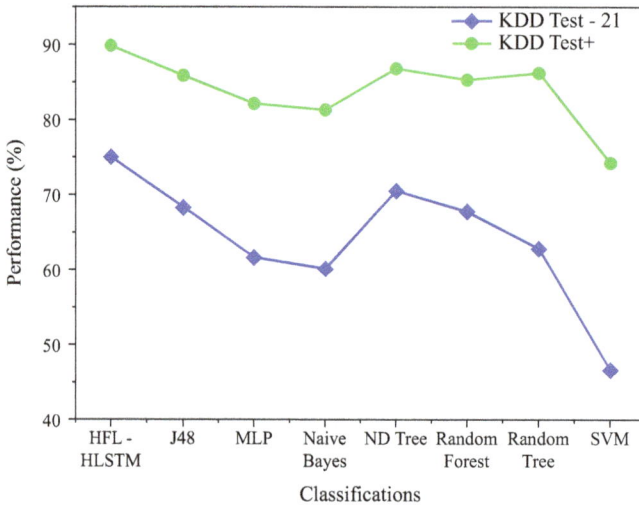

Figure 8.6   A binary-class comparison is used to evaluate the proposed model in relation to the performance of existing models

on the KDDTest+ dataset. If we compare this to KDDTest-21, we see that the suggested HFL−HLSTM model can reach an accuracy of over 75.07%.

When trained on the KDDTest+ dataset with 60 hidden layers, the HLSTM model achieves greater accuracy at a learning rate of 0.5.

Our comparisons also include binary classification, SVM, Naive Bayes, and MLP, as well as additional benchmarking algorithms. Figure 8.7 shows the outcomes of the existing model as well as the one that was proposed. On the KDDTest-21 dataset, the suggested model could reach an accuracy of 70.82%, while on the KDDTest+ dataset, it could reach an accuracy of 89.51%.

To sum up, when compared to the most recent and cutting-edge intrusion detection models, the suggested HFL−HLSTM model had many benefits. For mission-critical IoMT systems, it offers a solid option. The given architecture enables the HLSTM model to run with lowest deployment cost and great accuracy by combining Dew-cloud with well-developed computing technologies. The suggested model improves the availability while simultaneously increasing the privacy of users' data. For cyber-physical systems of the future to be effective, enterprises will need more control over their data and better threat detection capabilities. Over the course of the training period, we evaluated the model's accuracy, as well as its loss and accuracy. For the benefit of security enthusiasts, the proposed model is laid out in a way that is easy for them to understand and execute. Using the Dew-Cloud health institutions is a simple and inexpensive way for conventional healthcare facilities to improve. In addition, a newly created TON_IoT dataset and a benchmarked network security level-KDD dataset are used to validate the improved model. When compared to current models, the suggested one performs better in identifying breaches in medical data from various healthcare organizations.

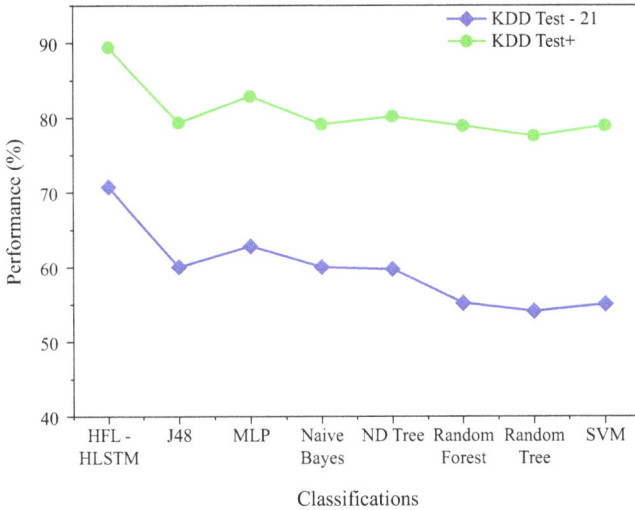

Figure 8.7   Assessment of the proposed model relative to the efficacy of current models across various categories

8.5   Summary and future directions

The integration of advanced artificial intelligence (AI) frameworks with scalable infrastructures such as cloud and edge computing is reshaping how healthcare systems manage and analyze large-scale data. In the context of the IoMT, this synergy holds particular promise, yet the domain remains in its formative stages and continues to face significant security challenges. The growing sophistication of cyberattacks poses a tangible risk to IoMT-enabled smart hospitals, necessitating proactive strategies to protect sensitive health data and maintain service continuity.

To address these issues, recent advancements have explored the use of HFL in conjunction with LSTM networks, deployed via Dew-Cloud architectures. This layered approach offers both scalability and enhanced privacy, enabling distributed intrusion detection without compromising data integrity. In this framework, HFL facilitates secure model training across decentralized nodes, while HLSTM enables the identification of temporal anomalies in network behavior.

Evaluation of the HFL–HLSTM model using established benchmark datasets, including TON_IoT and NSL-KDD, has demonstrated superior performance when compared to conventional intrusion detection methods. Reported metrics include an F1-score of 98.58%, accuracy of 99.31%, recall of 98.24%, and precision of 98.97%. In classification tasks, the model also achieved notable results—capturing 89.91% accuracy on the KDDTest+ (binary classification) dataset and 70.82% on KDDTest-21 (multi-class).

These outcomes highlight the model's robustness, efficiency, and suitability for deployment in mission-critical healthcare environments. Furthermore, its application enhances trust in IoMT systems by ensuring both predictive performance and operational reliability.

Looking ahead, future research may explore incorporating optimization algorithms such as Gurobi to further reduce latency and improve precision. Additional investigations could focus on the adaptability of IDS models in the face of heterogeneity, interoperability, and dynamic scalability, which are critical for supporting diverse healthcare infrastructures.

References

[1] C. Dhasaratha, M. K. Hasan, S. Islam, *et al.*, "Data privacy model using blockchain reinforcement federated learning approach for scalable internet of medical things," *CAAI Trans Intell Technol*, vol. 16, no. 1, 2024, doi:10.1049/cit2.12287.

[2] A. U. Haq, J. P. Li, I. Khan, *et al.*, "DEBCM: deep learning-based enhanced breast invasive ductal carcinoma classification model in IoMT healthcare systems," *IEEE J Biomed Health Inform*, vol. 28, no. 3, pp. 1207–1217, 2022.

[3] R. Mahalakshmi and N. Lalithamani, "Preventing COVID-19 using edge intelligence in internet of medical things," In *International Conference on Innovative Computing and Communications: Proceedings of ICICC 2022*, Vol. 1 (pp. 213–227). Singapore: Springer Nature Singapore.

[4] A. Akram, K. Zafar, A. N. Mian *et al.*, "On layout optimization of wireless sensor network using meta-heuristic approach," *Computer Systems Science and Engineering*, vol. 46, no. 3, pp. 1–17, 2023.

[5] Y. Otoum, Y. Wan, and A. Nayak, "Federated transfer learning-based IDS for the Internet of Medical Things (IoMT)," in 2021 *IEEE Globecom Workshops, GC Wkshps 2021 – Proceedings, 2021.* doi:10.1109/GCWkshps52748.2021.9682118.

[6] S. Saif, P. Das, S. Biswas, S. Khan, M. A. Haq, and V. Kovtun, "A secure data transmission framework for IoT enabled healthcare," *Heliyon*, vol. 10, no. 16, 2024.

[7] S. Mayakannan, R. Rathinam, R. Saminathan, R. Deepalakshmi, M. Gopal and J. J. M. Hillary, "Analysis of spectroscopic, morphological characterization and interaction of dye molecules for the surface modification of TiB2Nanoparticles," *J Nanomater*, vol. 2022, 2022, doi:10.1155/2022/1033216.

[8] B. L. Y. Agbley, J. P. Li, A. U. Haq, *et al.*, "Federated fusion of magnified histopathological images for breast tumor classification in the Internet of Medical Things," *IEEE J Biomed Health Inform*, 2023.

[9] I. S. N. V. R. Prasanth, P. Jeevanandam, P. Selvaraju, *et al.*, "Study of friction and wear behavior of graphene-reinforced AA7075 nanocomposites by machine learning," *J Nanomater*, vol. 2023, 2023, doi:10.1155/2023/5723730.

[10] M. Zubair, A. Ghubaish, D. Unal, *et al.*, "Secure Bluetooth communication in smart healthcare systems: a novel community dataset and intrusion detection system," *Sensors*, vol. 22, no. 21, 2022, doi:10.3390/s22218280.

[11] P. Muruganandhan, S. Jothilakshmi, R. Vivek, S. Nanthakumar, S. Sakthi and S. Mayakannan, "Investigation on silane modification and interfacial UV aging of flax fibre reinforced with polystyrene composite," *Mater Today Proc*, 2023, doi:10.1016/j.matpr.2023.03.272.

[12]   L. Chaari Fourati and S. Ayed, "Federated Learning toward Data Pre-processing: COVID-19 Context," in 2021 *IEEE International Conference on Communications Workshops, ICC Workshops 2021 – Proceedings, 2021.* doi:10.1109/ICCWorkshops50388.2021.9473590.

[13]   W. Jian, J. P Li, A. U. Haq, S. Khan, R. M. Alotaibi and S. A. Alajlan, "Feature elimination and stacking framework for accurate heart disease detection in IoT healthcare systems using clinical data." *Frontiers in Medicine* vol. 11, pp. 1362397, 2024.

[14]   R. Manikandan, P. Ponnusamy, S. Nanthakumar, A. Gowrishankar and V. Balambica, "Optimization and experimental investigation on AA6082/WC metal matrix composites by abrasive flow machining process," *Mater Today Proc*, 2023, doi:10.1016/j.matpr.2023.03.274.

[15]   R. Myrzashova, S. H. Alsamhi, A. V Shvetsov, A. Hawbani, and X. Wei, "Blockchain meets federated learning in healthcare: a systematic review with challenges and opportunities," *IEEE Internet Things J*, vol. 10, no. 16, pp. 14418–14437, 2023, doi:10.1109/JIOT.2023.3263598.

[16]   J. Almalki, "Intrusion detection and prevention model for blockchain based IoMT applications," *Computer Systems Science and Engineering*, vol. 48, no. 1, pp. 132–151, 2024, doi:10.32604/csse.2023.038085.

[17]   I. Keshta, M. Soni, M. W. Bhatt, *et al.*, "Energy efficient indoor localisation for narrowband internet of things," *CAAI Trans Intell Technol*, vol. 8, no. 4, 2023.

[18]   D. Dinesh Kumar, A. Balamurugan, K. C. Suresh, R. Suresh Kumar and N. Jayanthi, "Study of microstructure and wear resistance of AA5052/B4C nanocomposites as a function of volume fraction reinforcement to particle size ratio by ANN," *J Chem*, vol. 2023, 2023, doi:10.1155/2023/2554098.

[19]   M. Ibrahim, R. Elhafiz, H. Okasha, and A. Al-Wadi, "Securing the Internet of Medical Things: AI-Based Intrusion Detection," in *Proceedings of the 16th International Conference on Electronics, Computers and Artificial Intelligence, ECAI 2024*, 2024. doi:10.1109/ECAI61503.2024.10607556.

[20]   R. Srinivasan, S. Karunakaran, M. Hariprabhu, R. Arunbharathi, S. Suresh, S. Nanthakumar, "Investigation on the mechanical properties of powder metallurgy-manufactured AA7178/ZrSiO4 nanocomposites," *Advances in Materials Science and Engineering*, vol. 2023, 2023, doi:10.1155/2023/3085478.

[21]   A. Kumar, S. Jain, K. Kaushik, and R. Krishnamurthi, "Patient-centric smart health-care systems for handling COVID-19 variants and future pandemics: technological review, research challenges, and future directions," *The Internet of Medical Things: Enabling Technologies and Emerging Applications*, pp. 181–224, 2022. [Online]. Available: https://www.scopus.com/inward/record.uri?eid=2-s2.0-85142545514&partnerID=40&md5=23587ce079fafbc251ef321d6fc30654

[22]   O. Samuel, A. B. Omojo, A. M. Onuja, *et al.*, "IoMT: a COVID-19 healthcare system driven by federated learning and blockchain," *IEEE J Biomed Health Inform*, vol. 27, no. 2, pp. 823–834, 2023, doi:10.1109/JBHI.2022.3143576.

[23]   K. Gupta, D. K. Sharma, K. Datta Gupta, and A. Kumar, "A tree classifier based network intrusion detection model for Internet of Medical Things,"

Computers and Electrical Engineering, vol. 102, 2022, doi:10.1016/j.compeleceng.2022.108158.

[24] S. Almotairi, D. D. Rao, O. Alharbi, Z. Alzaid, Y. M. Hausawi, and J. Almutairi, "Efficient intrusion detection using OptCNN-LSTM model based on hybrid correlation-based feature selection in IoMT," *Fusion: Practice and Applications*, vol. 16, no. 1, pp. 171–194, 2024, doi:10.54216/FPA.160112.

[25] S. Khan and M. Alshara, "Development of Arabic evaluations in information retrieval," *International Journal of Advanced Applied Sciences*, vol. 6, no. 12, pp. 92–98, 2019.

[26] M. Fazil, S. Khan, B. M. Albahlal, R. M. Alotaibi, T. Siddiqui, and M. A. Shah, "Attentional multi-channel convolution with bidirectional LSTM cell toward hate speech prediction," *IEEE Access*, vol. 11, pp. 16801–16811, 2023.

[27] Khan, Shakir, and Amani Alfaifi. "Modeling of coronavirus behavior to predict it's spread." *International Journal of Advanced Computer Science and Applications* 11.5 vol. 11, no. 5, 394–399, 2020.

[28] D. Chowdhury, S. Banerjee, M. Sannigrahi, *et al.*, "Federated learning based Covid-19 detection," *Expert Syst*, vol. 40, no. 5, 2023, doi:10.1111/exsy.13173.

[29] D. Gupta, O. Kayode, S. Bhatt, M. Gupta, and A. S. Tosun, "Hierarchical federated learning based anomaly detection using digital twins for smart healthcare," in *Proceedings – 2021 IEEE 7th International Conference on Collaboration and Internet Computing, CIC 2021*, 2021, pp. 16–25. doi:10.1109/CIC52973.2021.00013.

[30] A. Khosrotabar and M. Kadoch, "Anomaly detection for Internet of Medical Things using chameleon optimization-based feature selection," in *Lecture Notes in Networks and Systems*, 2024, pp. 139–148. doi:10.1007/978-3-031-67447-1_10.

[31] L. AlSuwaidan, S. Khan, R. Almakki, A. R. Baig, P. Sarkar, and A. E. S. Ahmed, "Swarm intelligence algorithms for optimal scheduling for cloud-based fuzzy systems," *Math Probl Eng*, vol. 2022, no. 1, p. 4255835, 2022.

[32] D. C. Nguyen, Q. V. Pham, P. N. Pathirana, *et al.*, "Federated learning for smart healthcare: a survey," *ACM Comput Surv*, vol. 55, no. 3, 2022, doi:10.1145/3501296.

[33] M. Azrour, J. Mabrouki, A. Guezzaz, S. Ahmad, S. Khan, and S. Benkirane, *IoT, Machine Learning and Data Analytics for Smart Healthcare.* CRC Press, 2024.

[34] M. L. Hernandez-Jaimes, A. Martinez-Cruz, K. A. Ramírez-Gutiérrez, and C. Feregrino-Uribe, "Artificial intelligence for IoMT security: a review of intrusion detection systems, attacks, datasets and Cloud–Fog–Edge architectures," *Internet of Things (Netherlands)*, vol. 23, 2023, doi:10.1016/j.iot.2023.100887.

[35] Q. Abbas, M. E. Ibrahim, S. Khan, and A. R. Baig, "Hypo-driver: a multi-view driver fatigue and distraction level detection system." *CMC-Computers Mater Contin*, vol. 71, no. 1, pp. 1999–2017, 2022.

[36] K. Sattar, H. T. Ahmad, H. M. Abdulghani, S. Khan, J. John and S. A. Meo, "Social networking in medical schools: medical student's viewpoint." *Biomed Res.*, vol. 27, no. 4, pp.1378–1384, 2016.

Chapter 9

Revolutionizing pathological assessments with privacy-centric machine learning models

S. Shiva Prakash[1], Anu Tonk[2], Sailaja Manepalli[3], R. Revathi[4], K.V. Shahnaz[5], Roshan Nayak[6] and S. Mayakannan[7]

Abstract

This study aims to explore the possibility of using fully homomorphic encryption (FHE) in conjunction with machine learning for private pathological evaluation. Specifically, we will look at how support vector machine (SVM) inference phases might be used to classify sensitive medical data. To make SVM inference on encrypted datasets easier to implement, a system is presented that makes use of the Cheon–Kim–Kim–Song (CKKS) FHE protocol. This architecture eliminates the need to decode data before analysis and guarantees the confidentiality of patient records. Further, a method for efficiently extracting features from medical images for use in vector representations is introduced. The system's performance and usefulness are supported by its examination on different datasets. Classification accuracy and performance are comparable to that of conventional, non-encrypted SVM inference; however, the suggested technique protects the CKKS scheme from known cryptographic assaults with a 128-bit security level. In a matter of seconds, the secure inference procedure is carried out. Cardiology, oncology, and medical imaging are just a few of the areas that could gain from FHE's improved security and efficiency in bioinformatics analyses, according to these results. The significant future implications of this research for privacy-preserving machine learning can pave the way for improvements in diagnostic processes, individualized medical treatments, and clinical investigations.

[1]Department of Mechanical Engineering, New Horizon College of Engineering, India
[2]Department of Multidisciplinary Engineering, The NorthCap University, India
[3]Department of Mechanical Engineering, Anil Neerukonda Institute of Technology and Sciences, India
[4]Department of Data Science, K.S. Rangasamy College of Arts and Science, India
[5]Department of Electronics and Communication Engineering, Veltech Rangarajan Dr. Sagunthala R&D Institute of Science and Technology, India
[6]School of Electrical Engineering, Kalinga Institute of Industrial Technology (Deemed to be University), India
[7]Department of Mechanical Engineering, Rathinam Technical Campus, India

Keywords: Support vector machines; Cheon–Kim–Kim–Song (CKKS); homomorphic encryption; radial basis function (RBF); Wisconsin breast cancer (WBC)

9.1 Introduction

Bioinformatics is a dynamic area that is quickly developing to help analyze and understand complicated biological data by combining computational methods with biological principles [1]. Important medical applications include medical imaging diagnostics, the pathological evaluation of diseases (such as heart defects, cancer progression, and other similar problems), and the creation of therapeutic strategies for these diseases. Bioinformatics has tremendous promise, yet there are serious concerns about data privacy and security [2]. Medical records are sensitive and personal information, so it is critical that all necessary precautions be taken to protect them from prying eyes.

Fully homomorphic encryption (FHE), differential privacy (DP), and secure multi-party computation (MPC) are all examples of cryptographic countermeasures that have emerged as potential solutions to the problem outlined earlier [3]. Among these methods, FHE is the most encouraging one. By doing away with the requirement for pre-decryption, it permits the execution of arbitrary calculations on encrypted data [4]. This feature allows for the safe outsourcing of critical processes to untrusted cloud servers, ensuring that both the data and the insights obtained from them remain secure. Both general-purpose computing and bioinformatics stand to benefit greatly from this.

The mathematical structure that underpins FHE ensures the security of the two domains—ciphertext, which contains encrypted data, and plaintext, which contains unencrypted data. Without decryption, one feature of homomorphism is the capacity to directly conduct computational calculations on data encrypted. The plaintext space and the ciphertext space are directly mapped for every operation. This allows a server to do valuable computations and decrypt ciphertexts even when it does not have access to the original material. When these calculations are complete, the encrypted result is stored in the ciphertexts that are generated. The plaintext output, which is decrypted using the FHE scheme's secret key, is the same as the result produced from the original plaintext data after executing the same processes.

Nevertheless, it is crucial to recognize that FHE imposes computational overheads that make it impractical and inefficient for broad use. This inefficiency is caused by a number of things. To begin, FHE is dependent on mathematical objects, which are difficult to work with and incur a high processing cost. The second reason is that computational demands and bandwidth needs are already high due to the data size expansion that occurs during encryption. Lastly, although FHE is capable of handling simple arithmetic operations like adding and multiplying, it frequently has to rely on numerical approximations when assessing more complicated functions, such as exponentiation, which can lead to errors. Despite these obstacles, there has been a rise in the use of FHE to create safe medical applications, according to the literature [5].

By integrating the CKKS completely homomorphic encryption method with support vector machines (SVMs), this study expands upon previous work to examine the practicability of efficient and private pathological evaluation [6]. The aim of this work is to create an effective and safe system for pathological evaluation that functions on encrypted representations of sensitive medical data, protecting their confidentiality. Given its capacity to execute calculations on encrypted real or complex vectors, the CKKS–FHE technique stands out as an ideal contender for ML jobs. In contrast, SVMs are tried-and-true ML techniques for handling regression and classification problems reliably [7,8]. A more refined method for homomorphic assessment of the SVM prediction function on encrypted input is presented, along with a parallel attack on the inherent computational cost of FHE processes.

Here, we assess how well the suggested approach works using four open-source standard data: the text datasets Cleveland Heart Disease (CHD) dataset and WBC, as well as the BreastMNIST and PneumoniaMNIST medical imaging datasets from MedMNIST. The experimental results demonstrate that the system can encrypt medical data using the CKKS–FHE algorithm with a 128-bit security level in a matter of seconds without sacrificing performance. Furthermore, data privacy and usefulness are both preserved because the adoption of homomorphic encryption does not result in any noticeable loss of precision.

The main goal of bioinformatics system development was to create a way to quickly and safely analyze medical record pathology using the CKKS–FHE method in conjunction with support vector machines. During analysis, sensitive patient data is protected by framework, which follows the strict 128-bit security standard.

The method demonstrates considerable flexibility by enabling the construction of high-performance homomorphic SVM models that can utilize a wide range of kernels, including linear, polynomial, radial basis function (RBF), and Sigmoid functions. Because of this versatility, users can choose the kernel that works best with their data and the methods they use for analysis.

With performance comparable to open models, this approach scales quickly for tabular/imagery datasets and difficult issues. Through feature extraction and comprehensive optimizations, trials on four data—Breast modified national institute of standard and technologies, CHD, and pneumonia modified national institute of standard and technologies (MNIST), Wisconsin breast cancer (WBC)—prove this, attaining low latency and high accuracy.

Several machine learning models, such as logistic regression, convolutional neural networks (CNNs), and long short-term memory (LSTM), have privacy-preserving techniques created for them. Secure and private model training and inference are made possible by these algorithms by the use of several privacy-protecting approaches, including FHE and MPC [9,10]. One example is the use of FHE-based privacy-preserving genotype imputation techniques. Another is a commercial Genomewide Association Study (GWAS) that uses privacy-enhancing tools. Oncological data can be analyzed with several data owners using a hybrid method that combines MPC and FHE [11,12]. Use of CNNs on encrypted data and a FHE-based collective learning technique for LSTM that preserves privacy.

These developments make it possible to secure sensitive data in many contexts, such as healthcare and genomics [13].

We now focus on SVMs in this setting after investigating broad privacy-preserving ML approaches. With the introduction of a nonlinear kernel SVM for online medical prediagnosis, the framework safeguards patient privacy while classifying health data [14,15]. eDiag lets users secure their health records and enable prediagnosis on the server side without exposing their models or requiring data decryption. In order to improve performance, the scientists adjusted eDiag expression for the nonlinear SVM. They also used techniques like polynomial aggregation and multiparty random masking to reduce computation and communication overhead. On the Pima Indians Diabetes (PID) dataset, they achieved 94% accuracy using conventional elliptic curve cryptography. They evaluated their performance using sub-second SVM, which is encouraging. Being open-source also encourages community involvement and growth, which is great for eDiag. Further investigation is needed, though, because eDiag does run into some significant limits. The RBF kernel is its sole support at the moment, which limits its usefulness to certain use cases. In order to test its generalizability across varied medical data and disorders, it must be tested on the PID dataset. Solutions relying on conventional elliptic curve cryptography may no longer be the safest or most efficient option due to advancements in quantum computing. At last, the writers failed to provide a numerical value for the system's security.

Using the Okamoto–Uchiyama (OU) homomorphic encryption technique, the authors [16] suggested an SVM prediction on encrypted data that preserves privacy. Although their approach shows potential in several areas, it does have some serious drawbacks. One positive aspect of the system is its multi-class support, which allowed it to achieve a remarkable accurateness of 98.4% on the dermatologic dataset, which consisted of 366 samples [17]. This indicates that it may be able to accurately categorize specific medical domains. Its limited usefulness, however, is due to serious flaws [18]. Lack of open-source implementation limits community involvement and transparency. In addition, there is a significant computing cost associated with the system, which can hinder its usefulness in situations where time is of the essence, as the SVM prediction process can take anywhere from 5 to 48 s. It was concerned about its applicability for safeguarding very sensitive medical data because its resilience against adversarial assaults is obfuscated by the absence of a quantifiable security level [19].

An FHE-enabled homomorphic SVM designed for use outside of the medical field. The outcome is compatible with four SVM kernels: linear, polynomial, RBF, and sigmoid. To expedite computing, it makes use of GPUs and the CKKS technique to do approximation arithmetic on encrypted real values. But there are additional obstacles to this implementation. Not being open source makes it difficult to reproduce or verify, and its security level ranges from 80 to 100 bits. Its accuracy is also relatively low, ranging from 76.79 to 89.58%. In addition, classifying a single encrypted sample takes anywhere from 1.14 to 66.08 s, which results in a substantial computational cost even though the method makes use of low degree polynomials (5 to 18).

Using a variety of kernel functions—including sigmoid, RBF, linear, and polynomial—our method offers a comprehensive solution to the issue of privacy-preserving SVM inference. To put method to the test, we employ four real-world datasets: PneumoniaMNIST, CHD, WBC, and BreastMNIST. We show that it is accurate and efficient. For general-purpose central processing units, this system is freely downloadable as open-source software. In addition, solution of this work offers the highly recommended 128-bit security level and uses cutting-edge cryptographic techniques that are resistant to quantum assaults.

9.2   Background

To lay the groundwork for this system, we first give brief descriptions of SVMs and the CKKS homomorphic scheme [20].

9.2.1   Symbols and notations

In this chapter, the sets of integers and real numbers are referred to as Z and R, respectively. We can express the set $Z_q = Z \cap (-c/2, c/2)$ for any integer q. Put otherwise, for every positive integer z, the set [z]q represents the only integer in the interval $(-q/2, q/2)$ that is equal to z modulo q. The term q is used for each coordinate or coefficient in vectors and polynomials. Matrixes and vectors are represented by capital and lowercase bold characters, respectively. In vector space (u, v) denotes the dot product of two vectors u and v.

9.2.2   Support vector machine

SVMs are widely utilized for regression and classification in supervised machine learning. Decision boundaries in high-dimensional space, or hyperplanes, divide data points into several classifications in order to determine the ideal one. A well-defined training and inference strategy is essential for the effective use of SVMs. During training, a model is repeatedly adjusted with the help of a labeled dataset. Hence, in the inference step, this trained model is utilized to generate predictions on a distinct, unlabeled testing dataset.

9.2.3   Training SVM

Every data point in a labeled dataset is represented by an *n*-dimension features $x_i \in$ Rn and $y_i \in \{-1, 1\}$. X_i and y_i are the labels for the set, and the labels might be anything from zero to *m*. In SVM training, the feature vectors are expanded into a higher-dimensional space using the train dataset and a predefined function *K*. Afterwards, parameters are derived by resolving an optimization problem, either primal or dual.

This subgroup of the input consists of the support vectors designated as, $\{SV\}_0^{l-1}$. The support vectors can be seen as weight factors for the array of coefficients, denoted as, $\{\alpha\}_0^{l-1}$. The real numbers comprise the set to which the bias parameter $b \in R$ belongs.

The scope of this chapter does not permit a treatment of SVM optimization challenges. For efficient SVM model training and parameter acquisition, we make use of publicly available resources. The system uses the generated parameters to make predictions while ensuring user privacy is protected.

9.2.4   SVM inference

In inference, the y-coordinate of an input feature vector is predicted using the SVM model. This is accomplished by testing the choice function in (9.1), where i stands for the support vector i.

$$y = \text{sign}\left(\sum_{i=0}^{l-1} a_i y_i K(\mathbf{x}, \mathbf{x}_i) + b\right) \tag{9.1}$$

9.2.5   The CKKS scheme

An attractive component of homomorphic encryption, the CKKS method allows computations to be performed directly on encoded vectors of real-valued data [21]. As a foundational problem in cryptographically hard ring-learning with errors (RLWE), it provides strong security assurances. The parameterizability of CKKS allows it to achieve very secure levels, such as the generally acknowledged 128-bit threshold. The ring $RQ = ZQ[x]/(xN + 1)$ is used to implement the theory, where $Q \in Z$ is the coefficient modulus and $N \in Z$ is the ring dimension.

Users of CKKS have the option to select between leveled and bootstrapped computation modes, allowing them to customize the algorithm according to their needs. Bootstrapped mode permits evaluations of unlimited or unknown depth, whereas leveled mode is ideal for circuits with a defined depth. Due to the high computational burden associated with bootstrapped mode, leveled mode is usually preferable for circuits that are too deep. Nevertheless, this flexibility is not without its cost. Because of this efficiency/flexibility trade-off, choosing the right CKKS mode requires serious consideration of circuit complexity and performance needs. To precisely match the modest circuit depth requirements of this work, the leveled mode is employed in the implementation to improve efficiency and reduce computational cost.

9.2.6   Homomorphic operations in CKKS

In order for CKKS to provide computations on encrypted data, real vectors must first be encoded and then encrypted. One unique feature of CKKS is its ability to process long real number vectors simultaneously through the use of a single-instruction, multiple-data (SIMD) paradigm. This feature makes CKKS similar to a vector computer, which can process several data pieces within a single ciphertext at the same time.

Transferring the two vectors, 1 and 2, from $\in RN/2$ to the polynomial ring RQ, the internal representation inside CKKS, is the initial step in converting them into plaintext messages. With p1 = Encoding(v1) and p2 = Encoding(v2), we get plaintext messages. Both c1 and c2 are ciphertexts that are created when the plaintext communications are encrypted. The following homomorphic actions are performed by CKKS on top of these encrypted ciphertexts:

- By applying a homomorphic point-wise addition to the encrypted messages, the function EvalAdd(c1, c2) produces ciphertext cadd = Encryption(Encoding(v1·v2)). Just one parameter needs to be in plaintext for EvalAdd to generate an encrypted sum. One way to compute the encrypted total is by using the formula cadd = EvalAdd(c1, p2). Note that CKKS also makes subtraction easier.
- To create ciphertext, the function EvalMul(c1, c2) multiplies the encrypted messages homomorphically point-wise cmul = Encryption(Encoding(v1·v2)). The fact that EvalMul may produce a safe product with less computing burden even when given a parameter in plaintext is worth noting.
- This EvalRotate function iteratively rotates the encrypted message vector in the left-to-right direction by an amount a ∈ Z+.

The aforementioned basic techniques can be used to build more complicated homomorphic computations including polynomial evaluation, dot-product calculations, and vector–matrix operations.

9.3   Research methodology

Our method for efficient and secure bioinformatics analysis is detailed in this work, and it relies on homomorphic encryption. By doing so, vital computations can be performed on encrypted data while sensitive medical data remains protected. The foundational security assumptions and the threat model for the architecture are first established. We then reveal the system's components, detailing the technical details that allow efficient and secure bioscience calculations on encrypted data.

9.3.1   Threat model

The two primary components of the system are the client/inquirer and the server. The sensitive data is held by the client/inquirer, but the model used to produce privacy-preserving SVM predictions is stored on the server. If the server is "honest but curious," as the saying goes, it will adhere to all protocols without question. Nevertheless, it might be hiding something, like a desire to boost computational performance, that would allow it to gain an advantage by deciphering secret keys or encrypted data. By encrypting critical client data but leaving the less sensitive parameters of the SVM model unencrypted, we establish a balance between privacy and computational load. This ensures optimal performance while protecting customer privacy.

Priority is given to performance and the main objective of safeguarding client data privately, even if clients may be able to learn SVM model parameters through repeated requests [22]. As the model parameters in the threat model are not mission-critical, additional mitigation strategies that could increase computational overhead are deliberately avoided [23,24].

In summary, the proposed system guarantees the following security features under the defined assumptions:

Client data privacy: The server is guaranteed that no unencrypted data will ever reach it during the inference process. To guarantee that no private information is disclosed, all computations are carried out on encrypted data. This fulfills the fundamental need of privacy-preserving computing while also protecting the client's confidentiality.

Insights privacy: The server encrypts the computed result so that no one other than the client can decipher it and verify its accuracy. The customer might utilize the prediction to guide their future decisions and research when they receive it.

Insight accuracy: Using the encrypted data, the server successfully executes the SVM inference, resulting in a trustworthy and informative classification result for the client. While keeping data private, this keeps the calculation useful for the client.

Unlinkability: No client-specific inference requests can be associated with the server. Because the encrypted data and categorization result are unlinkable, the server has no way of knowing where a request came from or following a client's activity over time. This method is more effective in safeguarding the client's privacy.

9.3.2   Federated Homomorphic Support Vector Machine (FHSVM)

To tackle a wider range of kernel-based learning problems including homomorphic encryption, this technique builds on top of FHSVM and adds support for RBF and sigmoid kernels [25]. With the enhancements, FHSVM continues to support linear and polynomial kernels effectively, while now accommodating a broader range of nonlinear kernel functions, enabling more flexible and advanced data analysis. Figure 9.1 illustrates that the method aligns with FHSVM by assuming the server is already equipped with pre-qualified SVM techniques.

The method used to generate the pre-trained model is not essential to the system's functionality.

Public or synthetic datasets: In training, a dataset that closely resembles the client's samples is used, either one that is publically available or one that is synthetically constructed [25]. The method ensures model relevance without compromising client data privacy.

Collaborative training: In order to train a shared model, numerous clients use a secure federated learning protocol, which prevents the direct exposure of sensitive data [26]. A paradigm that is specific to the distribution of collective data and protects individual privacy is the result of this joint effort.

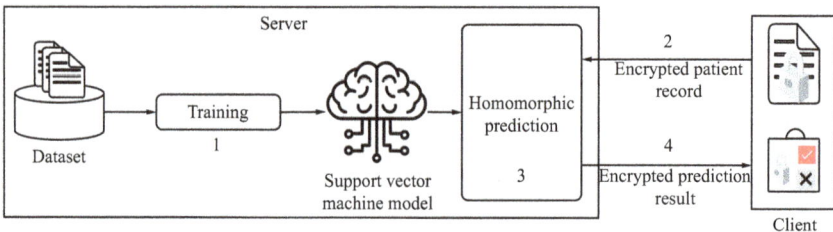

Figure 9.1   Data low-level structure in support vector machines.

Given that the variables of the SVM model are not encrypted, great care must be exercised. Although the server retains access to unencrypted SVM parameters, a key aspect of the solution that safeguards privacy is its capability to perform homomorphic computations on encrypted client requests. Because the server cannot access the sensitive data itself, the risk of privacy breaches due to unencrypted model parameters is reduced. It is worth noting that there may be situations where encrypting the parameters of SVM models could provide extra security benefits. Nevertheless, we have chosen to encrypt solely client requests in order to optimize efficiency and utilize resources within the limitations of structure of the system. The benefits and drawbacks of encrypting model parameters could be investigated in future studies.

After the server has trained an SVM model, it can securely and privately analyze encrypted client requests.

Query encryption: The customer employs a robust CKKS solution to safeguard their query data. This cryptographic method encrypts data so that it cannot be decrypted, protecting sensitive information but still allowing for useful computations on encrypted data.

Transmission of encrypted query: Confidentiality is maintained throughout transmission by sending the encrypted query to the server.

Homomorphic evaluation: The server starts directly assessing the encrypted query after receiving the trained SVM model. The server can perform model actions and generate an encrypted response without accessing the raw data of the query due to homomorphic encryption.

Return encrypt result: By returning the encrypted SVM evaluation result to the client, we ensure that the query and insight remain confidential throughout.

Decryption client-side: The client can view the prediction or classification made by the SVM model in plaintext after decrypting the encrypted output using the correct decryption key.

The technology integrates encryption, secure evaluation, and decryption to enable the server to process sensitive client data using SVM-based computations without compromising privacy in any form.

9.3.3 Non-linear kernel functions

The original concentration of FHSVM on polynomial and linear kernels laid a strong groundwork, but we saw the necessity to include non-linear to make it more versatile. To tackle this, we painstakingly navigated the intricacies of the homomorphic environment while effectively integrating RBF and sigmoid kernel implementations.

$$\text{RBF}(x, x_i) = \exp\left(-\gamma \|x - x_i\|^2\right) \tag{9.2}$$

$$\text{Sigmoid}(x, x_i) = \tanh\left(\alpha x^\top x_i + \beta\right) \tag{9.3}$$

Here α, β, and γ are constants.

Since transcendental functions such as exp and tanh are not natively supported by CKKS, we employ Chebyshev polynomial approximations as a workaround. This method, extensively recognized for its efficacy and accuracy in homomorphic settings, allows us to evaluate non-polynomial smooth kernel functions like RBF and sigmoid in the encrypted domain [27].

Determining two critical parameters is necessary for an accurate approximation of f(x) using Chebyshev polynomials.

Input range: Specifies the range [a, b] where the approximation holds. In order to find this range, it is necessary to experimentally check that the testing dataset distribution includes input values that are representative. An appropriate range strikes a balance between precision and computing efficiency.

Polynomial degree: In FHE, the computing cost increases as the degree of the approximation increases, but in general, higher degrees offer more precision. It is a common practice to conduct experiments in order to determine the ideal degree, which requires a compromise between efficiency and accuracy.

The following limitations should be considered when adjusting these parameters:

Improving performance: Processing polynomial operations in FHE settings can be a computational load. Because of this, choosing a lower-degree polynomial can facilitate faster computations, but accuracy may suffer as a result.

While higher-degree polynomials sometimes produce more precise approximations, they also carry the risk of introducing noise and mistakes into FHE computations, which is known as approximation error. It is critical to thoroughly assess how approximation errors affect the overall performance of the system.

9.3.4 Computational infrastructure

Two key computing libraries, scikit-learn and OpenFHE, form the foundation of the system.

9.3.4.1 Scikit-learn

The main SVM model is built and trained using scikit-learn, a widely recognized Python library known for its comprehensive machine learning toolkit. Support for various kernel types (linear, polynomial, RBF, and sigmoid), along with scikit-learn's robust implementation of SVM algorithms, meets the system's requirements. The fact that its API is both easy to understand and use adds to its allure [28].

9.3.4.2 Open FHE library

Open FHE, a framework designed for developing FHE applications, serves as the foundation for the system's homomorphic prediction capabilities. The following key components contribute to achieving the system's privacy protection objectives:

Effective CKKS scheme execution: Provides a highly optimized implementation of the CKKS scheme, satisfying computational requirements for both security and performance.

Algorithm Toolbox for FHE: A wide variety of difficult computations, such as Chebyshev polynomial evaluation, vector element reduction, dot-product, and many more, can be performed quickly using the algorithms provided, which are optimized for the FHE domain [29].

9.3.5 Execution

To clearly highlight their characteristics, the two main components of the system are described in detail:

1. Training the support vector machine structure and
2. Conducting homomorphic prediction on encrypted data obtained from customers.

9.3.6 Datasets

The Cleveland Heart Disease dataset and the WBC dataset are two widely recognized bioinformatics resources utilized in the evaluation of the system.

Cleveland Heart Disease dataset: For machine learning purposes, this dataset serves as a gold standard for identifying cardiac problems from patient records. This open-source dataset provides useful information about cardiovascular health and is used to compare and contrast different illness prediction algorithms. The Cleveland Clinic Foundation was the site of its collection. Age, sex, type of chest pain, resting blood pressure, serum cholesterol, fasting blood sugar, electrocardiogram (ECG) results, numbers of major vessel defects, the angle of the highest ST segment of exercise, peak heart rate, and thallium stress test results are among the numerous pieces of information included in the dataset.

WBC dataset: The WBC dataset is a valuable resource for machine learning algorithms that identify breast cancer [30]. It contains a plethora of information derived from digitalized fine-needle aspiration (FNA) photographs. Through the use of 30 features, including properties of cell nuclei, model training enables precise categorization of benign and malignant instances.

MedMNIST: Based on the famous MNIST dataset for handwritten digits, MedMNIST is a massive collection of standardized biological photographs. Among its many primary data modalities are 12 2D datasets and 6 3D datasets, all of which are commonplace in biomedical imaging. Every image is first resized to a standard 2D or 3D size of 28 x 28 pixel or voxel dimensions before classification. Classification applications using 100–100,000 samples and complexity levels spanning multi-label issues, ordinal regression, binary, and multi-class analysis might benefit from the lightweight 2D and 3D pictures offered by MedMNIST. Machine learning, computer vision, and biomedical image analysis all benefit from this. Around 10,000 3D images and 708,000 2D pictures make up the collection, which is a great resource for various academic and scientific projects. For this research, we used the BreastMNIST and PneumoniaMNIST databases.

An example of a binary classification benchmark, the BreastMNIST dataset contains 780 breast ultrasound pictures that have been classified as either normal or cancerous [31]. The dataset is divided into three parts: validation, training, and test. The ratio of the parts is 7:1, 1:2. Preprocessing involves downsizing the 150 x 150 pixel source images to 28 x 28 pixel size to ensure consistency with the MNIST dataset.

Among the 5862 chest X-ray images included in the PneumoniaMNIST dataset are those taken from children. From 384 x 127 to 2916 x 2713 pixels in original size, each image is grayscale. In order to achieve uniformity, each image is first enlarged to 28 x 28 pixels, and then center-cropped with a square window that is the similar size as the source image's shorter dimension. The goal of the dataset's binary-class classification architecture is to differentiate between chest X-rays showing pneumonia and those showing normal anatomy. Training accounts for 90% of the data, while validation and testing account for 10%.

The primary statistical features of the two datasets are displayed in Table 9.1. Even though the two data sets are moderate in size, we observe that there are substantially more specimens in the WBC dataset. The CHD dataset has more complicated features and may have data restrictions because of missing values, while both datasets contain solely numerical attributes and binary classification targets. Additionally, there is a class imbalance in both datasets; the WBC dataset contains more benign cases and the CHD dataset does not contain any cardiac illness.

The BreastMNIST and PneumoniaMNIST are compared in Table 9.2. The original image sizes of the two datasets varies greatly; BreastMNIST images are 150 x 150 pixels, whereas PneumoniaMNIST images are 384 x 127 to 2916 x 2713 pixels. The preprocessing steps ensure that the datasets are uniformly 28 x 28 pixels

Table 9.1   Datasets from CHD and WBC compared statistically

Characteristic	Wisconsin breast cancer	Cleveland heart disease
Target variable	Binary (malignant/benign)	Binary (disease absence/presence)
No. of classes	2	2
No. of features	9	13
No. of samples	569	303
Feature types	Numerical	Numerical
Missing values	None	Present
Class distribution	unstable	Unstable

Table 9.2   Analysis of the BreastMNIST and PneumoniaMNIST datasets from MedMNIST

Specifications	PneumoniaMNIST	BreastMNIST
Source data	5862 chest X-ray images	780 Breast ultrasound images
Original size range	384 x 127 2916 x 2713 pixels	150 x 150 pixels
Classes	Normal, pneumonia	Normal, malignant
Pre-processing	28 x 28 pixels	28 x 28 pixels
Image modality	Grayscale	Grayscale
Train, validate, test	90%, 10%, 0% (same as validation set)	70%, 10%, 20%

in size. Another difference between BreastMNIST and PneumoniaMNIST is the data split used for validation, training, and testing. BreastMNIST employs a 70%–10%–20% split, in contrast to PneumoniaMNIST's 90%–10% split.

9.3.7 SVM training

Following well-known procedures for SVM, we employ a systematic pipeline that includes the steps outlined below. By following this thorough and conventional workflow, we guarantee the creation of a strong SVM model that can reliably and accurately classify data.

9.3.8 Data preprocessing

To promote model convergence and stability, we normalize the data using min–max (eq. (9.4)) to make sure that the feature values are evenly distributed. Because numerical stability requires keeping intermediate values relatively modest throughout computations, this step is very helpful inside the CKKS framework.

$$\mathbf{x}_{norm} = \frac{\mathbf{x} - x_{min}}{x_{max} - x_{min}} \tag{9.4}$$

9.3.9 Feature extraction

Feature engineering was not performed on either the CHD or WBC tabular datasets. Prior to classification, the PneumoniaMNIST and BreastMNIST medical imaging datasets are preprocessed to extract relevant numerical features for input into the SVM classifier (Table 9.3).

Autoencoders are now widely used for feature extraction from various types of data, such as photos, message, and video. This ability stalks from their latent representation learning capacity, which allows them to grasp the essential structure and content of data. In order to get ready for the next machine learning assignment, autoencoders neatly compress the original data into a latent space with fewer dimensions and try to reconstruct the input while decoding, keeping the most important attributes. In the Autoencoder design, a network of encoders and decoders is constructed. The spatial dimensions of the input image are gradually reduced by the encoder using max-pooling2D layers, convolutional layers (Conv2D) activated with ReLU, and so on. A 4 x 4 x 8 tensor is generated from the flattened feature maps. Using upsampling2D and Conv2D layers driven by ReLU, decoders can upsample implicit representations. The final step is to restore the initial image using a convolutional layer activated with a sigmoid function.

9.3.10 Data splitting

At random, we divide the input dataset into two sets that cannot be joined:

Training set: For use in training models and estimating parameters.

Testing set: Set aside for objective examination of models and performance reviews.

*Table 9.3   Automatic feature extraction using the BreastMNIST and
PneumoniaMNIST datasets using an autoencoder architecture*

Type	Param #	Output shape
UpSampling2D	0	(None, 8, 8, 8)
Conv2D	584	(None, 8, 8, 8)
Conv2D	1160	(None, 13, 13, 8)
Conv2D	160	(None, 26, 26, 16)
MaxPooling2D	0	(None, 13, 13, 16)
Conv2D	584	(None, 4, 4, 8)
Reshape	0	(None, 4, 4, 8)
Flat	0	(None, 128)
UpSampling2D	0	(None, 16, 16, 8)
Conv2D	584	(None, 4, 4, 8)
MaxPooling2D	0	(None, 7, 7, 8)
Conv2D	73	(None, 28, 28, 1)

As shown in Table 9.2, we used the MedMNIST datasets that had been split into training and testing sets. Eighty percent of the data in the CHD dataset and 20% in the WBC dataset are reserved for training purposes.

9.3.11   Model training

There are four distinct kernel functions that we use to train SVM algorithms: polynomial, linear, RBF, and sigmoid. These functions differ for each dataset.

9.3.12   Model assessment

Recall, precision, and F1-score are appropriate measures for evaluating the model's predictive power on the unknown testing dataset.

9.3.13   Model parameters extraction

After the training is complete, the calibrated selection boundary variables, which are the model's predictive heart, can be extracted. The following homomorphic prediction component is designed around these parameters.

9.3.14   Homomorphic prediction

To ensure reliable model inference while safeguarding data security, the second component of the system is the homomorphic SVM prediction. Using CKKS, an encrypted sample, and a pre-trained SVM model, we may assess the decision function, as shown in (9.1). The following model parameters are utilized in this.

Support vectors (*SV*): The samples are chosen to define the decision boundary of the model.

Dual coefficients ($ai \cdot yi$): The relative importance of each support vector as measured by its numerical weights, which are used to make classification judgments.

Table 9.4   The parameters of the support vector machine models used for Chebyshev polynomial approximation

	Function	*min*	*max*	**Poly degree**
BreastMNIST and *PneumoniaMNIST*	exp	−10	0	13
CHD, WBC	exp	−100	0	120
	tanh	−60	60	496

Kernel function (*K*): In charge of translating the information into a 3D space so that non-linear correlations can be more easily identified.

Kernel parameters: Controlling the variables that make up the selected kernel function, honing its pattern-detecting capabilities.

By approximating the kernel function (K) with a Chebyshev polynomial, the compatibility issue of CKKS with non-polynomial kernels is resolved. This method allows for the homomorphic evaluation of CKKS by successfully aligning its supported operations with the decision function operations. Simplifying (9.1) to only include dot products, additions, and multiplications can help prevent computational issues and guarantee that CKKS will run smoothly. Table 9.4 displays the parameters linked to the SVM model's Chebyshev polynomial approximations. The RBF and sigmoid kernels were used to achieve these approximations on the MedMNIST, WBC, and CHD datasets.

Finally, it is important to note that the system does not compute the sign function directly. The rationale behind this decision is that efficient homomorphic domain approximation is more challenging for non-smooth functions (such as the sign function) compared to smooth kernels. Rather than relying on the computationally intensive part of the protocol to perform the sign function, we move its computation to the client side in order to avoid this problem. This optimization, however, is predicated on the important premise that the parameters of the Support vector machine model do not contain critical confidentiality information and do not require protection from any leakage attacks initiated by the client, even when there are several queries. Because homomorphic calculations are no longer burdened with the difficulties of sign function approximation, we attain a tremendous efficiency boost under this optimization. Notably, Chebyshev polynomials can still be used as approximants for the sign function if the user prefers to compute it on the server.

9.3.15   Factors of CKKS

This article describes the cryptographic variables of the CKKS approach, which is utilized to do the homomorphic SVM prediction.

The effectiveness of the system's SVM prediction algorithms are governed by a carefully chosen set of cryptographic parameters within the CKKS scheme. Critical parameters include N, the ring's diameter, and Q, the encrypted text coefficient modulus bit-width log2. How many integers may be encapsulated in a single

Table 9.5   The 128-bit cryptographic parameters of each kernel and dataset used by the CKKS scheme.

	Kernel			
Dataset	**Linear**	**RBF**	**Polynomial**	**Sigmoid**
CHD	(258,16k)	(804,32k)	($32k$, 756)	($65k$, 892)
		($\log_2 Q$, N)		
WBC	(287, 16k)	(804, 32k)	(644, $32k$)	(996, $65k$)
PneumoniaMNIST	–	(640, 64k)	–	–
BreastMNIST	–	(640, 64k)	–	–

ciphertext is determined by the ring dimension N, which in turn affects the computing efficiency of homomorphic processes. In addition, N is critical for deciding the scheme's security level; higher values typically provide better defense against adversarial assaults. Permitted multiplicative depth, noise accumulation, and plaintext precision are all controlled by the ciphertext coefficient modulus Q. With a big Q, we can examine deeper circuits, but at the cost of greater computer requirements and ciphertext growth. Thus, security, efficiency, and accuracy are all factors to consider while determining the best CKKS parameters.

Table 9.5 details the parameters that were used to calibrate the N and log2 Q values of all SVM implementations. This includes those with linear, polynomial, RBF, and sigmoid kernel functions, and those that used the Breast MNIST, Pneumonia MNIST, and WBC datasets. Every setup strictly follows the industry-standard 128-bit CKKS security standard. The exact computing requirements of the particular SVM, including the desired computational precision and the multiplicative depth needed for its circuit, determine the choice of Q. Keep in mind that the RBF kernel was the sole one used for the medical images because of how well it performed.

The chart further proves that the ring diameter is intricately related to the ciphertext modulus. In particular, it shows that keeping the desired level of security requires an equal and opposite increase in N for every log2 Q that is increased.

9.4   Experimental results

We will go over the experimental setup, which includes all the settings and evaluation metrics. We proceed to assess the SVM models' precision. The study concludes with an analysis of the computational demands and efficiency of homomorphic prediction.

9.4.1   Experimental framework

Our evaluation laptop featured a 12th Generation Intel(R) Core (TM) i7-12700H CPU, Ubuntu (v. 22.04.2 LTS) and 64 GB of RAM installed. The tools used included scikit-learn (v.1.3.0) for SVM training, OpenFHE (v.1.1.1) for homomorphic SVM inference, gcc (v.11.4.0) for C++ compilation, and Python (v.3.10.12) for scripting. To guarantee the greatest performance possible, OpenFHE was developed with multi-threading enabled.

9.4.2   Assessment of the SVM formulations

Recall, Precision, and F1 score are commonly used assessment metrics for data-driven classifiers, and we employed them to systematically evaluate the built SVM models' efficacy. To provide a precise assessment of the model's efficacy, these measures were computed solely using the testing dataset.

The proportion of expected positive results that actually occur is known as the accuracy rate or precision. This demonstrates that the model is capable of avoiding false positives. Recall is the proportion of actual positive results to the overall number of positive results. As a balanced measure of performance, the F1 score takes precision and memory into account in a harmonic manner. An assessment of the model's false positive detection capability is the end outcome.

Table 9.6 shows the outcomes of four SVM models trained on the Breast MNIST, WBC, Pneumonia MNIST, and CHD datasets, utilizing distinct kernels: linear, polynomial, RBF, and sigmoid. Important to SVMs are kernels, which influence the decision boundary and define the sample similarity function.

Examining the WBC and CHD tabular datasets. Importantly, regardless of the metric or kernel used, the WBC dataset consistently outperforms the CHD dataset when comparing the two datasets. This could be because SVM models work better with the WBC dataset or because the WBC dataset is easier to classify overall. Additionally, the sigmoid kernel and the RBF kernel are the two best performers on both datasets. The performance of linear and polynomial kernels is either poorer or about the same. This trend may point to underfitting or overfitting problems with

Table 9.6   SVM model quality as measured by F1, recall, and accuracy

		Kernel			
Dataset	**Measurement**	**Linear**	**Polynomial**	**Radial basis function**	**Sigmoid**
WBC	Recall	0.96	0.93	0.96	0.96
	Precision	0.97	0.91	0.98	0.98
	F1-score	0.96	0.92	0.97	0.97
CHD	Recall	0.74	0.71	0.74	0.76
	Precision	0.74	0.71	0.74	0.76
	F1-score	0.74	0.71	0.74	0.76
PneumoniaMNIST	Recall	–	–	0.86	–
	Precision	–	–	0.88	–
	F1-score	–	–	0.86	–
BreastMNIST	Recall	–	–	0.82	–
	Precision	–	–	0.81	–
	F1-score	–	–	0.81	–

linear or polynomial kernels, or it could show that sigmoid and RBF kernels are more flexible and adaptive to the data distribution. Last but not least, the WBC dataset shows much more noticeable performance disparities between kernels compared to the CHD dataset. One possible explanation is that the WBC dataset contains more complicated and nonlinear patterns, whereas the CHD dataset contains more noise and outliers.

Our SVM models trained with RBF kernels outperformed other kernel configurations on the medical imaging datasets Pneumonia MNIST and Breast MNIST, showing encouraging results. The RBF kernel achieved an F1 of 0.81 on Breast MNIST and an F1-score of 0.86 on Pneumonia MNIST. When it comes to medical picture classification, the results demonstrate that the chosen kernel and learning approach are effective. The method remains competitive, as the original authors of MedMNIST also employed Auto-sklearn and achieved comparable results, with accuracies of 0.81 on BreastMNIST and 0.86 on PneumoniaMNIST.

9.4.3   Analyzing the performance of homomorphic prediction

Following the prior demonstration that encrypted and unencrypted SVM inference achieve statistically identical predictive accuracy, the performance of the homomorphic prediction phase during runtime will be examined. The CKKS technique enables high-fidelity computations, as demonstrated by this finding. We use different benchmarking methods to measure the time needed on the server side to evaluate the homomorphic SVM prediction in order to put a number on the prediction delay that comes with homomorphic execution. Its applicability to real-world deployment circumstances and the possible overhead of homomorphic computations can be determined by examining the fine-grainedness of these runtime measurements.

Using the BreastMNIST, WBC, CHD, and PneumoniaMNIST datasets, researchers examined the latency properties of homomorphic SVM inference. Table 9.7 summarizes these findings. Runtime performance, measured in seconds, allows one to quantify the time it takes for the server to generate encrypted predictions across various kernel SVM types.

Table 9.7   Average delay (second) for homomorphic support vector machine inference on datasets

Dataset	Kernal			
	Linear	RBF	Poly	Sigmoid
WBC	0.52	5.73	4.41	12.80
PneumoniaMNIST	–	2.03	–	–
BreastMNIST	–	2.09	–	–
CHD	0.39	4.81	3.83	11.23

The latency and complexity of the kernel are shown to be correlated. In both the CHD and WBC datasets, the linear kernel—named for its direct structure—repeatedly obtains the lowest latency. The RBF and sigmoid kernels have the longest latencies, followed by the polynomial kernel. The innate computing complexity of each kernel is reflected in this hierarchy. Nonlinear kernels necessitate more complex circuits because of kernel evaluation, in contrast to the linear kernel's straightforward form that just requires a dot product computation.

Additionally, latency is affected by the size and dimensionality of the collection. Regardless of the kernel, the WBC dataset always has greater latency than the CHD dataset, which has more samples and support vectors. This lines up with the prediction that more support vectors are required for bigger datasets, which in turn increases processing cost.

There is a huge difference between linear and nonlinear kernels, as shown by the latency spectrum analysis. With latencies as low as 0.39 to 0.52 s, linear kernels routinely beat the competition when it comes to practicality. Based on these promising results, it seems that homomorphic prediction using linear kernels could have some real-world applications. The sigmoid kernel and other nonlinear kernels, on the other hand, display far larger latencies, between 3.83 and 12.80 s.

To achieve optimal results, it is crucial to select kernels with care, balancing latency, accuracy, and as this clear contrast shows, overall performance on specific tasks.

According to the statistics, the average latency for the two healthcare imaging data is 2.09 s for BreastMNIST and 2.03 s for PneumoniaMNIST [32,33]. Results on both medical imaging datasets are consistent when using the suggested homomorphic SVM learning methodology with the RBF kernel.

1. It is hard to make direct comparisons because different technologies are used to enhance privacy.
2. There is a lack of access to specific information about the methodologies used since most implementations are not open-source.
3. The application of different datasets can affect the efficiency and effectiveness of privacy-preserving techniques.
4. The findings can be affected by the computing platform used, whether it's a single-thread CPU or a multi-thread GPU.

Table 9.8 offers an analysis of pertinent state-of-the-art solutions to illuminate the delay linked to privacy-preserving SVM inference. Instead, rather than advocating for a single approach that works in every situation, we want to let the reader understand how each technique performs under different problem setups. The approach consistently outperforms previous methods across various datasets, with CPU latencies ranging from 2.03 to 5.73 s. It employs SVMs with an RBF kernel and privacy-preserving FHE. A GPU dataset with 13 features were delayed by 21 and four-half seconds. Technologies that enhance privacy using random masking and aggregate polynomial achieve better performance with a 1.5 s CPU latency [34,35].

Table 9.8 Time required for SVM inference (in seconds)

Dataset	# Features	Support vector machine kernel	Platform	Privacy- enhancing technology	Duration (s)
CHD	14				4.82
Breast MNIST	129				2.10
Pneumonia MNIST	129				2.04
WBC	10	RBF	CPU	CKKS	5.74

Notable results also include solutions that use the CKKS FHE approach rather than the Paillier partial homomorphic encryption algorithm. Paillier is inefficient and inconvenient for users because it only allows homomorphic additions and requires additional steps for multiplication [36].

9.5 Discussion

This section summarizes the research, emphasizing the main contributions and their significance. It also highlights existing limitations and explores potential directions for future work. Using SVMs and homomorphic encryption, we assess the viability of providing bioinformatics that preserves user privacy. The CHD, WBC, BreastMNIST, and PneumoniaMNIST datasets serve as real-world examples of tabular and imaging data used to evaluate and refine the proposed method. The results show that, while preserving the anonymity of input samples, the method achieves accuracy comparable to that of unencrypted SVM prediction. This demonstrates the method's feasibility and effectiveness for real-world applications requiring secure and accurate SVM-based inference.

To begin, unlike methods based on secure multi-party computation (MPC) or garbled circuits (GC), the proposed solution removes the need for client–server interaction or communication during the prediction phase. As a result, the prediction process experiences less delay and less network overhead. Second, unlike attribute-based encryption (ABE) techniques, the proposed methodology does not rely on complex cryptographic assumptions or trusted third parties. This simplifies both the security model and the implementation compared to such methods. Unfortunately, we haven't thought of a way to prevent hostile actors from tampering with the model parameters or encrypted data. The solution is based on a semi-honest, or honest-but-curious, threat model in which the server follows the protocol but may attempt to infer information by decrypting the data. Additionally, the use of polynomial approximations for RBF and sigmoid kernel functions introduces some approximation errors in the method. The CKKS technique makes these kinds of errors because it can't do exact mathematical calculations on encrypted data. However, by employing improved polynomial approximation techniques and carefully selecting appropriate CKKS parameters, these inaccuracies are minimized. As a result, the method maintains a high level of accuracy and reliability in its predictions.

The approach offers several advantages and has yielded promising results; however, certain limitations and challenges remain that should be addressed in future research. Currently, the system is limited to binary classification tasks and does not support multi-class or multi-label SVM prediction. Future efforts should aim to extend the technique to accommodate more complex and realistic classification scenarios. Secondly, the method has not yet been evaluated on high-dimensional or large-scale datasets, which may pose significant challenges in terms of memory usage and computational efficiency. Future work could involve testing the approach on more complex and diverse datasets and benchmarking it against leading privacy-preserving SVM prediction techniques. Thirdly, hardware-accelerated CKKS implementations offer performance improvements of two to three orders of magnitude over CPU-based versions, presenting an opportunity to further enhance the efficiency of the proposed technique. For decryption and encryption accelerators based on CKKS on FPGA, GPU, and ASIC platforms, several works have suggested suitable hardware solutions. All of these studies show that CKKS-based apps can be made faster and more scalable by taking advantage of hardware's high data rates and huge parallelism. The software-based approach utilizing the OpenFHE library, along with other available implementations, could be compared to assess the performance and efficiency of the system. Possible implications for understanding the pros and cons of software vs hardware approaches to privacy conserving support vector machine prediction are worth exploring.

9.6 Conclusions

The possibility of doing efficient and secure bioinformatics analysis utilizing FHE is explored in this work. We offer a strong solution that combines the CKKS–FHE approach with SVMs to allow the secure pathological evaluation of encrypted medical data. Users can be assured that their input samples remain secure when using homomorphic SVM inference on encrypted data, as the solution has been tested and validated on real-world datasets, including two tabular datasets (CHD and WBC) and two medical imaging datasets (PneumoniaMNIST and BreastMNIST). In comparison to open SVM inference, the proposed architecture achieves high precision by utilizing many kernel types. Depending on the dataset size and kernel selected, the framework's execution duration can range from 0.39 s to 12.80 s, indicating its efficiency. To protect CKKS from known dangers, the framework offers a 128-bit security level. This chapter highlights the significance of selecting the right kernel according to data properties and application limitations, and how effective SVMs are for medical data classification. Moreover, the CKKS scheme's homomorphic prediction capability is proven, providing encouraging accuracy but calling for additional research into latency improvement in practical settings.

References

[1]   R. Leon, B. Martinez-Vega, H. Fabelo, *et al.*, "Non-invasive skin cancer diagnosis using hyperspectral imaging for in-situ clinical support," *J Clin Med*, vol. 9, no. 6, 2020, doi:10.3390/jcm9061662.

[2]   Y. Terada, M. Isaka, T. Kawata, *et al.*, "The efficacy of a machine learning algorithm for assessing tumour components as a prognostic marker of surgically resected stage IA lung adenocarcinoma," *Jpn J Clin Oncol*, vol. 53, no. 2, pp. 161–167, 2023, doi:10.1093/jjco/hyac176.

[3]   W. Alghamdi, S. Mayakannan, G. A. Sivasankar, J. Singh, B. Ravi Naik, and C. Venkata Krishna Reddy, "Turbulence Modeling Through Deep Learning: An In-Depth Study of Wasserstein GANs," in *Proceedings of the 4th International Conference on Smart Electronics and Communication, ICOSEC 2023*, 2023, pp. 793–797. doi:10.1109/ICOSEC58147.2023.10275878.

[4]   A. U. Haq, J. P. Li, I. Khan, "DEBCM: deep learning-based enhanced breast invasive ductal carcinoma classification model in IoMT healthcare systems," *IEEE J Biomed Health Inform*, vol. 28, no. 3, 2022.

[5]   J. Shu, X. Ren, H. Cheng, *et al.*, "Beneficial or detrimental: recruiting more types of benign cases for cancer diagnosis based on salivary glycopatterns," *Int J Biol Macromol*, vol. 252, 2023, doi: 10.1016/j.ijbiomac.2023.126354.

[6]   A. M. Gejea, S. Mayakannan, R. M. Palacios, A. A. Hamad, B. Sundaram, and W. Alghamdi, "A Novel Approach to Grover's Quantum Algorithm Simulation: Cloud-Based Parallel Computing Enhancements," in *Proceedings of the 4th International Conference on Smart Electronics and Communication, ICOSEC 2023*, 2023, pp. 1740–1745. doi:10.1109/ICOSEC58147.2023.10276383.

[7]   P. Nikolaidis, M. Ismail, L. Shuib, S. Khan, and G. Dhiman, "Predicting student attrition in higher education through the determinants of learning progress: a structural equation modelling approach," *Sustainability*, vol. 14, no. 20, p. 13584, 2022.

[8]   Q. Abbas, M. E. A. Ibrahim, S. Khan and A. R. Baig, "Hypo-driver: a multiview driver fatigue and distraction level detection system." *CMC-computers Mater Contin*, vol. 71, no. 1, pp. 1999—2017, 2022.

[9]   M. D. Rajab, T. Taketa, S. B. Wharton, and D. Wang, "Ranking and filtering of neuropathology features in the machine learning evaluation of dementia studies," *Brain Pathology*, vol. 34, no. 4, 2024, doi:10.1111/bpa.13247.

[10]   J. Ren, R. Lax, J. G. Krueger, *et al.*, "Detecting nodular basal cell carcinoma in pathology imaging using deep learning image segmentation," in *Progress in Biomedical Optics and Imaging –Proceedings of SPIE*, 2020. doi:10.1117/12.2549950.

[11]   C. Yu and E. J. Helwig, "The role of AI technology in prediction, diagnosis and treatment of colorectal cancer," *Artif Intell Rev*, vol. 55, no. 1, pp. 323–343, 2022, doi:10.1007/s10462-021-10034-y.

[12]   R. AlShboul, F. Thabtah, A. J. Walter Scott, and Y. Wang, "The application of intelligent data models for dementia classification," *Applied Sciences (Switzerland)*, vol. 13, no. 6, 2023, doi:10.3390/app13063612.

[13]   S. Mayakannan, M. Saravanan, R. Arunbharathi, V. P. Srinivasan, S. V Prabhu, and R. K. Maurya, *Navigating Ethical and Legal Challenges in Smart Agriculture: Insights from Farmers. In Predictive Analytics in Smart Agriculture* (pp. 175–190). CRC Press.

[14]   A. ul Haq, J. P. Li, S. Khan, M. A. Alshara, R. M. Alotaibi, and C. Mawuli, "DACBT: deep learning approach for classification of brain tumors using MRI data in IoT healthcare environment," *Sci Rep*, vol. 12, no. 1, p. 15331, 2022.

[15]   F. A. Alrashed, A. M. Alsubiheen, H. Alshammari, S. I. Mazi, S. A. Al-Saud and S. Alayoubi, "Stress, anxiety, and depression in pre-clinical medical students: prevalence and association with sleep disorders," *Sustainability*, vol. 14, no. 18, p. 11320, 2022.

[16]   S. Mayakannan, J. B. Raj, V. L. Raja, and M. Nagaraj, "Effectiveness of silicon nanoparticles on the mechanical, wear, and physical characteristics of PALF/ sisal fiber–based polymer hybrid nanocomposites," *Biomass Convers Biorefin*, vol. 13, no. 14, pp. 13291–13305, 2023, doi:10.1007/s13399-023-04654-3.

[17]   S. Khan and L. AlSuwaidan, "Agricultural monitoring system in video surveillance object detection using feature extraction and classification by deep learning techniques," *Computers and Electrical Engineering*, vol. 102, p. 108201, 2022.

[18]   R. Girimurugan, C. Shilaja, A. Ranjithkumar, R. Karthikeyan, and S. Mayakannan, "Numerical Analysis of Exhaust Gases Characteristics in Three-Way Catalytic Convertor Using CFD," in *AIP Conference Proceedings*, 2023, doi:10.1063/5.0150561.

[19]   W. Su, Dachao Zheng, Jiacheng Zhou, *et al.*, "Rapid and precise multifocal cutaneous tumor margin assessment using fluorescence lifetime detection and machine learning," *APL Photonics*, vol. 9, no. 9, 2024, doi:10.1063/5.0224181.

[20]   S. Mayakannan, N. Krishnamurthy, K. V Devi, R. Deepalakshmi, S. Rani, and A. Jose Anand, *Navigating the Complexity of Macro-Tasks: Federated Learning as a Catalyst for Effective Crowd Coordination. In Handbook on Federated Learning* (pp. 308–332). CRC Press.

[21]   A. T. Francis, B. Manifold, E. C. Carlson, *et al.*, "In vivo simultaneous nonlinear absorption Raman and fluorescence (SNARF) imaging of mouse brain cortical structures," *Commun Biol*, vol. 5, no. 1, 2022, doi:10.1038/ s42003-022-03166-6.

[22]   T. Lindsey and J.-J. Lee, "Automated cardiovascular pathology assessment using semantic segmentation and ensemble learning," *J Digit Imaging*, vol. 33, no. 3, pp. 607–612, 2020, doi:10.1007/s10278-019-00197-0.

[23]   R. Wu, B. Wang, and Z. Zhao, "Privacy-preserving medical diagnosis system with Gaussian kernel-based support vector machine," *Peer Peer Netw Appl*, vol. 17, no. 5, pp. 3094–3109, 2024, doi:10.1007/s12083-024-01743-6.

[24]   M. Gupta, D. Pantola, I. Budhiraja, and R. Chaudhary, "P-HrDPS: Security-Aware Heart Disorder Prediction Support Model Using Ensemble Learning Technique," in *2023 IEEE International Conference on Communications Workshops: Sustainable Communications for Renaissance, ICC Workshops*

2023, 2023, pp. 654–659. doi:10.1109/ICCWorkshops57953.2023. 10283624.

[25]   S. Khan and M. Alshara, "Development of Arabic evaluations in information retrieval," *International Journal of Advanced Applied Sciences*, vol. 6, no. 12, pp. 92–98, 2019.

[26]   Y. Zhan, Q. Jin, T. Y. E. Yousif, M. Soni, Y. Ren, and S. Liu, "Predicting pediatric Crohn's disease based on six mRNA-constructed risk signature using comprehensive bioinformatic approaches," *Open Life Sci*, vol. 18, no. 1, 2023, doi: 10.1515/biol-2022-0731.

[27]   C.-H. Pham, T. Huynh-The, E. Sedgh-Gooya, M. El-Bouz, and A. Alfalou, "Extension of physical activity recognition with 3D CNN using encrypted multiple sensory data to federated learning based on multi-key homomorphic encryption," *Comput Methods Programs Biomed*, vol. 243, 2024, doi:10. 1016/j.cmpb.2023.107854.

[28]   J.-C. Bajard, P. Martins, L. Sousa, and V. Zucca, "Improving the efficiency of SVM classification with FHE," *IEEE Transactions on Information Forensics and Security*, vol. 15, pp. 1709–1722, 2020, doi:10.1109/TIFS.2019. 2946097.

[29]   Y. Chen, Q. Mao, B. Wang, P. Duan, B. Zhang, and Z. Hong, "Privacy-preserving multi-class support vector machine model on medical diagnosis," *IEEE J Biomed Health Inform*, vol. 26, no. 7, pp. 3342–3353, 2022, doi:10. 1109/JBHI.2022.3157592.

[30]   L. AlSuwaidan, S. Khan, R. Almakki, A. R. Baig, P. Sarkar, and A. E. S. Ahmed, "Swarm intelligence algorithms for optimal scheduling for cloud-based fuzzy systems," *Math Probl Eng*, vol. 2022, no. 1, p. 4255835, 2022.

[31]   M. Azrour, J. Mabrouki, A. Guezzaz, S. Ahmad, S. Khan, and S. Benkirane, *IoT, Machine Learning and Data Analytics for Smart Healthcare*. CRC Press, 2024.

[32]   M. D. Rajab, M. Azrour, J. Mabrouki, *et al.*, "Assessment of Alzheimer-related pathologies of dementia using machine learning feature selection," *Alzheimers Res Ther*, vol. 15, no. 1, 2023, doi:10.1186/s13195-023-01195-9.

[33]   S. Saif, P. Das, S. Biswas, S. Khan, M. A. Haq, and V. Kovtun, "A secure data transmission framework for IoT enabled healthcare," *Heliyon*, vol. 10, no. 16, 2024.

[34]   K. Sattar, T. Ahmad, H. M. Abdulghani, S. Khan, J. John and S. A. Meo, "Social networking in medical schools: medical student's viewpoint." *Biomed Res* vol. 27, no. 4, pp.1378–1384, 2016.

[35]   S. Khan and A. Alfaifi, "Modeling of coronavirus behavior to predict it's spread." *International Journal of Advanced Computer Science and Applications,* vol. 11, no. 5, pp.394–399, 2020.

[36]   S. Khan and S. Alqahtani, "Hybrid machine learning models to detect signs of depression." *Multimedia Tools and Applications,* vol. 83, no. 13, pp. 38819–38837, 2024.

Chapter 10

Exploring the synergy of SSL and IoMT in breast cancer detection: a theoretical framework

Ahmad Waleed Salehi[1] and Abdulahi Mahammed Adem[1]

Abstract

The global prevalence of breast cancer remains a significant health challenge and thus warrants the development of advanced diagnostic tools. This chapter studies the integration of self-supervised learning (SSL) with Internet of Medical Things (IoMT) to enhance breast cancer diagnosis. The proposed system makes use of IoMT devices for continuous data collection and uses SSL algorithms for efficient image analysis. It aims to enhance diagnostic accuracy and efficiency. The chapter describes a comprehensive state-of-the-art overview of IoMT technologies and SSL methods applied to medical imaging, along with their advantages and drawbacks. We discuss different data integration approaches and give several concrete examples that illustrate the practical benefit of this approach in clinical practice. It also shows technical, ethical, and operational challenges with some proposed approaches for smooth integration. The key findings support the combined use of IoMT and SSL for substantial improvement in early medical condition identification and patient treatment personalization. These technologies should be generalized in future research to clinical contexts and then demonstrated to be effective in real-world settings. This chapter opens the doors for any innovative theory with extensive insight and practical direction on integration of SSL with IoMT into breast cancer diagnosis to the existing diagnostic techniques with the best possible outcomes for patients.

Keywords: Self-supervised learning; Internet of Medical Things; machine learning; breast cancer; medical imaging

10.1 Introduction

Breast cancer is one of the most frequently occurring cancers around the world and is an important issue for women's health. The World Health Organization

[1]Yogananda School of AI Computers and Data Sciences, Shoolini University, India

asserts it to be the most common form of cancer among females; in 2020, more than 2.3 million new cases were diagnosed, which amounts to approximately 11.7% of all newly diagnosed cancer cases around the world [1]. This has not only had a health implication on the patients' body and psyche but has also created fiscal pressure on health systems through expenditure on treatment and care. The burden of breast cancer varies very significantly across various parts of the world. Incidence rates in more developed countries are higher because of better screening programs, high awareness, and better diagnostic tools [2].

The early identification and diagnosis of breast cancer could improve the outcomes for patients. In fact, early detection of breast cancer has a very high potential for successful treatment and survival. For example, the 5-year survival rate for breast cancer that has not spread beyond the breast is more than 99%, whereas when the cancer reaches other parts of the body as distant cancer, then the survival rate is just 27% [3]. This large difference underlines the necessity of early detection of the disease by improving the accuracy and precision of screening techniques.

Mammography is still the best screening method for breast cancer, able to identify tumours before they become palpable [4]. It is quite limited, such as decreased sensitivity for the examination of dense breast tissue and high false-positive rates, which generate unnecessary biopsies and worry for patients [5,6]. In this regard, the rising demand for breast cancer diagnosis by supplementary technologies and more accurate and reliable diagnostic tools in breast cancer detection has been in sight. Advances in imaging techniques, combined with sophisticated computational approaches such as machine learning, have been showing potential for the success of efforts in tackling such challenges.

In fact, the introduction of machine learning, more particularly deep learning methods, revolutionized medical imaging. This would develop advanced algorithms for the accurate analysis of intricate medical data [7]. These technologies can help with the identification of patterns and irregularities in imaging data that may turn out to be very elusive for human observers to perceive. Self-supervised learning (SSL) is a class of machine learning that uses high amounts of unlabelled data in the training models [8], improving their performance for specific specified tasks and needing only a small amount of labelled data. This will turn out to be of particular importance in medical imaging, where getting annotated data is time-consuming and expensive. It is also fast becoming a game-changer in healthcare: the Internet of Medical Things (IoMT). IoMT brings together medical devices and applications that connect to healthcare IT systems via online computer networks. IoMT supports real-time patient data capture for immediate analysis, leading to quicker and more accurate diagnosis. Such diagnostics tool integration with imaging techniques in IoMT would aid in getting a bird's-eye view for breast cancer detection. Therefore, this interlinked approach can improve accuracy in screening and diagnosis of breast cancer, reduce false positives, and make a better outcome for the patient by allowing treatment plans that are personally engineered based on full data analyses.

10.1.1   Objectives

- Summarize current state of breast cancer detection and the limitations of existing methods.
- Discuss the principles and applications of SSL and IoMT in medical imaging.
- Present a conceptual framework for integrating these technologies in breast cancer detection.
- Highlight potential real-world applications.
- Identify the challenges and future opportunities in this field.

10.1.2   Motivation of the chapter

The increasing cases of breast cancer globally thus call for better diagnosis methodologies. Traditional diagnostic processes may be useful, but they usually are not sensitive and effective enough for early detection and follow-up with the patient. Recent advancements on IoMT and SSL could resolve pending challenges. Data collection and real-time monitoring can be very beneficial with IoMT sensors while SSL algorithms are supportive in image processing and anomaly detection. All these technologies, when integrated together, might offer an improved way of diagnosis in breast cancer by increasing its accuracy, speed, and personalization.

10.1.3   Contributions

- To present a theoretical framework for integrating SSL with IoMT.
- Provides an extensive review of current state of IoMT technologies and self-supervised learning.
- Examines several approaches to integrate IoMT data with SSL algorithms.
- Highlights potential avenues of future research in addition to what is expected about the integration of IoMT with self-supervised learning for future effectiveness.

To provide better guidance in this chapter, the following research questions have been formulated:

1. What are the present techniques in use in SSL and IoMT for the diagnosis of breast cancer?
2. How might the integration of IoMT-enabled devices with that of SSL algorithms enhance diagnostic accuracy and efficiency in the detection of breast cancer?
3. What challenges and limitations are arising because of the integration of SSL with IoMT in the breast cancer screening and diagnosis context?
4. Which are the possible future research lines that would be highly sought in the integration of SSL with IoMT for breast cancer diagnosis and treatment?

The structure of this chapter is organized as follows: Section 10.2 covers the use of SSL techniques for breast cancer detection, including an overview of the key methods and their applications in breast cancer diagnostics. Section 10.3 discusses IoMT applications in breast cancer detection, showing how IoMT technologies facilitate real-time monitoring and integration of diagnostic tools. Section 10.4 looks into the possible synergy between SSL and IoMT in breast cancer diagnosis improvement and patient outcome enhancement. Section 10.5 will elaborate on an hypothetical framework outlining the integration of SSL and IoMT for breast cancer detection with specifications of structure and workflow that could be proposed. Section 10.6 addresses challenges in implementing SSL and IoMT for breast cancer detection. It also identifies challenges and how those can be mitigated. Section 10.7 concludes with a summary of the findings and implications of integrating SSL with IoMT in breast cancer detection, with emphasis on the possible benefits and future research directions.

10.2   SSL techniques for breast cancer detection

10.2.1   *Pretext tasks for representation learning*

Pretext tasks are specially designed auxiliary tasks for producing useful feature representations without requiring manual annotations [9]. Models learn from the vast quantity of unlabelled data in tasks of breast cancer detection; hence, they improve their ability in the identification of complex patterns in medical images.

- *Image rotation prediction*

The model is trained to predict the extent of rotation imparted to the input image accurately in an image rotation prediction task. This method commonly entails rotating images by a few angles, say 0°, 90°, 180°, and 270°, as shown in Figure 10.1, and then training the model for the accurate categorization of the specific rotation angle [10]. It learns identify the various rotations which in turn enhance its geometric and structural knowledge of an images.

This is an important challenge, especially in medical imaging, as the orientation of breast tissue might vary considerably from one image to another. For instance, mammograms can be performed in several angles, such as craniocaudal and mediolateral oblique views [11]. Training a model to predict rotation accurately will enable it to understand and normalize differences in order to robustly detect abnormalities, such as tumors or calcifications.

- *Jigsaw puzzle solving*

Solution of the jigsaw puzzle involves breaking an image into patches, reordering the patches, and training the model to put them back together in the correct order [12]. This is because the model should learn spatial relationships and the persistent existence of objects in the image.

It can be used in solving jigsaw puzzles for identifying spatial patterns of anomalies in medical images like mammograms, ultrasounds, and histopathology slides to detect breast cancer [13]. Specifically, the model learns to recognize that specific tissue structures or patterns must be connected even though they appear separated in the rearranged patches. This understanding is key in the identification of tumours that are spread out or irregularly shaped and, therefore, take no form of a pattern.

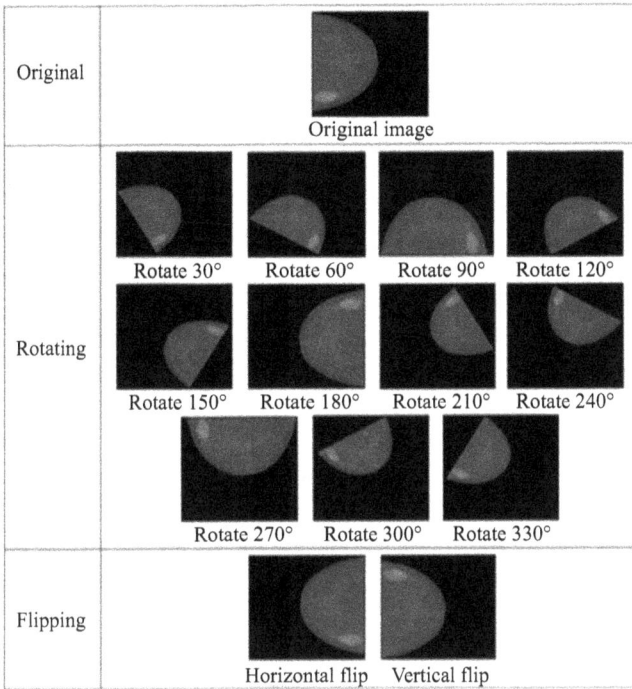

Figure 10.1   Image rotation example

- **Image colorization**

Image colorization is a task whereby a model tries to predict what the original colors of a grayscale image are [14]. This challenge will encourage the model to learn complex characteristics and textures, since estimation of accurate colors relies on understanding the content and context of the image. Colorization activities can provide large benefits for the analysis of images in the context of breast cancer detection. By training a model to colorize grayscale histopathology images, it can learn the very slight variations in texture and structure that identify either cancerous or non-cancerous tissue, as indicated in Figure 10.2.

10.2.2   Contrastive learning approaches

Contrastive learning strategies train the model to contrast similar image pairs with dissimilar image pairs, and as a result, a model is learned that draws on the intrinsic structure of the data to learn meaningful representations [15]. These methods are quite effective in fine-grained distinction medical image analysis tasks.

- **SimCLR**

A Simple Framework for Contrastive Learning of Visual Representations (SimCLR). SimCLR uses data augmentation to form positive pairs, consisting of

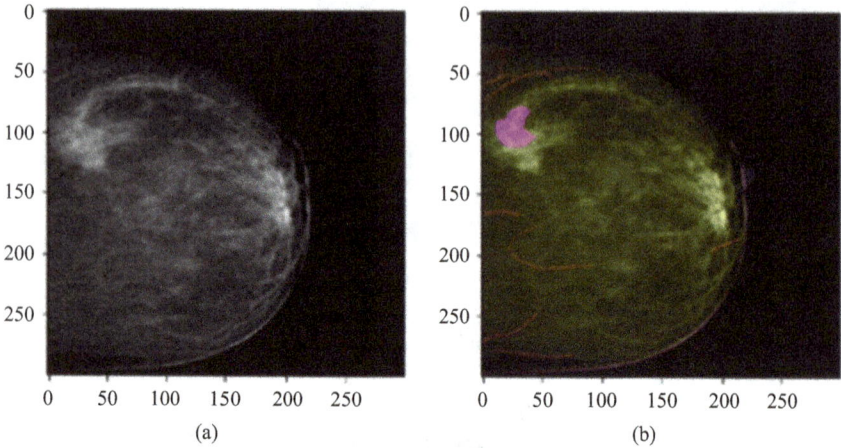

Figure 10.2   Image colorization

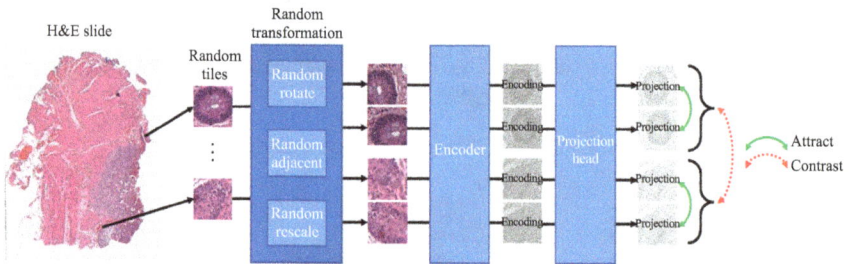

Figure 10.3   Effective tile representations learned by SimCLR from WSIs.

two augmented copies of the same image, and negative pairs, consisting of augmented versions of different images. The model learns to maximize the similarity of the positive pair while minimizing that of the negative pairs. This framework will force the model to learn robust and discriminative features that are invariant to common transformations such as rotation, scaling, and cropping [16].

One domain in which SimCLR can be applied is breast cancer detection. There it can enhance the model's capacity to capture tiny differences in medical imaging, hence differentiating between benign and malignant tumors. For example, in Figure 10.3 [17], there are two random transformations done on every input tile. These operations include rotation, selection of a neighbor tile, and rescaling. This approach transforms each input tile twice. All the augmented tiles are then encoded by feature extractors, projected by a fully connected layer, and contrasted with every other input tile's transformation while being attracted to their own. SimCLR builds a robust feature representation by training the model to recognize augmentations as corresponding to the same underlying tissue structures, hence diminishing the effect of imaging artifacts and variabilities [16].

- **MoCo (Momentum Contrast)**

Momentum contrast (MoCo) uses momentum-based updates to maintain a dynamic dictionary of the encoded representations. The method allows one to learn stable and resilient representations for a long time. This is because the momentum encoder has a slower update rate compared to the online encoder, ensuring that the obtained representations are stable and continuous [18]. The two basic modules of MoCoV3 are according to (10.1), query encoder, which processes the current batch of images, and the key encoder, kept up to date by momentum-based moving average of the parameters of the query encoder:

$$\theta_k = m\theta_k + (1 - m)\theta_q \tag{10.1}$$

The parameters of the key encoder are denoted as θ_k, whereas the momentum coefficient is represented by θ_q. The model's capacity to promote agreement across distinct augmented views of the same image (positive pairs) and limit similarity with other images (negative pairs) is enhanced since the key encoder can keep a queue of encoded keys representing previously observed images, which is also shown in Figure 10.4. This structure is based on the InfoNCE loss, given in (10.2):

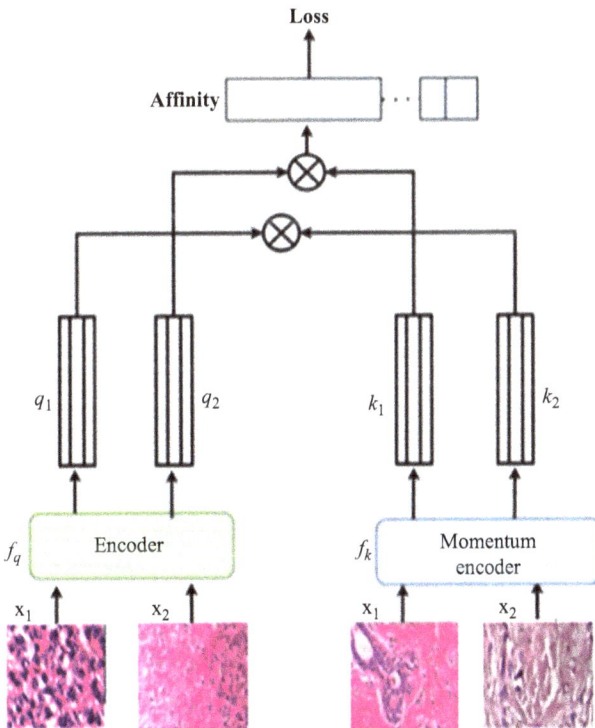

Figure 10.4   Architecture of MoCoV3.

$$\mathcal{L}_{InfoNCE} = -\log \frac{\exp(q.k^+/\tau)}{\sum_{i=0}^{k} \exp(q.k^+/\tau)} \tag{10.2}$$

q and k^+ denote the query and positive key feature vectors, k and i denote the negative key vectors, K denotes the number of negative keys, and τ denotes the temperature parameter.

- **BYOL (bootstrap your own latent)**

In *bootstrap your own latent* (BYOL), there are two networks: an online network and a target network. The online network is trained to predict the target network's representations. The parameters of the target network are updated with the running average of the parameters of the online network. This self-supervised learning approach improves the model's potential to learn rich and diverse representations. For this reason, in breast cancer diagnosis, BYOL enables efficient learning of representations without the hassle of negative sampling [19]. BYOL may be particularly appropriate for medical imaging because, on the one hand, the strategy is simple, and it does not require defining negative samples that may be complex and challenging due to the variable and often subtle nature of abnormalities [20].

The ability of BYOL to learn robust features without requiring explicit negative pairs can be useful when labelled data is scarce. Histopathological images often contain very complex patterns and textures, which make it difficult to annotate them appropriately [21]. BYOL uses this intrinsic organization of histopathological images to learn meaningful representations, thereby enabling the identification and classification of different types of breast lesions.

10.2.3   Predictive coding networks

Predictive coding networks make predictions of their future states given input data [22]. The working is achieved through generating predictions about future sensory information and updating the internal models of the network according to errors between these predictions and the actual inputs. Thus, the predictive coding network can be used to predict future frames in a medical imaging sequence or propagate identified anomalies for breast cancer diagnosis [23]. During dynamic contrast-enhanced MRI, early frames can be used to train a model in predicting forthcoming patterns of contrast enhancement. This feature helps in early diagnosis of those types of cancerous growth that show marked changes in their enhancing features over some time.

For instance, by making use of robust autoregressive models, constructive predictive coding (CPC) can learn self-supervised representations through future prediction in latent space. To get the most out of the latent space and make the best predictions for future samples, it employs a probabilistic contrastive loss. An initial step involves using g_{enc} a non-linear encoder to convert x_t the input observational sequence into a latent representational sequence $\mathcal{Z}_t = g_{enc}(x_t)$, which may have a lower temporal resolution. Afterwards, a context latent representation $c_t = g_{ar}(\mathcal{Z} \leq t)$ is produced by an autoregressive model g_{ar}, which summarizes all

$\mathcal{Z} \leq t$ in the latent space. A density ratio is represented in the following way given in (10.3), which maintains the mutual information between x_{t+k} and c_t:

$$f_k(x_{t+k}, c_t) \propto \frac{p(x_{t+k}|c_t)}{p(x_{t+k})} \qquad (10.3)$$

Predictive coding networks can simulate the temporal variations of ultrasound or mammography image and thus developmental patterns of tumors can be understood [24]. These networks learn how to predict future changes in medical images, enhancing the potentials of early breast cancer identification and monitoring, consequently leading to early and more effective therapies. Table 10.1 summarizes the research done in SSL with respect to breast cancer disease.

10.2.4 Other self-supervised techniques

Many innovative self-supervised techniques in breast cancer detection are emerging beyond the already presented methods. These techniques use innovative ways of augmenting the process of knowledge acquisition about specific features and improving accuracy for diagnosis.

- **Masked image modelling**

Masked image modeling trains the model to reconstruct correctly the missing or incomplete parts of the image [35]. This exercise is intended to force the model to learn contextual features while learning to understand relationships between what

Table 10.1 Summary of studies done on SSL and breast cancer disease.

Paper	Year of publication	Method/Technique	Application
[25]	2024	Variational autoencoder	Mammogram classification
[26]	2024	Contrastive learning, Weakly Supervised learning	Immunotherapy prediction
[27]	2021	Domain adaptation, Contrastive learning	Surgical margin detection
[28]	2021	Contrastive learning	Classification
[29]	2023	Domain Adaptation, Contrastive learning	Segmentation
[30]	2019	Image context restoration	Medical image analysis
[31]	2023	Pretext tasks, Multi-instance Learning	Molecular subtype prediction
[32]	2023	Multi-graph encoder	Classification
[33]	2024	Contrastive learning, Masked autoencoders	Classification
[34]	2022	Masked video Modelling, contrastive learning	Diagnosis

is visible and what is masked. Masked image modelling will be useful in breast cancer detection since a model may learn intricate details and contextual information from the images. These are important for factors in identifying small abnormalities within a medical image [35]. A formal definition of the masked image modeling problem is as follows: Multiple patches $x \in \mathbb{R}^{N \times (p^2C)}, x = |x_i|_{i=1}^N$ are used to divide an image $X \in \mathbb{R}^{H \times W \times C}$, where N is the number of patches. One way to represent a masking sequence is as $x \odot M$. By utilizing an encoder $f_\theta(\cdot)$ and a decoder $g_\theta(\cdot)$, the original pixel can be recreated using the remaining unmasked patches $x^\sim$. m_i is the hidden layer at the masked region in natural language processing, and its formula is $m_i = f_\theta(x^\sim)$ The learning is given in (10.4):

$$\mathscr{L}_{MIM} = \frac{1}{||M||} \sum_{i \in N} M_{i=1} ||m_i - x_i||^2 \tag{10.4}$$

For instance, reconstruction of occluded parts in mammograms or histological images may be done, where the model learns to highlight the patterns and structures that characterize malignant cell growths. This task increases the resilience of a model towards correct breast cancer detection and localization, especially those highly challenging cases with very small or diffuse aberrations.

- *Cluster-based self-supervision*

Cluster-based self-supervision is a method that involves grouping or categorizing images based on their feature representations [36]. These clusters are used to learn features that distinguish effectively among different classes or categories. The cluster-based techniques for breast cancer detection make it easier for the model to identify unique patterns and structures correlated with various kinds of breast lesions [37]. For instance, images of benign and malignant tumors can be grouped into two different clusters and then the model could be trained to learn the difference between them. This method will help in enhancing the model's capabilities in capturing fine-grained differences across various subtypes of BC, improving diagnostic accuracy and reliability.

- *Adversarial learning*

Adversarial methods, predominantly Generative Adversarial Network (GAN), involve training a generator to create realistic medical images and a discriminator to tell between the real and generated images [38]. The application of adversarial learning in detecting breast cancer is by generating synthetic images of breast cancer, which can be used to supplement the training data of the model [39].

The ability of adversarial learning to model complex patterns and anomalies in medical images is much more enhanced by enabling the model to differentiate between real and generated images. In this way, the approach will be able to achieve a higher model accuracy for the identification of rare or atypical cases of breast cancer. This can offer comprehensive and accurate diagnostic abilities [39]. With such sophisticated SSL techniques, researchers will significantly improve the performance of models in breast cancer detection. These methods diminish their dependence on labelled data, enhance the features for representation, and permit robust and accurate breast cancer detection in multiple settings clinically.

10.3   IoMT applications in breast cancer detection

10.3.1   Overview of IoMT architecture and components

The IoMT refers to the network of medical devices and apps communicating with each other through the internet to improve the healthcare delivery process. IoMT facilitates the integration of different diagnosis equipment, data collection systems, and analytical platforms in the domain of diagnosis of breast cancer to ensure complete and timely treatment of the patient [40]. The IoMT architecture for diagnosis of breast cancer includes the following vital components:

Smart devices: These encompass sophisticated imaging instruments such as mammography systems, ultrasound machines, MRI scanners, and histopathology analysers [41]. These gadgets are outfitted with sensors and networking capabilities to gather and transfer data.

Connectivity solutions: The success of IoMT hinges on the utilization of resilient and fortified communication networks to guarantee dependable data transfer between devices and central systems [42]. These technologies encompass Wi-Fi, Bluetooth, and cellular networks.

Cloud computing and data storage: Centralized cloud systems are necessary for storing the immense volumes of data produced by IoMT devices [43]. These platforms offer flexible storage solutions and facilitate immediate access to data from any location worldwide.

Data analytics and machine learning platforms: These platforms utilize complex algorithms, such as SSL models, to process and analyze the collected data. They assist in recognizing patterns, detecting deviations from the norm, and producing diagnostic insights [44].

User interfaces and applications: Interfaces such as mobile applications and online portals enable healthcare practitioners to retrieve, display, and analyze the data. Furthermore, these applications enable and support patient involvement and the ability to monitor patients from a distance [45].

The incorporation of these components into IoMT results in the establishment of a full ecosystem that facilitates the diagnosis of breast cancer. This ecosystem enhances the precision of data, makes it possible to monitor in real time, and offers personalized diagnostic insights. Figure 10.5 shows architecture and components of IOMT in breast cancer diagnosis.

10.3.2   IoMT-enabled devices for breast cancer screening

Breast cancer screening utilizes IoMT-enabled devices due to their advanced imaging and real-time data communication capabilities. Some of these devices include:

- **Smart mammography systems:** Intelligent mammography devices with advanced sensors and connectivity features take very fine detail images of the breasts [46]. These intelligent systems employ digital mammography techniques to identify microcalcifications and masses, which could signal the

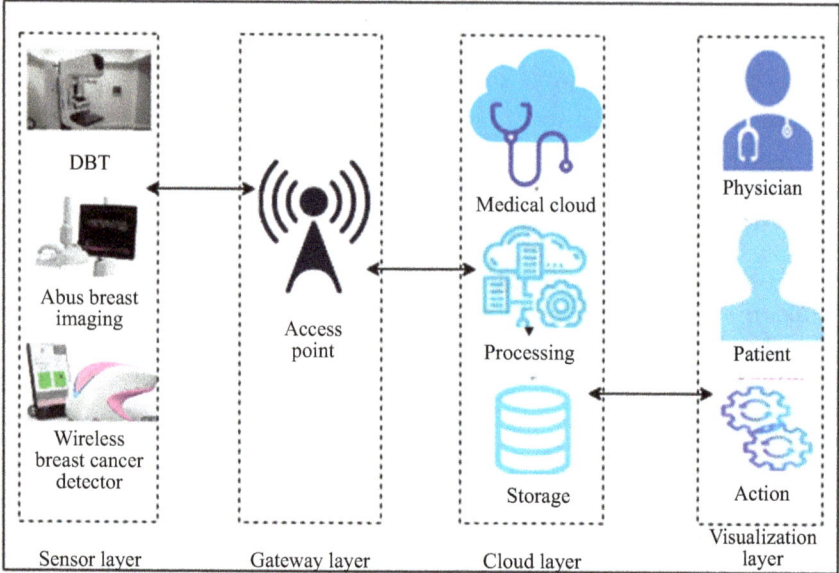

Figure 10.5   IOMT architecture in breast cancer diagnosis

presence of cancer in the breasts. IoMT integration makes it possible to send images directly to the cloud for storing and processing. This would grant radiologists and oncologists instant access to mammograms anywhere in the world. Moreover, the system is able to implement SSL models for analysis of such images in detection of patterns and anomalies, improving the screening for breast cancer both in accuracy and efficiency [47].

- *Ultrasound devices:* In breast cancer screening, ultrasound devices are extremely important. This is particularly true for dense breast tissue patients for whom mammography may not be very effective. IoMT-enabled ultrasound devices have several advantages associated with them; these include real-time image transmission and features for image processing [48]. These high-frequency sound wave devices come up with detailed images of breast tissue, pointing out cysts, solid masses, and other irregularities. IoMT connectivity facilitates real-time ultrasound image transmission to the cloud for investigation by SSL models to detect any cancerous cell growth [49]. In addition, portable and handheld IoMT-enabled ultrasound equipment offers great flexibility for implementation across different healthcare settings, including remote and underdeveloped areas. This ensures that the maximum number of patients get quality breast cancer screening.
- *MRI systems:* MRI systems have excellent differentiation of soft tissues and have been pretty effective in breast cancer detection, more so in high-risk populations and in women with dense breasts. IoMT-enabled MRI devices integrated into practice enhance the diagnosis process by instant transmission

of imaging data to centralized platforms [50]. These machines use very strong magnets and radio waves to generate detailed images of the breasts, which can highlight problems that may not appear on mammograms or ultrasound images. Integration with IoMT allows MRI data to be shared efficiently among radiologists and oncologists for collaboration in diagnosis and planning treatment. Next-generation SSL algorithms can even read MRI data for breast cancer patterns, such as abnormal outlines of tumors and changes in contrast enhancement over time. Incorporation of MRI into the process increases the accuracy and reliability of the detection process of breast cancer [51].

- *Histopathology scanners:* Histopathological examination is indispensable for confirming the diagnosis of breast cancer. IoMT-enabled histopathology scanners digitize tissue samples and send high-resolution images to cloud platforms for analysis. These scanners capture detailed images of tissue sections that are stained with various dyes, hence revealing crucial cellular and structural information critical for diagnosing cancers. Connectivity in IoMT ensures that these images can easily be accessed by pathologists for analysis using sophisticated machine learning algorithms [52]. The SSL methods can be used on the histopathology images to find tiny characteristics that indicate cancer, such as changes in cell structure and organization of tissue. This enhances the accuracy of the histological diagnosis and reduces the time required for manual analysis [53]. Table 10.2 summarizes studies done relating to IOMT and breast cancer.

10.3.3 Real-time data collection and monitoring

Two of the key features associated with IoMT applications in breast cancer identification are real-time data collection and monitoring [67]. Devices enabled with IoMT continuously collect the data from patients and forward it to a central platform for analysis and monitoring purposes. Real-time data collection in breast cancer screening involves image acquisition from mammography, ultrasound, MRI, and histopathology scanners. These photos are then instantaneously transferred to cloud platforms, where processing and analysis can be done with complex machine learning algorithms. This would have enormous value in the clinical setting: health care practitioners could identify and act on potential abnormalities in real-time. In this way, radiologists would be able to view the mammogram or ultrasound image immediately after they have been captured, not missing any area of potential concern [68]. Moreover, real-time data collection will allow continuous monitoring of patients' medication prescribed for in breast cancer and enable the introduction of the latest imaging data in medication.

10.3.4 Remote diagnosis and telemedicine

IoMT played a substantial role in improving remote diagnosis and telemedicine in breast cancer detection. These are technologies that facilitate health providers offering diagnosis and consultation services to patients in spite of their location [69]. IoMT-enabled devices facilitate the electronic transmission of medical images

Table 10.2   Summary of studies on IOMT related to breast cancer

Paper	Year of publication	Method/Technique	Application
[30]	2019	IoMT, image context restoration	Medical image analysis
[50]	2022	IoMT, wireless sensor networks	Detection and monitoring
[55]	2020	IoMT, AI-radiologist collaboration	Screening mammogram interpretation
[56]	2022	IoMT, fluorescence-guided surgery	Surgical guidance
[57]	2024	IoMT, longitudinal cognitive study	Cognitive impact analysis
[58]	2023	IoMT, roll-to-roll printing	Radiative cooling
[59]	2024	IoMT, inhibitor resistance analysis	Treatment resistance analysis
[60]	2024	IoMT, post-surgical monitoring	Post-mastectomy assessment
[61]	2024	IoMT, BREAST-Q questionnaire	Patient outcome measurement
[62]	2024	IoMT, survival analysis	Survival rate comparison
[63]	2024	IoMT, genomic profiling	Genomic differences analysis
[64]	2024	IoMT, immune profiling	Prognostic value analysis
[65]	2024	IoMT, adaptive control systems	System security analysis
[66]	2024	IoMT, optical solitons	Phase modulation analysis

and data of patients to experts for analysis, after which diagnostic opinions can be generated. This becomes more advantageous in rural and backward regions of the world where access to specialized medical care may not be easily available [70]. For instance, a patient in a remote region can have a mammogram test done in a local clinic with an IoMT-enabled mammography machine. The images are then sent for examination by a radiologist at another location. Thereafter, he or she can make a diagnosis, and further investigation, such as supplementary imaging or biopsy, may be recommended without the patient having to travel long distances.

10.4   Potential synergies between SSL and IoMT

10.4.1   *Advantages of SSL for medical data analysis*

SSL offers substantial advantages for medical data analysis, namely in the field of breast cancer detection. It tackles the difficulties posed by a scarcity of labelled data and allows for the extraction of valuable characteristics from extensive amounts of unlabelled data [71].

Reduced dependence on labelled data: Probably one of the main benefits derived from SSL is the capability of using unlabelled data, which is abundant in the medical field. Medical images are time-consuming to label and require special expertise; this makes labelled datasets relatively scarce and expensive [71]. SSL is becoming more and more independent of labelled data by learning through pretext tasks from unlabelled data and then fine-tuning on much smaller labelled datasets. In addition to saving resources, this approach enables the use of larger, more diverse datasets and therefore enhances model robustness and generalization ability.

Learning rich data representations: Among them, especially, contrastive learning, autoencoders, and generative models are very strong SSL methods for extracting complex patterns and representations from medical images. Only through this approach will models learn inherent characteristics that may be independent but important for a myriad of diagnostic tasks [71]. If the SSL models can successfully accomplish tasks like image inpainting or rotation prediction, they would truly understand the structure and context of an image.

Better generalization: Models trained with SSL exhibit better generalization to new, unseen data, an especially important property in medical applications known for extreme variability across patient demographics and imaging conditions [71]. Especially, through learning more resilient and invariant features, an SSL model could retain high performance across different groups of patients or imaging machines, hence establishing more reliable diagnostic tools. This generality is also very useful in therapy situations, where consistency and accuracy are paramount.

10.4.2   Benefits of IoMT in data collection and real-time monitoring

The IoMT improves healthcare by enabling ongoing data gathering, immediate monitoring, and seamless integration with healthcare systems, all of which are crucial for early detection and efficient management of breast cancer [72].

Continuous monitoring: IoMT devices allow for the ongoing monitoring of patients' health, which helps to identify anomalies at an early stage. For example, wearable technologies and intelligent breast patches have the capability to monitor alterations in breast tissue and notify healthcare professionals about possible concerns. Continuous monitoring enables prompt interventions, enhancing patient outcomes by detecting cancer at earlier, more manageable stages [73].

Real-time data collection: The real-time data collecting capabilities of IoMT devices guarantees that healthcare providers can promptly obtain the most current patient information [74]. The real-time component is essential for making immediate and well-informed judgements, especially in rapidly changing clinical settings. IoMT-enabled mammography devices offer real-time assessment of image quality and can identify abnormalities during the screening procedure, enabling prompt re-imaging if needed.

Integration with healthcare systems: The integration of IoMT devices with hospital information systems, electronic health records (EHRs), and cloud-based analytics platforms is smooth and seamless [75]. This integration guarantees that patient data is easily accessible for thorough analysis and individualized therapy. By connecting IoMT devices to centralized healthcare databases, practitioners can get comprehensive patient histories and thereby make more informed decisions [76]. In addition, this connectivity enables remote monitoring and telemedicine, thereby expanding the accessibility of healthcare services to locations that lack sufficient access.

10.4.3 Benefits of integrating SSL with IoMT in breast cancer detection

Integration of SSL within IoMT has several advantages in breast cancer identification. These algorithms can develop complex feature representations using large amounts of unmarked data that can make the resulting diagnosis model more accurate [77]. This minimizes the reliance on labelled data and decreases the burden of needing large, labelled datasets, which is usually expensive and time-consuming in medical imaging. Improved generalization, data augmentation, and self-supervised pretraining provide better adaptation to diverse and new data, hence enhancing robustness in the clinical real-world scenario. Scalability: being continuously connected, IoMT devices can collect and transmit data continuously, aggregating huge amounts of unlabeled data for training SSL models [78]. This enables scalable and continuous learning. This might help in the reduction of costs within the healthcare sector by an early diagnosis, hence eliminating the repetition of diagnosis procedures. SSL and IoMT show much potential for improving the diagnosis of breast cancer. Therefore, this integration could revolutionize diagnostic techniques and bring about improved outcomes should the effect of better accuracy and efficiency be achieved.

10.5 Hypothetical framework for integration

10.5.1 IoMT data collection and transmission

Informed consent is highly dependent on IoMT capabilities in terms of data collection and transmission. IoMT devices include high-end mammography machines, ultrasound machines, MRI machines, and wearable health monitoring devices. These devices are smartly installed to keep recording good-quality images and physiological data constantly. The information obtained is then securely transmitted over encrypted communication channels to some central storage systems, ensuring that it meets all standards of data protection, such as Health Insurance Portability and Accountability Act (HIPAA) and General Data Protection Regulation (GDPR). Secure transmission and central storage provide for handling voluminous medical data, highly relevant to further analytic stages.

10.5.2 Self-supervised model pre-training and fine-tuning

In this approach, SSL models are pre-trained with large amounts of unlabelled medical data obtained from IoMT devices. The pretext tasks within the pre-training phase include image in-painting, rotation prediction, and contrastive learning. All these tasks enable meaningful representations to be learnt from the raw data itself and enhance their ability for pattern and anomaly detection in medical images.

After pre-training, they get fine-tuned with labelled datasets designed for breast cancer diagnosis. Fine-tuning adjusts the models to perform relevant, accurate diagnostic tasks in tumor detection, segmentation, and classification. As a result of the dual-phase training—first with unlabeled data, then with labeled data—the accuracy and resiliency in the prediction outputs from the models are very high [79].

10.5.3 Integration workflow from data acquisition to diagnosis

The workflow goes all the way from data collection to diagnosis. First, imaging and physiological data are collected by IoMT devices, followed by its preprocessing for quality enhancement and removal of artefacts. This preprocessing will ensure that the data fed into the SSL models are of the best possible quality.

Following preprocessing, the data is fed into the pre-trained and finely tuned SSL models. These models incorporate state-of-the-art techniques in the identification and extraction of relevant characteristics to generate diagnostic results. Radiologists and clinicians interpret the AI-generated results, combining their class-specific expertise with the information provided by the models to arrive at definitive diagnostic decisions. IoMT devices produce feedback in a continuous manner in real time; this is helpful in promptly initiating follow-up actions and in creating custom care plans. This workflow in Figure 10.6 for the integration of both SSL and IoMT technologies could provide a proper and efficient breast cancer detection approach, taking advantage of the two technologies.

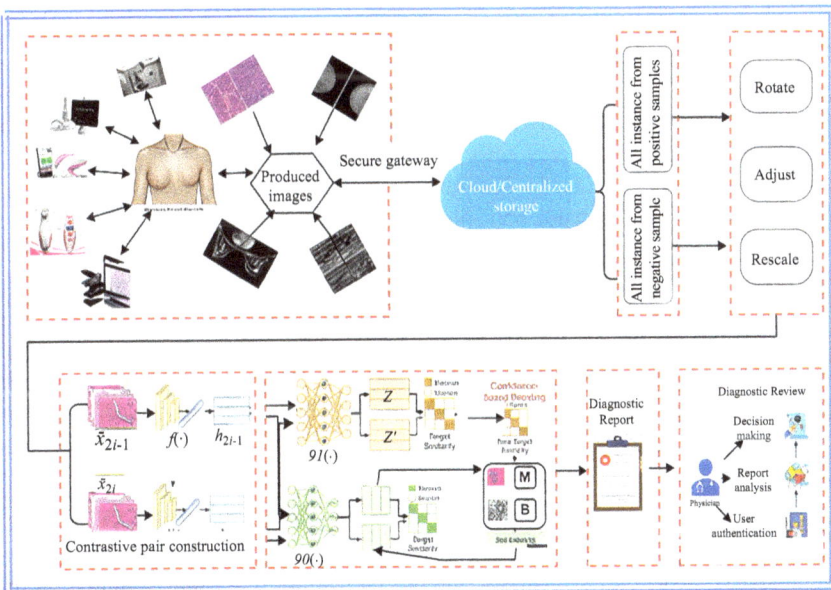

Figure 10.6 Proposed hypothetical integration

10.6   Challenges and considerations

10.6.1   *Data standardization and interoperability*

The integration of SSL with IoMT devices in breast cancer detection involves the first and major challenge of ensuring data standardization and interoperability. The medical data is derivative from several sources and equipment that show distinct formats, resolutions, and protocols. Heterogeneity may be present and can complicate the data integration and analysis [80].

Data standardization: This is a very critical process that would ensure that uniform formats and protocols for data are adopted [81]. This would mean that the same SSL models can handle data from the different IoMT devices. It simply calls for the development of standardized protocols for collecting, classifying, and storing data. For instance, the Digital Imaging and Communications in Medicine (DICOM) standard is widely used in medical imaging.

Interoperability: The development of robust frameworks that enable smooth data exchange between IoMT devices and SSL systems would require assertions on compatibility. This would be realized by creating software solutions working together in a manner prescribed by international standards, such as HL7 (Health Level Seven International) [82]. This also involves data integration from heterogeneous healthcare systems, which requires the application of APIs and middleware solutions during integration and harmonization of multiple systems.

10.6.2   *Privacy and security concerns*

IoMT and SSL integration into healthcare poses severe privacy and security concerns because medical information involves very confidential and sensitive data.

Data privacy: It is very critical as far as patients' data protection is concerned. It is thus imperative to observe laws such as the Health Insurance Portability and Accountability Act in the United States and the General Data Protection Regulation in Europe [83]. These policies enforce very strict regulations that govern access, storage, and exchange of data. This will involve the development of strong data encryption algorithms, access controls, and anonymization within IoMT devices and SSL systems to safeguard patient identities.

Data security: IoMT devices are susceptible to cyber security threats such as data leakage, hacking, and unauthorized access [84]. In order to reduce these threats, sophisticated security mechanisms at every level are required, including end-to-end encryption, secure communication protocols, and regular security auditing procedures. Moreover, the development of secure firmware and software updates for IoMT devices will actually alleviate the exploitation of their vulnerabilities, thus improving the resilience of the whole system [85].

10.6.3   *Technical and infrastructure challenges*

Breast cancer detection with IoMT and SSL integration is both technique-bound and infrastructurally very challenging.

Technical challenges: The development and deployment of models for SSL in medical data analysis are computationally resource-intensive and demanding in terms of machine learning expertise [86]. High-performance computing infrastructure and good data management facilities are needed to train SSL models on large-scale medical datasets. Besides, it is always important to carefully validate and finely adjust these SSL models so that they generalize well to unseen patient populations and imaging modalities.

Infrastructural issues: The IoMT devices implemented in the clinical environment require quite a robust infrastructure, including reliable network connectivity, efficient data storage systems, and good maintenance of IoMT equipment [87]. This is likely that inadequate healthcare infrastructure will retard IoMT and SSL implementations in many areas, especially in low- and middle-income countries. These challenges could be overcome through committing resources to strengthening hospital infrastructure, establishing reliable internet access, and training and supporting health workers.

10.6.4   Clinical adoption, trust, regulatory and ethical issues

Successful integration of SSL and IoMT into breast cancer detection requires examination of the clinical adoption, trust, and regulatory and ethical issues associated with these technologies. Demonstrating that these technologies will work in clinical practice includes establishing their reliability and therapeutic benefits, sorting out issues of regulatory compliance, and addressing ethical concerns related to bias, fairness, and patient autonomy. Table 10.3 identifies key considerations for clinical integration, trust-building, and regulatory and ethical frameworks in SSL and IoMT technologies adoption in breast cancer detection.

Table 10.3   Integration of SSL and IoMT: clinical adoption, building trust, regulatory and ethical considerations

Key aspects	Description	Objective	Impact	Challenge
Clinical adoption	Prove efficiency and reliability through rigorous clinical trials	Ensure therapeutic benefits and user-friendly	Improved clinical confidence and seamless work-flow integration	Designing comprehensive trials and studies, integrating into existing workflow
Building trust	Foster transparency in AI decision making,	Enhance doctor and patient confidence	Increased acceptance and validation by health care professionals	Developing explainable AI solutions, physician engagement
Regulatory compliance	Ensure SSL models and IoT devices comply with regulatory requirements	Validate and certify safety	Safe and effective implementation in clinical practice	Keeping regulation up to date with AI and IoMT innovations
Ethical considerations	Address biases and fairness in AI, ensure variability in training data and obtain informed consent	Protect patient groups from unfair treatment	Fair and unbiased AI outcomes, respect for patient autonomy	Managing biases in datasets, ensuring informed consent

10.6.5   Overview of analogous research in related field

10.6.5.1   SSL in medical imaging

In medical imaging, the topic of SSL gained intense interest fairly recently because it efficiently acquires knowledge from data that have no labels. According to a number of recent publications, SSL demonstrated efficiency in many different imaging applications such as segmentation, classification, and anomaly detection. For instance, SSL methods have been used to analyze MRI scans to segment brain tumors and achieve a drastic performance improvement by leveraging large amounts of unlabelled data. Application of SSL methods in chest X-ray analysis provides good results when pre-training models allow higher precision in anomaly detection as compared to models trained from scratch.

10.6.5.2   IoMT applications in healthcare

The IoMT is the network of interconnected devices used in the collection, transmission, and analysis of data related to health information. IoMT has been at the forefront of increasing patient monitoring, improving diagnostic accuracy, and allowing telemedicine. Wearable devices and smart sensors are, therefore, vastly used in the constant monitoring of vital signs, providing real-time data necessary for the management of a chronic condition. The IoMT technologies, such as smart mammography systems and ultrasonic probes, help collect quality imaging data for cancer detection, hence its accurate diagnosis. Besides, IoMT platforms have been integrated with EHRs to ease data movement and support clinical decision making.

10.7   Conclusion

This chapter analyzed the integration of SSL with IoMT in diagnosing breast cancer. Major conclusions emphasized that the integration can boost diagnostic accuracy, detection at the earliest stage possible, and the monitoring of patients. At the same time, SSL algorithms can enhance the analysis of medical images using an enormous amount of unlabelled medical data while IoMT devices allow for data collection and transmission in real time. These technologies could irrevocably transform breast cancer detection and treatment to be more efficient, accurate, and patient centered. IoMT and SSL are likely to become promising methods of screening for breast cancer in the future. Growing algorithms of AI and IoMT technology will increase diagnostic instruments advanced and reliable in nature. Further research should focus on creating multi-modal AI models, durable enough and that can assimilate data coming from mammography, ultrasound, and MRI, quite easily. Clinical validation and established methods are required to apply these technologies in real life. AI models must be more understandable and transparent to build physician and patient trust. Notably, the combination of SSL and IoMT advances breast cancer diagnosis. It is one such technique which can revolutionize diagnostic practice through the convergence of AI with IoMT, leading to faster identification, more accurate diagnosis, and better outcomes for the patient. This sector will grow in the future, and more research, collaboration, and clinical validation will be needed for more innovations to benefit appropriately. More efficient, effective, and patient-centered healthcare solutions are bright. This integration of SSL will drive this improvement.

References

[1]   M. Arnold, V. C. Morgan, M. Rumgay, J. Mafra, C. Singh, and I. Soerjomataram, "Current and future burden of breast cancer: global statistics for 2020 and 2040," *The Breast*, vol. 66, pp. 15–23, 2022, doi: 10.1016/j. breast.2022.08.010.

[2]   V. Sopik, "International variation in breast cancer incidence and mortality in young women," *Breast Cancer Res Treat*, vol. 186, no. 2, pp. 497–507, 2021, doi: 10.1007/S10549-020-06003-8/METRICS.

[3]   B. Smolarz, A. Zadrożna Nowak, and H. Romanowicz, "Breast cancer— epidemiology, classification, pathogenesis and treatment (Review of Literature)," *Cancers* vol. 14, no. 10, p. 2569, 2022, doi: 10.3390/ CANCERS14102569.

[4]   R. Vijayarajeswari, P. Parthasarathy, S. Vivekanandan, and A. A. Basha, "Classification of mammogram for early detection of breast cancer using SVM classifier and Hough transform," *Measurement*, vol. 146, pp. 800–805, 2019, doi: 10.1016/J.MEASUREMENT.2019.05.083.

[5]   D. Abdelhafiz, C. Yang, R. Ammar, and S. Nabavi, "Deep convolutional neural networks for mammography: advances, challenges and applications," *BMC Bioinformatics*, vol. 20, no. 11, pp. 1–20, 2019, doi: 10.1186/S12859-019-2823-4/FIGURES/3.

[6]   T. Habuza, A. N. Navaz, F. Hashim, *et al.*, "AI applications in robotics, diagnostic image analysis and precision medicine: current limitations, future trends, guidelines on CAD systems for medicine," *Informatics in Medicine Unlocked*, vol. 24, p. 100596, 2021, doi: 10.1016/j.imu.2021.100596.

[7]   M. Rana and M. Bhushan, "Machine learning and deep learning approach for medical image analysis: diagnosis to detection," *Multimed Tools Appl*, vol. 82, no. 17, pp. 26731–26769, 2023, doi: 10.1007/S11042-022-14305-W/TABLES/5.

[8]   R. Krishnan, P. Rajpurkar, and E. J. Topol, "Self-supervised learning in medicine and healthcare," *Nature Biomedical Engineering* vol. 6, no. 12, pp. 1346–1352, 2022, doi: 10.1038/s41551-022-00914-1.

[9]   J. Xu, L. Xiao, and A. M. Lopez, "Self-supervised domain adaptation for computer vision tasks," *IEEE Access*, vol. 7, pp. 156694–156706, 2019, doi: 10. 1109/ACCESS.2019.2949697.

[10] K. Ohri and M. Kumar, "Review on self-supervised image recognition using deep neural networks," *Knowl Based Syst*, vol. 224, p. 107090, 2021, doi: 10. 1016/J.KNOSYS.2021.107090.

[11] A. Jouirou, A. Baâzaoui, and W. Barhoumi, "Multi-view information fusion in mammograms: a comprehensive overview," *Information Fusion*, vol. 52, pp. 308–321, 2019, doi: 10.1016/J.INFFUS.2019.05.001.

[12] S. C. Huang, A. Pareek, M. Jensen, M. P. Lungren, S. Yeung, and A. S. Chaudhari, "Self-supervised learning for medical image classification: a systematic review and implementation guidelines," *npj Digital Medicine*, vol. 6, no. 1, pp. 1–16, 2023, doi: 10.1038/s41746-023-00811-0.

[13] F. Behrad and M. Saniee Abadeh, "An overview of deep learning methods for multimodal medical data mining," *Expert Syst Appl*, vol. 200, p. 117006, 2022, doi: 10.1016/J.ESWA.2022.117006.

[14] L. Fan, A. Sowmya, E. Meijering, and Y. Song, "Cancer survival prediction from whole slide images with self-supervised learning and slide consistency," *IEEE Trans Med Imaging*, vol. 42, no. 5, pp. 1401–1412, 2023, doi: 10.1109/TMI.2022.3228275.

[15] H. Zheng, Y. Zhou, and X. Huang, "Improving cancer metastasis detection via effective contrastive learning," *Mathematics* vol. 10, no. 14, p. 2404, 2022, doi: 10.3390/MATH10142404.

[16] R. Gong, L. Wang, J. Wang, B. Ge, H. Yu, and J. Shi, "Self-distilled supervised contrastive learning for diagnosis of breast cancers with histopathological images," *Comput Biol Med*, vol. 146, p. 105641, 2022, doi: 10.1016/J.COMPBIOMED.2022.105641.

[17] T. E. Tavolara, M. N. Gurcan, and M. K. K. Niazi, "Contrastive multiple instance learning: an unsupervised framework for learning slide-level representations of whole slide histopathology images without labels," *Cancers* vol. 14, no. 23, p. 5778, 2022, doi: 10.3390/CANCERS14235778.

[18] Y. Xie, J. Long, J. Hou, D. Chen, and G. Guan, "Weakly supervised pathological whole slide image classification based on contrastive learning," *Multimed Tools Appl*, vol. 83, no. 21, pp. 60809–60831, 2024, doi: 10.1007/S11042-023-17988-X/METRICS.

[19] K. Pandaram, P. R. Genssler, and H. Amrouch, "WaSSaBi: wafer selection with self-supervised representations and brain-inspired active learning," *IEEE Transactions on Circuits and Systems I: Regular Papers*, vol. 71, no. 4, pp. 1808–1818, 2024, doi: 10.1109/TCSI.2024.3357975.

[20] K. Pani and I. Chawla, "Examining the quality of learned representations in self-supervised medical image analysis: a comprehensive review and empirical study," *Multimed Tools Appl*, vol. 83, no. 10, pp. 1–31, 2024, doi: 10.1007/S11042-024-19072-4/METRICS.

[21] R. Yan, Y. Shen, X. Zhang, *et al.*, "Histopathological bladder cancer gene mutation prediction with hierarchical deep multiple-instance learning," *Med Image Anal*, vol. 87, p. 102824, 2023, doi: 10.1016/J.MEDIA.2023.102824.

[22] L. Li, Y. Liang, M. Shao, S. Lu, S. Liao, and D. Ouyang, "Self-supervised learning-based Multi-Scale feature Fusion Network for survival analysis from whole slide images," *Comput Biol Med*, vol. 153, p. 106482, 2023, doi: 10.1016/J.COMPBIOMED.2022.106482.

[23] A. S. Panayides, A. Amini, N. D. Filipovic, *et al.*, "AI in medical imaging informatics: current challenges and future directions," *IEEE J Biomed Health Inform*, vol. 24, no. 7, pp. 1837–1857, 2020, doi: 10.1109/JBHI.2020.2991043.

[24] L. Luo, X. Wang, Y. Lin, *et al.*, "Deep learning in breast cancer imaging: a decade of progress and future directions," *IEEE Rev Biomed Eng*, vol. 17, no. 4, pp. 312–332, 2024, doi: 10.1109/RBME.2024.3357877.

[25] M. A. Karagoz and O. U. Nalbantoglu, "A self-supervised learning model based on variational autoencoder for limited-sample mammogram classification," *Applied Intelligence*, vol. 54, no. 4, pp. 3448–3463, 2024, doi: 10. 1007/S10489-024-05358-5/TABLES/8.

[26] G. Yu, Y. Zuo, B. Wang, and H. Liu, "Prediction of tumor-associated macrophages and immunotherapy benefits using weakly supervised contrastive learning in breast cancer pathology images," *Journal of Imaging Informatics in Medicine*, vol. 8, no. 1, pp. 1–11, 2024, doi: 10.1007/S10278-024-01166-Y.

[27] A. M. L. Santilli, L. G. Bruzzone, F. C. De Luca, *et al.*, "Domain adaptation and self-supervised learning for surgical margin detection," *Int J Comput Assist Radiol Surg*, vol. 16, no. 5, pp. 861–869, 2021, doi: 10.1007/S11548-021-02381-6/METRICS.

[28] S. Perek, M. Amit, and E. Hexter, "Self Supervised Contrastive Learning on Multiple Breast Modalities Boosts Classification Performance," *Lecture Notes in Computer Science (including subseries Lecture Notes in Artificial Intelligence and Lecture Notes in Bioinformatics)*, vol. 12928, LNCS, pp. 117–127, 2021, doi: 10.1007/978-3-030-87602-9_11.

[29] K. Lee, H. Lee, G. El Fakhri, J. Woo, and J. Y. Hwang, "Self-Supervised Domain Adaptive Segmentation of Breast Cancer via Test-Time Fine-Tuning," *Lecture Notes in Computer Science (including subseries Lecture Notes in Artificial Intelligence and Lecture Notes in Bioinformatics)*, vol. 14220, LNCS, pp. 539–550, 2023, doi: 10.1007/978-3-031-43907-0_52.

[30] L. Chen, P. Bentley, K. Mori, K. Misawa, M. Fujiwara, and D. Rueckert, "Self-supervised learning for medical image analysis using image context restoration," *Med Image Anal*, vol. 58, p. 101539, 2019, doi: 10.1016/J. MEDIA.2019.101539.

[31] Z. Shang, H. Liu, K. Wang, and X. Wang, "BM-SMIL: A Breast Cancer Molecular Subtype Prediction Framework from H&E Slides with Self-supervised Pretraining and Multi-instance Learning," *Lecture Notes in Computer Science (including subseries Lecture Notes in Artificial Intelligence and Lecture Notes in Bioinformatics)*, vol. 14243, LNCS, pp. 81–90, 2023, doi: 10.1007/978-3-031-45087-7_9.

[32] M. Ibrahim, S. Henna, and G. Cullen, "Multi-graph convolutional neural network for breast cancer multi-task classification," *Communications in Computer and Information Science*, vol. 1662, pp. 40–54, 2023, doi: 10. 1007/978-3-031-26438-2_4/TABLES/1.

[33] Q. Wu, L. Zhang, M. Wang, *et al.*, "Cross-view Contrastive Mutual Learning Across Masked Autoencoders for Mammography Diagnosis," *Lecture Notes in Computer Science*, vol. 14349, pp. 74–83, 2024, doi: 10.1007/978-3-031-45676-3_8.

[34] Z. Lin, R. Huang, D. Ni, J. Wu, and B. Luo, "Masked Video Modeling with Correlation-Aware Contrastive Learning for Breast Cancer Diagnosis in

Ultrasound," *Lecture Notes in Computer Science (including subseries Lecture Notes in Artificial Intelligence and Lecture Notes in Bioinformatics)*, vol. 13543, LNCS, pp. 105–114, 2022, doi: 10.1007/978-3-031-16876-5_11.

[35]   D. Muhtar, X. Zhang, P. Xiao, Z. Li, and F. Gu, "CMID: a unified self-supervised learning framework for remote sensing image understanding," *IEEE Transactions on Geoscience and Remote Sensing*, vol. 61, 2023, doi: 10.1109/TGRS.2023.3268232.

[36]   J. Bai, R. Posner, T. Wang, C. Yang, and S. Nabavi, "Applying deep learning in digital breast tomosynthesis for automatic breast cancer detection: a review," *Med Image Anal*, vol. 71, p. 102049, 2021, doi: 10.1016/J.MEDIA.2021.102049.

[37]   X. Liu, Y. Chen, Z. Wang, *et al.*, "Self-supervised learning: generative or contrastive," *IEEE Trans Knowl Data Eng*, vol. 35, no. 1, pp. 857–876, 2023, doi: 10.1109/TKDE.2021.3090866.

[38]   Y. Zhao, J. Zhang, D. Hu, H. Qu, Y. Tian, and X. Cui, "Application of deep learning in histopathology images of breast cancer: a review," *Micromachines* vol. 13, no. 13, p. 2197, 2022, doi: 10.3390/MI13122197.

[39]   X. Zhang, J. Mu, X. Zhang, H. Liu, L. Zong, and Y. Li, "Deep anomaly detection with self-supervised learning and adversarial training," *Pattern Recognit*, vol. 121, p. 108234, 2022, doi: 10.1016/J.PATCOG.2021.108234.

[40]   A. Swiecicki, N. Konz, M. Buda, and M. A. Mazurowski, "A generative adversarial network-based abnormality detection using only normal images for model training with application to digital breast tomosynthesis," *Scientific Reports*, vol. 11, no. 1, pp. 1–13, 2021, doi: 10.1038/s41598-021-89626-1.

[41]   J. B. Awotunde, S. O. Folorunso, S. A. Ajagbe, J. Garg, and G. J. Ajamu, "AiIoMT: IoMT-based system-enabled artificial intelligence for enhanced smart healthcare systems," *Machine Learning for Critical Internet of Medical Things: Applications and Use Cases*, pp. 229–254, 2022, doi: 10.1007/978-3-030-80928-7_10.

[42]   L. Pinto-Coelho, "How artificial intelligence is shaping medical imaging technology: a survey of innovations and applications," *Bioengineering* vol. 10, no. 12, p. 1435, 2023, doi: 10.3390/BIOENGINEERING10121435.

[43]   S. U. Khan, N. Islam, Z. Jan, I. U. Din, A. Khan, and Y. Faheem, "An e-Health care services framework for the detection and classification of breast cancer in breast cytology images as an IoMT application," *Future Generation Computer Systems*, vol. 98, pp. 286–296, 2019, doi: 10.1016/J.FUTURE.2019.01.033.

[44]   R. Kumar and R. Tripathi, "Towards design and implementation of security and privacy framework for Internet of Medical Things (IoMT) by leveraging blockchain and IPFS technology," *Journal of Supercomputing*, vol. 77, no. 8, pp. 7916–7955, 2021, doi: 10.1007/S11227-020-03570-X/METRICS.

[45]   G. S. P. Ghantasala, N. V. Kumari, and R. Patan, "Cancer prediction and diagnosis hinged on HCML in IOMT environment," *Machine Learning and the Internet of Medical Things in Healthcare*, vol. 1, no. 1, pp. 179–207, 2021, doi: 10.1016/B978-0-12-821229-5.00004-5.

[46]   R. O. Ogundokun, S. Misra, A. O. Akinrotimi, and H. Ogul, "MobileNet-SVM: a lightweight deep transfer learning model to diagnose BCH scans for IoMT-based imaging sensors," *Sensors*, vol. 23, no. 2, p. 656, 2023, doi: 10. 3390/S23020656.

[47]   A. R. Khan, T. Saba, T. Sadad, H. Nobanee, and S. A. Bahaj, "Identification of anomalies in mammograms through Internet of Medical Things (IoMT) diagnosis system," *Comput Intell Neurosci*, vol. 2022, no. 1, p. 1100775, 2022, doi: 10.1155/2022/1100775.

[48]   A. N. Edmund, C. A. Alabi, O. O. Tooki, A. L. Imoizeand, and T. D. Salka, "Artificial intelligence-assisted Internet of Medical Things enabling medical image processing," *Handbook of Security and Privacy of AI-Enabled Healthcare Systems and Internet of Medical Things*, pp. 309–334, 2023, doi: 10.1201/9781003370321-13/ARTIFICIAL-INTELLIGENCE-ASSIS-TED-INTERNET-MEDICAL-THINGS-ENABLING-MEDICAL-IMAGE-PROCESSING-AJIMAH-NNABUEZE-EDMUND-CHRISTOPHER-AKI-NYEMI-ALABI-OLUWASEUN-OLAYINKA-TOOKI-AGBOTINAME-LUCKY-IMOIZE-TANKO-DANIEL-SALKA.

[49]   P. Manickam, R. V. Venkatesan, A. P. Kumar, *et al.*, "Artificial intelligence (AI) and Internet of Medical Things (IoMT) assisted biomedical systems for intelligent healthcare," *Biosensors*, vol. 12, no. 8, p. 562, 2022, doi: 10.3390/BIOS12080562.

[50]   C. Kaushal, M. K. Islam, A. Singla, and M. Al Amin, "An IoMT-based smart remote monitoring system for healthcare," *IoT-enabled Smart Healthcare Systems, Services and Applications*, vol. 1, no. 1, pp. 177–198, 2022, doi: 10. 1002/9781119816829.ch8.

[51]   Z. Tang, Z. H. Sun, E. Q. Wu, C. F. Wei, D. Ming, and S. Di Chen, "MRCG: A MRI retrieval framework with convolutional and graph neural networks for secure and private IoMT," *IEEE J Biomed Health Inform*, vol. 27, no. 2, pp. 814–822, 2023, doi: 10.1109/JBHI.2021.3130028.

[52]   T. A. Khan, S. K. Mohammed, A. R. Shaikh, *et al.*, "Secure IoMT for disease prediction empowered with transfer learning in Healthcare 5.0, the concept and case study," *IEEE Access*, vol. 11, pp. 39418–39430, 2023, doi: 10.1109/ACCESS.2023.3266156.

[53]   B. L. Y. Agbley, M. K. Agbley, R. Adu-Gyamfi, *et al.*, "Federated fusion of magnified histopathological images for breast tumor classification in the Internet of Medical Things," *IEEE J Biomed Health Inform,* vol. 28, no. 6, pp. 3389–3400, 2024, doi: 10.1109/JBHI.2023.3256974.

[54]   R. O. Ogundokun, S. Misra, M. Douglas, R. Damaševičius, and R. Maskeliūnas, "Medical Internet-of-Things based breast cancer diagnosis using hyperparameter-optimized neural networks," *Future Internet*, vol. 14, no. 5, p. 153, 2022, doi: 10.3390/FI14050153.

[55]   T. Schaffter, M. N. Catley, S. Bakker, *et al.*, "Evaluation of combined arti-ficial intelligence and radiologist assessment to interpret screening mam-mograms," *JAMA Netw Open*, vol. 3, no. 3, pp. e200265–e200265, 2020, doi: 10.1001/JAMANETWORKOPEN.2020.0265.

[56] J. Li, M. Xu, C. Zhang, *et al.*, "Indocyanine green fluorescence imaging-guided laparoscopic right posterior hepatectomy," *Surg Endosc*, vol. 36, no. 2, pp. 1293–1301, 2022, doi: 10.1007/S00464-021-08404-2/ METRICS.

[57] E. M. Martin, J. V. Lopez, R. Brown, *et al.*, "Persistent cognitive slowing in post-COVID patients: longitudinal study over 6 months," *J Neurol*, vol. 271, no. 1, pp. 46–58, 2024, doi: 10.1007/S00415-023-12069-3/FIGURES/5.

[58] K. Te Lin, T. Ho, M. Chen, *et al.*, "Highly efficient flexible structured meta-surface by roll-to-roll printing for diurnal radiative cooling," *eLight*, vol. 3, no. 1, pp. 1–12, 2023, doi: 10.1186/S43593-023-00053-3/FIGURES/5.

[59] S. Kumar, A. S. Mehta, P. S. Kaur, *et al.*, "Resistance to FOXM1 inhibitors in breast cancer is accompanied by impeding ferroptosis and apoptotic cell death," *Breast Cancer Research and Treatment*, vol. 203, no. 1, pp. 1–14, 2024, doi: 10.1007/S10549-024-07420-9.

[60] O. Kaidar-Person, R. D. Schneps, P. I. Fernandez, *et al.*, "A BRILLIANT-BRCA study: residual breast tissue after mastectomy and reconstruction," *Breast Cancer Research and Treatment*, vol. 203, no. 2, pp. 1–9, 2024, doi: 10.1007/S10549-024-07425-4.

[61] R. Rampal, J. K. Patel, S. V. Narayan, *et al.*, "Three and twelve-month analysis of the PROM-Q study: comparison of patient-reported outcome measures using the BREAST-Q questionnaire," *Breast Cancer Research and Treatment*, vol. 203, no. 3, pp. 1–8, 2024, doi: 10.1007/S10549-024-07416-5.

[62] K. Sakashita, T. Muto, and S. Sato, "ASO visual abstract: clinical significance of primary tumor resection in perihilar cholangiocarcinoma with positive peritoneal lavage cytology," *Annals of Surgical Oncology*, vol. 31, no. 8, pp. 5407–5408, 2024, doi: 10.1245/S10434-024-15599-W.

[63] C.-Y. Huang, F. Chang, J. H. Wu, *et al.*, "Study of sex-biased differences in genomic profiles in East Asian hepatocellular carcinoma," *Discover Oncology*, vol. 15, no. 1, pp. 1–14, 2024, doi: 10.1007/S12672-024-01131-9.

[64] H. Xu, D. Xu, Y. Zheng, H. Wang, A. Li, and Xiaofei Zheng, "Investigation of prognostic values of immune infiltration and LGMN expression in the microenvironment of osteosarcoma," *Discover Oncology*, vol. 15, no. 1, pp. 1–22, 2024, doi: 10.1007/S12672-024-01123-9.

[65] Y. Liu and Y. Chen, "Fast finite-time secure control for nonlinear systems under dynamic event-triggered mechanism," *Nonlinear Dyn*, vol. 112, no. 7, pp. 5405–5419, 2024, doi: 10.1007/S11071-024-09319-Y/METRICS.

[66] E. M. E. Zayed, M. R. Abdel-Aty, and H. A. Sayed, "Optical solitons for dispersive concatenation model with power-law of self-phase modulation: a sub-ODE approach," *Journal of Optics (India)*, vol. 1, pp. 1–11, 2024, doi: 10.1007/S12596-024-01728-X/FIGURES/1.

[67] M. Elhoseny, G. Bin Bian, S. K. Lakshmanaprabu, K. Shankar, A. K. Singh, and W. Wu, "Effective features to classify ovarian cancer data in internet of medical things," *Computer Networks*, vol. 159, pp. 147–156, 2019, doi: 10. 1016/J.COMNET.2019.04.016.

[68]   S. Bharati, P. Podder, M. R. H. Mondal, and P. K. Paul, "Applications and challenges of cloud integrated IoMT," *Studies in Systems, Decision and Control*, vol. 311, no. 4, pp. 67–85, 2021, doi: 10.1007/978-3-030-55833-8_4.

[69]   L. Syed, S. Jabeen, and S. Manimala, "Telemammography: a novel approach for early detection of breast cancer through wavelets based image processing and machine learning techniques," *Studies in Computational Intelligence*, vol. 730, pp. 149–183, 2018, doi: 10.1007/978-3-319-63754-9_8.

[70]   S. Chaudhury and K. Sau, "A blockchain-enabled internet of medical things system for breast cancer detection in healthcare," *Healthcare Analytics*, vol. 4, p. 100221, 2023, doi: 10.1016/J.HEALTH.2023.100221.

[71]   B. VanBerlo, J. Hoey, and A. Wong, "A survey of the impact of self-supervised pretraining for diagnostic tasks in medical X-ray, CT, MRI, and ultrasound," *BMC Med Imaging*, vol. 24, no. 1, pp. 1–24, 2024, doi: 10.1186/S12880-024-01253-0/TABLES/9.

[72]   C. Huang, J. Wang, S. Wang, and Y. Zhang, "Internet of medical things: a systematic review," *Neurocomputing*, vol. 557, p. 126719, 2023, doi: 10.1016/J.NEUCOM.2023.126719.

[73]   K. Kakhi, R. Alizadehsani, H. M. D. Kabir, A. Khosravi, S. Nahavandi, and U. R. Acharya, "The internet of medical things and artificial intelligence: trends, challenges, and opportunities," *Biocybern Biomed Eng*, vol. 42, no. 3, pp. 749–771, 2022, doi: 10.1016/J.BBE.2022.05.008.

[74]   S. F. Ahmed, M. S. Bin Alam, S. Afrin, S. J. Rafa, N. Rafa, and A. H. Gandomi, "Insights into Internet of Medical Things (IoMT): data fusion, security issues and potential solutions," *Information Fusion*, vol. 102, p. 102060, 2024, doi: 10.1016/J.INFFUS.2023.102060.

[75]   J. B. Awotunde, A. E. Adeniyi, R. O. Ogundokun, G. J. Ajamu, and P. O. Adebayo, "MIoT-based big data analytics architecture, opportunities and challenges for enhanced telemedicine systems," *Studies in Fuzziness and Soft Computing*, vol. 410, pp. 199–220, 2021, doi: 10.1007/978-3-030-70111-6_10.

[76]   A. Abbas, R. Alroobaea, M. Krichen, S. Rubaiee, S. Vimal, and F. M. Almansour, "Blockchain-assisted secured data management framework for health information analysis based on Internet of Medical Things," *Pers Ubiquitous Comput*, vol. 28, no. 1, pp. 59–72, 2024, doi: 10.1007/S00779-021-01583-8/METRICS.

[77]   X. Li, M. Jia, M. T. Islam, L. Yu, and L. Xing, "Self-supervised feature learning via exploiting multi-modal data for retinal disease diagnosis," *IEEE Trans Med Imaging*, vol. 39, no. 12, pp. 4023–4033, 2020, doi: 10.1109/TMI.2020.3008871.

[78]   S. Baker and W. Xiang, "Artificial intelligence of things for smarter healthcare: a survey of advancements, challenges, and opportunities," *IEEE Communications Surveys and Tutorials*, vol. 25, no. 2, pp. 1261–1293, 2023, doi: 10.1109/COMST.2023.3256323.

[79]   X. Liu, J. Zhang, and Z. Pei, "Machine learning for high-entropy alloys: progress, challenges and opportunities," *Prog Mater Sci*, vol. 131, no. 2, p. 101018, 2023, doi: 10.1016/J.PMATSCI.2022.101018.

[80] J. C. Caicedo, C. McQuin, A. Goodman, *et al.*, "Data-analysis strategies for image-based cell profiling," *Nature Methods*, vol. 14, no. 9, pp. 849–863, 2017, doi: 10.1038/nmeth.4397.

[81] S. Studer, M. Baur, B. Schneider, *et al.*, "Towards CRISP-ML(Q): a machine learning process model with quality assurance methodology," *Machine Learning and Knowledge Extraction*, vol. 3, no. 2, pp. 392–413, 2021, doi: 10.3390/MAKE3020020.

[82] R. Saripalle, C. Runyan, and M. Russell, "Using HL7 FHIR to achieve interoperability in patient health record," *J Biomed Inform*, vol. 94, p. 103188, 2019, doi: 10.1016/J.JBI.2019.103188.

[83] M. Phillips, "International data-sharing norms: from the OECD to the General Data Protection Regulation (GDPR)," *Hum Genet*, vol. 137, no. 8, pp. 575–582, 2018, doi: 10.1007/S00439-018-1919-7/METRICS.

[84] G. Hatzivasilis, O. Soultatos, S. Ioannidis, C. Verikoukis, G. Demetriou, and C. Tsatsoulis, "Review of security and privacy for the internet of medical things (IoMT): resolving the protection concerns for the novel circular economy bioinformatics," *Proceedings - 15th Annual International Conference on Distributed Computing in Sensor Systems, DCOSS 2019*, pp. 457–464, 2019, doi: 10.1109/DCOSS.2019.00091.

[85] D. Koutras, G. Stergiopoulos, T. Dasaklis, P. Kotzanikolaou, D. Glynos, and C. Douligeris, "Security in IoMT communications: a survey," *Sensors*, vol. 20, no. 17, p. 4828, 2020, doi: 10.3390/S20174828.

[86] J. Qiu, Y. Zhang, X. Liu, *et al.*, "Large AI models in health informatics: applications, challenges, and the future," *IEEE J Biomed Health Inform*, vol. 27, no. 12, pp. 6074–6087, 2023, doi: 10.1109/JBHI.2023.3316750.

[87] J. Indumathi, R. R. Naik, S. Gupta, *et al.*, "Block chain based Internet of Medical Things for uninterrupted, ubiquitous, user-friendly, unflappable, unblemished, unlimited health care services (BC IoMT U6HCS)," *IEEE Access*, vol. 8, pp. 216856–216872, 2020, doi: 10.1109/ACCESS.2020.3040240.

Chapter 11

Proteomics techniques for characterizing microbial proteins

Sheilina Choudhary[1], Vandana Guleria[1] and Rupak Nagraik[2]

Abstract

The large-scale study of proteins known as proteomics has transformed our knowledge of microbial life by shedding light on the dynamics, structure, and function of microbial proteins. With an emphasis on extremophiles—microorganisms that flourish in harsh settings including high temperatures, high salinity, acidic environments, or strong radiation—this chapter explores sophisticated proteomics techniques. For the identification and measurement of microbial proteins, post-translational modifications, and protein-protein interactions in these organisms, essential techniques such as tandem mass spectrometry (MS/MS), liquid chromatography-mass spectrometry (LC-MS), and two-dimensional gel electrophoresis (2-DE) are essential. Furthermore, shotgun proteomics allows for the thorough identification of complicated microbial proteomes, while methods like stable isotope labeling by amino acids in cell culture (SILAC) and isobaric tags for relative and absolute quantitation (iTRAQ) provide improved protein quantification.

Proteomics has shown the distinct adaptations in extremophiles, such as metabolic pathways, stress-response proteins, and specialized enzymes, that enable survival in hostile environments. Proteomics, when combined with other omics techniques like transcriptomics and genomes, provides a systems-level understanding of the biology of extremophiles, demonstrating how their regulatory networks are tailored for harsh conditions. Rebuilding metabolic circuits is another benefit of this multi-omics integration, which also provides information about possible biotechnological uses like enzyme engineering for industrial operations.

Keywords: Microbial proteomics; Mass spectrometry; Two-dimensional gel electrophoresis; Chromatography techniques; Post-translational modifications; Protein–protein interactions

[1]School of Bioengineering and Food Technology, Shoolini University, Solan, Himachal Pradesh, India
[2]Department of Biotechnology, Graphic Era (Deemed to Be University), Dehradun, Uttarakhand, India

11.1 Introduction

Understanding the diverse roles that microorganisms play in different ecosystems, industries, and human health requires an understanding of the complex biological processes that underpin their existence. Deciphering these processes relies heavily on proteomics, the large-scale study of proteins. Proteomics offers significant insights into the functional dynamics, relationships, and environmental adaptations of microbial cells by examining their complete protein complement.

In microorganisms, proteins are vital macromolecules that carry out and control almost every cellular function, from signaling and metabolism to replication and structural upkeep. Researchers can understand how microbial proteins work and interact within biological networks by using proteomics to investigate these molecules. Applications such as drug development, biotechnology, and environmental sustainability all depend on this knowledge [1].

Advanced analytical methods including mass spectrometry, chromatography, and gel electrophoresis are essential to proteomics because they have transformed the study of microbial proteins. A key instrument in proteomics, mass spectrometry (MS) enables the accurate identification and measurement of proteins as well as the identification of minute alterations such as post-translational modifications (PTMs). PTMs, including phosphorylation, glycosylation, and acetylation, are important for controlling the function of proteins and are frequently necessary for microbes to adapt to changes in their environment [2].

Studying metabolic pathways and how proteins regulate them are among the major objectives of microbial proteomics. The number and activity of proteins, the molecular workhorses that catalyze biological events, frequently dictate a cell's metabolic status. Proteomic research, for example, has been crucial in finding the enzymes that microorganisms use for carbon fixation, nitrogen cycling, and the synthesis of secondary metabolites. These discoveries have significant ramifications for industrial biotechnology, which involves engineering microorganisms to generate useful substances such as medications or biofuels.

The development of high-throughput proteomic technology has also made it possible to research microbiomes or microbial communities. Understanding the collective protein expression of microbial consortia in complex environments, including soil, oceans, or the human gut, is the focus of microbial proteomics. The functional capacities of microbial communities are revealed by this method, called metaproteomics, which also provides insight into the functions that these communities play in processes such as disease progression, nutrient cycling, and ecosystem stability. We now know more about how microbial proteins interact with host systems to affect health and disease, for instance, thanks to the use of proteomics in the study of the gut microbiome [3].

The study of microbial reactions to antimicrobial drugs and the creation of countermeasures against antimicrobial resistance (AMR) both heavily rely on proteomics. Researchers can find resistance mechanisms and possible treatment targets by identifying proteins that are altered or increased in response to antibiotics. Proteomics also makes it possible to characterize the proteins on the surface of bacteria, which are frequently the first to come into contact with antibiotics or human immune systems. The development of novel medications or vaccines to combat the escalating AMR epidemic can be guided by these investigations [4].

Protein separation and purification are made easier by chromatography techniques like liquid chromatography, which guarantees that complicated combinations may be examined with high specificity. On the basis of their molecular weight and isoelectric point, proteins can be resolved using gel electrophoresis, including two-dimensional gel electrophoresis. When combined, these approaches allow scientists to create thorough protein profiles and learn more about microbial proteomes [5].

Studying extremophilic organisms—microbes that flourish in harsh conditions like high salinity, acidic pH, or extremely high temperatures—is another area in which proteomics is important. The distinct protein adaptations that enable these creatures to live and function in such environments are clarified by proteomics. Gaining insight into these adaptations advances our understanding of microbial biology and has real-world applications, such as the identification of enzymes that can be used in hostile environments.

By mapping protein–protein interactions and establishing connections between genotype and phenotype, microbial proteomics seeks to uncover proteins and their changes. Proteomics data can be combined with transcriptomics and genomes to create detailed models of the biology of microbial systems. This holistic approach is particularly important in understanding pathogenic bacteria, as it can offer possible targets for therapeutic intervention [6].

Proteomics, in summary, provides comprehensive insights into the structure, function, and regulation of microbial proteins, making it an essential tool in the study of microbial biology. The analytical basis for protein identification and the investigation of post-translational changes is provided by methods such as mass spectrometry, chromatography, and gel electrophoresis. Proteomics' investigation of extremophilic organisms not only advances our knowledge of life in harsh environments but also presents novel approaches to biotechnology and medicine. Proteomics has enormous promise to improve our knowledge of microbial life and its implications in solving global issues as it develops further.

11.1.1 Unlocking life's potential in extreme environments with extremophilic organisms

Extremophilic organisms, often known as extremophiles, flourish in settings that were once thought to be unsuitable for life. These microbes have evolved to withstand harsh environments, including high salinity, acidity, alkalinity, high pressure, severe radiation, and extremely high temperatures. In addition to enhancing our knowledge of the diversity of life on Earth, their capacity to thrive in environments where most other living forms fail offers enormous potential for a range of biotechnological uses. The tenacity of life is demonstrated by extremophiles, which are now the focus of studies on their biological processes and possible applications.

The use of biological agents to eliminate or neutralize environmental contaminants is known as bioremediation, and it is one of the most promising uses of extremophiles. Thermophilic bacteria and halophilic archaea are examples of extremophiles that can break down harmful substances in environments that would denature most enzymes, including high salt concentrations or temperatures. For instance, halophiles can detoxify high-salinity environments like salt mines or hypersaline lakes, whereas thermophilic bacteria can be used to break down industrial contaminants in hot wastewater systems. Their resilience renders them perfect for tackling environmental issues that demand resilience that is beyond the capabilities of traditional organisms [7].

Enzyme manufacturing for industrial processes is another important application area. Extremophiles produce extremozymes, which are enzymes that are stable and active in harsh environments. These enzymes are essential in sectors including biofuels, food processing, and pharmaceuticals. Examples of these include salt-tolerant lipases from halophiles and thermostable DNA polymerases from thermophiles. For example, a key component of molecular biology, polymerase chain reaction (PCR) technology for DNA amplification, heavily relies on thermostable enzymes. Similar to this, psychrophilic organisms' cold-adapted enzymes are useful in food preservation and laundry detergents because they work well at low temperatures [8].

Extremophiles' capacity for adaptation to harsh environments offers valuable insights into evolutionary biology. Their genomes frequently contain distinct stress-resistance genes, like those involved in membrane stability, protein folding, or DNA repair, which provide hints about extinct life forms and the development of cellular processes. Researchers can discover the molecular advancements that have enabled extremophiles to endure in severe habitats for billions of years by examining these genetic adaptations. The creation of synthetic biology applications, which involve introducing extremophilic features into other species to improve their functionality, is facilitated by this knowledge, which also advances evolutionary biology [9].

To sum up, the potential of extremophilic organisms is enormous from a scientific and industrial standpoint. Their extraordinary environmental adaptations, as shown in Figure 11.1, have applications in bioremediation, enzyme synthesis, and even the hunt for extraterrestrial life. Their biology also helps us better comprehend the evolutionary strategies and durability of life. Extremophiles' contributions to science and technology are expected to be as remarkable as their surroundings, as long as research into them keeps expanding.

11.1.2   *Mass spectrometry-based techniques in proteomics*

In proteomics, MS, as shown in Figure 11.2(a), has become a key technology that allows for the very sensitive and specific identification and measurement of proteins. MS offers vital information about the molecular makeup, structure, and changes of proteins by examining the mass-to-charge ratio (m/z) of ionized molecules. Protein research has been transformed by cutting-edge methods like matrix-assisted laser desorption/ionization-time of flight mass spectrometry (MALDI-TOF MS) and liquid chromatography-electrospray ionization-tandem mass spectrometry (LC-ESI-MS/MS), which provide unmatched capabilities for examining complex proteomes [10].

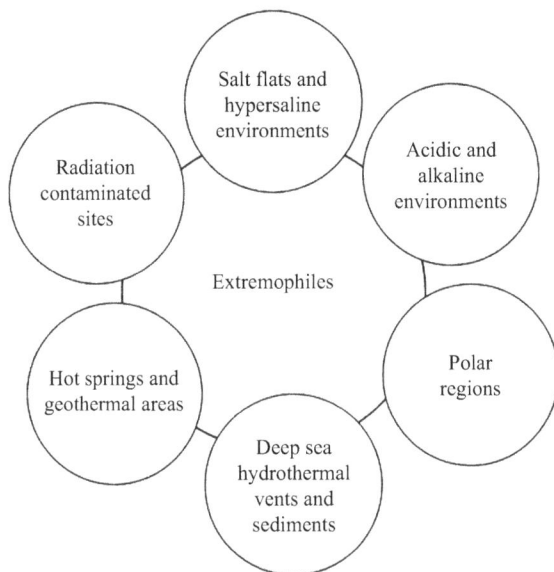

Figure 11.1   Diagrammatic representation of the primary habitats for extremophilic microbes

Separating, ionizing, and analyzing peptides produced from protein digestion is accomplished using LC-ESI-MS/MS as shown in Figure 11.2(c), which combines liquid chromatography (LC), electrospray ionisation (ESI), and tandem mass spectrometry (MS/MS). Peptide mixtures are resolved in the liquid chromatography step, guaranteeing that complicated samples are sent to the mass spectrometer sequentially. Peptides undergo ionization and are then moved into the gas phase for m/z analysis during electrospray ionisation. The peptides are then broken up into smaller ions by tandem mass spectrometry, which is subsequently used to determine the amino acid sequences of the ions. This methodical methodology makes LC-ESI-MS/MS highly accurate in identifying thousands of proteins from a single sample, which makes it essential for proteomics research, including medication development and biomarker identification [10].

MALDI-TOF MS, as shown in Figure 11.2(b), provides an additional method for protein analysis. This method involves ionizing proteins or peptides with a laser and co-crystallizing them with a matrix material. The resultant ions are accelerated in a vacuum, and their m/z ratio is determined by measuring their time of flight (TOF) to a detector. Because MALDI-TOF MS yields quick findings with little sample preparation, it is especially well-suited for high-throughput analysis and protein identification. Its capacity to produce peptide mass fingerprinting (PMFs) gives scientists a strong tool for identifying proteins in intricate mixtures by enabling them to compare experimental data with theoretical protein databases [11].

The detection of PTMs, such as phosphorylation, glycosylation, and acetylation, which are essential for controlling protein activity, is a strength of both

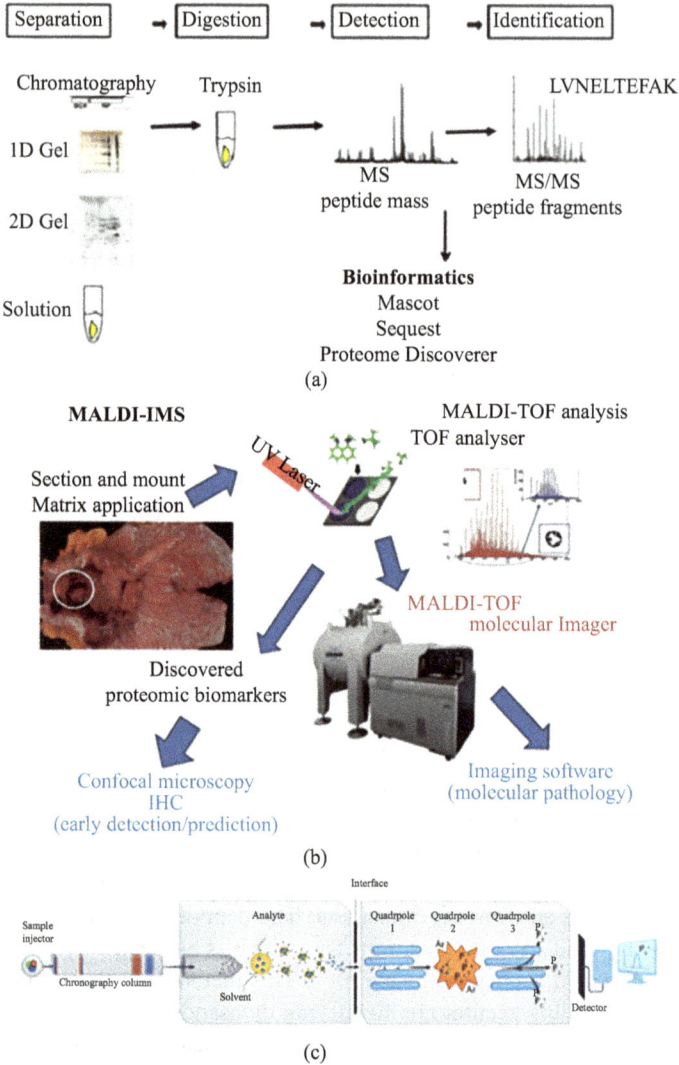

Figure 11.2   *Workflow of mass spectrometry techniques. (a) Mass spectrometry,*
(b) MALDI-TOF MS, and (c) LC-ESI-MS/MS.

LC-ESI-MS/MS and MALDI-TOF MS. For instance, by precisely locating a protein's phosphorylation site, LC-ESI-MS/MS might reveal information about signalling pathways. Similarly, MALDI-TOF MS may identify and examine glycan structures linked to proteins when paired with particular chemical derivatization methods. Understanding cellular functions and disease mechanisms—where PTMs frequently play as important modulators of protein activity—requires these talents [12].

Quantitative proteomics, which measures the relative or absolute abundance of proteins across various samples, is another important use of mass spectrometry-based techniques. To accomplish accurate quantification, LC-ESI-MS/MS is combined with methods such as label-free quantification, tandem mass tag (TMT) labelling, and stable isotope labelling by amino acids in cell culture (SILAC). Comparative proteomics research, such as determining which proteins exhibit differential expression in sickness and healthy conditions, has benefited greatly from these techniques. The identification of potential proteins for therapeutic targeting or diagnostics is made possible by quantitative proteomics, which has a significant impact on biomarker discovery [13].

Because of their accuracy and adaptability, LC-ESI-MS/MS and MALDI-TOF MS are vital tools for expanding our knowledge of the proteome. These methods offer a thorough understanding of protein dynamics in biological systems, from determining protein sequences and changes to measuring protein expression levels. It is anticipated that mass spectrometry's uses in proteomics will grow even more as the technology advances and incorporates new developments like ion mobility spectrometry and ultra-high-resolution analyzers, leading to breakthroughs in basic biology, biotechnology, and medicine.

11.1.3 Chromatography techniques in microbial proteomics

Chromatography methods play a significant part within the examination of microbial proteins by empowering the partition of complex protein blends, as shown in Figure 11.3. Among these, high-performance fluid chromatography (HPLC) has demonstrated to be an irreplaceable instrument. HPLC encourages the productive partition, refinement, and examination of proteins and peptides, permitting analysts to dismember the complex proteome of microbial cells. Its capacity to handle complex natural tests with high determination and reproducibility

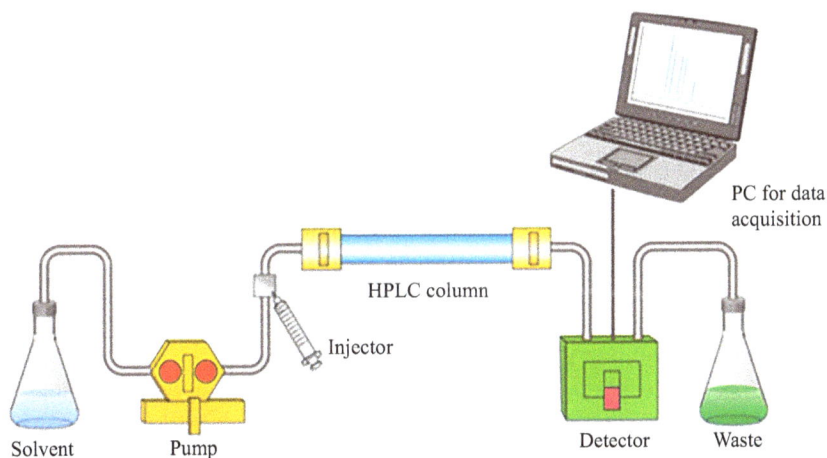

Figure 11.3   HPLC chromatography technique

makes it a foundation in proteomics workflows, particularly when combined with downstream methods like MS [14].

In HPLC, proteins or peptides are isolated based on their physical or chemical properties, such as hydrophobicity, charge, or mass. The foremost commonly utilized modes in protein examination are reverse-phase HPLC (RP-HPLC) and ion-exchange HPLC. RP-HPLC, which utilizes a hydrophobic stationary stage and a polar versatile stage, is especially successful for isolating peptides and proteins based on their hydrophobicity. This method is broadly utilized in conjunction with mass spectrometry, because it conveys peptides in a dissolvable framework congruous with electrospray ionization, guaranteeing consistent integration into proteomics ponders [15].

Ion-exchange HPLC (IEX-HPLC) is another capable device in microbial protein examination. This method isolates proteins based on their net charge, which is affected by their amino corrosive composition and the pH of the buffer framework. IEX-HPLC can be assist partitioned into cation-exchange and anion-exchange chromatography, depending on the charge of the stationary stage utilized. This mode is especially valuable for settling proteins with inconspicuous contrasts in charge, empowering the confinement of isoforms or proteins with post-translational alterations. Such exact division is basic for examining the useful differing qualities of microbial proteins.

Another eminent application of HPLC in proteomics is its utilization in affinity chromatography, a method that utilizes a particular interaction between a protein of intrigued and a ligand joined to the stationary phase. For occurrence, fondness chromatography is utilized to separate proteins that bind to specific substrates, co-factors, or antibodies. This focused approach is important in examining proteins that play parts in microbial signaling, digestion system, or pathogenesis because it permits analysts to center on particular components of the proteome.

HPLC is especially viable when analyzing microbial tests due to its capacity to handle complex blends, such as those inferred from lysed microbial cells. These tests frequently contain thousands of proteins with a wide run of physicochemical properties, requiring vigorous and flexible division procedures. Advances in HPLC innovation, such as ultra-high-performance fluid chromatography (UHPLC), have encouraged upgraded the determination and speed of protein partitions. When combined with division collection frameworks, HPLC permits the confinement of particular proteins or peptides for advanced examination, such as auxiliary characterization or useful measures [16].

The integration of HPLC with other procedures, such as MS and gel electrophoresis, increases its utility in proteomics workflows. For occurrence, proteins isolated by HPLC can be specifically analyzed by pair MS for recognizable proof and measurement, or they can be subjected to encourage preparation, such as enzymatic assimilation. This flexibility underscores the significance of chromatography strategies within the comprehensive examination of microbial proteins. As HPLC innovation proceeds to advance, its applications in microbial proteomics will extend, advertising more profound experiences into the atomic instruments basic microbial life and their intuitive with the environment [17].

11.1.4   Gel electrophoresis techniques in protein analysis

Gel electrophoresis could be a foundational method in proteomics that empowers the determination and investigation of proteins based on their estimate and charge. One-dimensional gel electrophoresis (1D-GE) and two-dimensional gel electrophoresis (2D-GE) are broadly utilized strategies that give a visual representation of protein tests, encouraging their methods in different organic settings. These methods are especially profitable for analyzing microbial proteins, as they permit analysts to resolve complex protein blends and pick up experiences into protein expression, structure, and work.

1D-GE isolates proteins fundamentally based on their atomic weight employing a polyacrylamide gel network. In this method, as shown in Figure 11.4, proteins are treated with a denaturing specialist, such as sodium dodecyl sulfate (SDS), which gives a uniform negative charge to the proteins, permitting them to emigrate through the gel exclusively based on estimate. Littler proteins travel speedier through the gel lattice, while larger proteins move more gradually. 1D-GE may be a clear and reproducible strategy that gives an introductory diagram of the protein

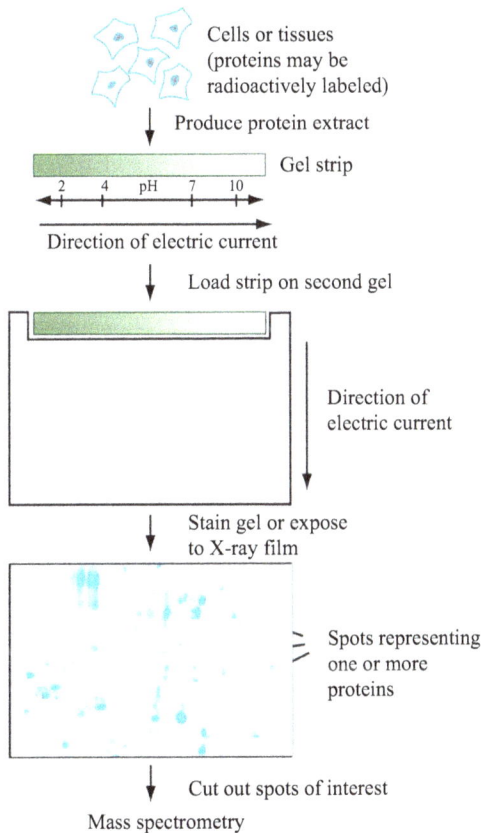

Figure 11.4   1D-Gel electrophoresis overview

composition in a test, making it reasonable for analyzing particular protein divisions or checking protein virtue in microbial considers [18].

2D-GE, on the other hand, offers a more comprehensive determination of proteins by separating them based on two particular properties: isoelectric point (pI) and atomic weight, as shown in Figure 11.5. Within the begin with measurement, proteins are settled by isoelectric centering (IEF), which isolates them based on their pI employing a pH slope. Proteins relocate inside the angle until they reach a point where their net charge is zero. Within the moment measurement, the proteins are encouraged isolated by SDS-PAGE, giving a visual representation of proteins in a two-dimensional outline. This procedure is profoundly successful for analyzing complex microbial proteomes because it can resolve thousands of proteins in a single try [19].

One of the noteworthy applications of gel electrophoresis, especially 2D-GE, is in considering protein expression profiles. By comparing the 2D protein maps of tests beneath distinctive conditions, analysts can recognize proteins that are differentially communicated in reaction to natural changes, push, or sedate medications. This capability is significant for understanding the versatile components of organisms, especially extremophiles or pathogens, in different situations. Moreover, proteins of intrigued can be extracted from the gel and subjected to MS for exact recognizable proof and characterization [20].

Another vital include of gel electrophoresis methods is their capacity to distinguish post-translational alterations (PTMs). Adjusted proteins regularly display shifts in their isoelectric point or atomic weight, which can be visualized as changes in their position on a gel. For illustration, phosphorylation ordinarily decreases a protein's pI, causing it to emigrate in an unexpected way within the to begin with measurement of 2D-GE. This makes gel electrophoresis a fundamental instrument for examining PTMs, which play basic parts in microbial signaling and control.

Despite its focal points, gel electrophoresis, as shown in Figure 11.6, has impediments, such as trouble in settling exceptionally huge or hydrophobic

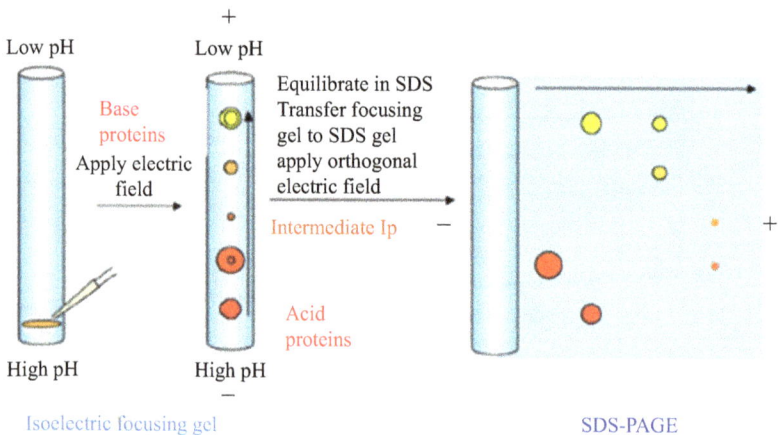

Figure 11.5   2D-Gel electrophoresis overview

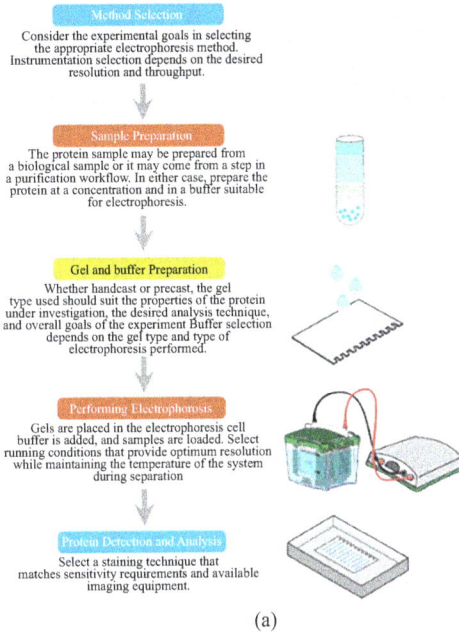

Method Selection

Consider the experimental goals in selecting
the appropriate electrophoresis method.
Instrumentation selection depends on the desired
resolution and throughput.

Sample Preparation

The protein sample may be prepared from
a biological sample or it may come from a step in
a purification workflow. In either case, prepare the
protein at a concentration and in a buffer suitable
for electrophoresis.

Gel and buffer Preparation

Whether handcast or precast, the gel
type used should suit the properties of the protein
under investigation, the desired analysis technique,
and overall goals of the experiment Buffer selection
depends on the gel type and type of
electrophoresis performed.

Performing Electrophoresis

Gels are placed in the electrophoresis cell
buffer is added, and samples are loaded. Select
running conditions that provide optimum resolution
while maintaining the temperature of the system
during separation

Protein Detection and Analysis

Select a staining technique that
matches sensitivity requirements and available
imaging equipment.

(a)

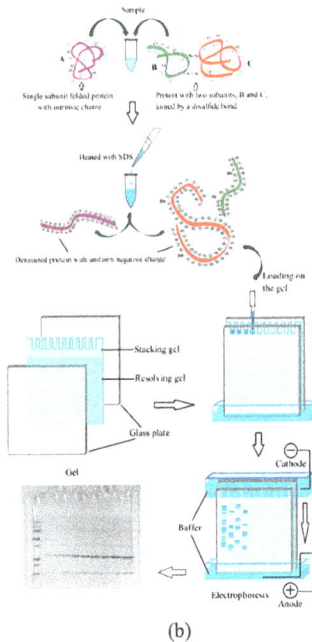

(b)

Figure 11.6   Gel electrophoresis overview

proteins and lower affectability compared to fluid chromatography-mass spectrometry (LC-MS). Be that as it may, it remains a capable and cost-effective method for subjective and semi-quantitative proteomic ponders. Progresses in gel electrophoresis, such as fluorescence-based recoloring and robotized gel imaging frameworks, have assisted upgraded its utility. As a result, 1D-GE and 2D-GE proceed to be broadly utilized in microbial proteomics for their capacity to supply nitty gritty bits of knowledge into protein composition, expression, and alterations [21].

11.1.5   *Protein identification and quantification in proteomics*

Protein distinguishing proof and measurement are central goals in proteomics, giving fundamental data for understanding cellular forms and atomic instruments. The exact characterization of proteins includes distinguishing their arrangements and deciding their wealth. MS plays an urgent part in accomplishing these objectives, and its integration with database look calculations has revolutionized the field. By coordinating exploratory MS information with hypothetical information from protein databases, analysts can recognize proteins with tall certainty and measure their relative or outright expression levels [22].

The method of protein recognizable, as shown in Figure 11.7, proof starts with the assimilation of proteins into smaller peptides utilizing proteins like trypsin. These peptides are analyzed by MS, creating spectra that speak to their mass-to-charge (m/z) proportions. Pair mass spectrometry (MS/MS) encourages parts peptides into smaller particles, giving data approximately their amino corrosive arrangements. Database look calculations, such as Mascot, Sequest, and X! Tandem, compare the test spectra with hypothetical spectra produced from protein databases. These calculations utilize scoring systems to decide the leading matches, empowering the recognizable proof of proteins based on their peptide groupings [23].

In addition to identifying proteins, MS encourages protein measurement, which is fundamental for comparing protein expression levels over diverse tests or conditions. Measurement can be accomplished utilizing label-free or labeling strategies. In label-free measurement, the escalation of peptide particles in MS

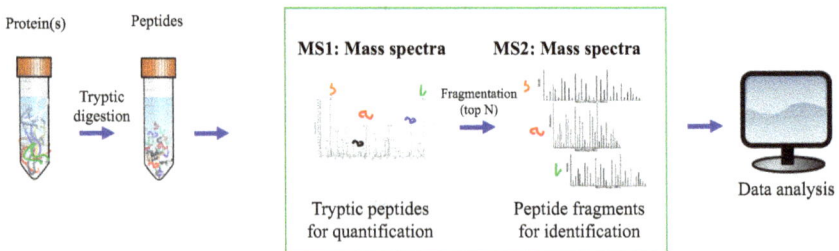

Figure 11.7   Protein identification and quantification

spectra is utilized to gauge protein plenitude. This approach is straightforward and cost-effective but requires high reproducibility in test planning and MS investigation. Labeling procedures, such as isobaric labels for relative and outright measurement (iTRAQ), as shown in Figure 11.8(a), or pair mass labels (TMT), as shown in Figure 11.8(b), include chemically labeling peptides with isotopic names. These labels permit concurrent evaluation of different tests in a single MS run, giving higher precision and throughput [24].

Database search algorithms moreover play a basic part in recognizing posttranslational adjustments (PTMs), which are significant for controlling protein work. Adjusted peptides frequently create special fracture designs in MS/MS, and specialized calculations, such as PTMap and MODa, are outlined to distinguish these designs. For this case, phosphorylation or glycosylation destinations can be pinpointed with high accuracy, advertising experiences into signaling pathways and cellular control. The capacity to recognize PTMs upgrades our understanding of protein movement and interaction systems, especially in complex frameworks like microbial communities [25].

Measurement approaches are advanced and improved by joining MS information with bioinformatics apparatuses and factual examination. For cases, calculations like MaxQuant and Horizon robotize the preparation of MS information, guaranteeing exact evaluation and decreasing manual mistakes. These instruments moreover empower normalization and factual comparison of protein plenitudes over diverse exploratory conditions. Such integration is basic for large-scale proteomics thinks about, such as recognizing biomarkers in maladies or checking microbial reactions to natural changes [26].

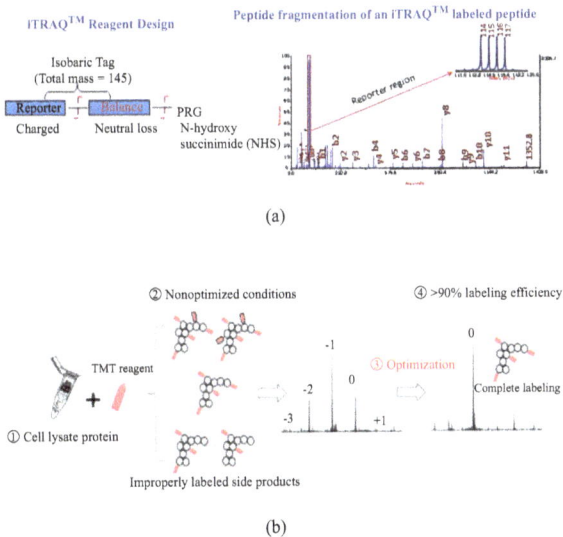

Figure 11.8   Protein identification and quantification using (a) iTRAQ method and (b) TMT method

In conclusion, database look calculations and progressed measurement strategies have changed protein distinguishing proof and measurement, making them more precise and proficient. These strategies are basic for characterizing proteomes, distinguishing utilitarian proteins, and investigating protein elements in different organic frameworks. As MS advances and computational instruments proceed to advance, the scope and profundity of proteomics inquiries about will extend, empowering more comprehensive bits of knowledge into the atomic components that administer life.

11.1.6   Post-translational modifications in microbial cells

Post-translational alterations (PTMs) are chemical changes that proteins experience after their blend, altogether affecting their structure, work, and interaction systems. Among these, phosphorylation is one of the foremost urgent PTMs in microbial cells. It includes the expansion of a phosphate gather, ordinarily to serine, threonine, or tyrosine buildups, catalyzed by chemicals called kinases. Phosphorylation acts as an atomic switch, changing protein action, localization, soundness, or intuition, and is basic for directing cellular forms such as the digestion system, flag transduction, and stretch reactions in organisms [27].

In microbial signaling pathways, phosphorylation plays a central part in two-component frameworks (TCS), which are key to natural detection and adjustment. TCS ordinarily comprises of a sensor histidine kinase and a reaction controller. Upon identifying outside boosts, the sensor kinase experiences autophosphorylation at a preserved histidine buildup, exchanging the phosphate gathered for an aspartate buildup on the reaction controller. This phosphorylation occasion actuates the reaction controller, empowering it to balance quality expression and microbial behavior. This instrument underscores the significance of phosphorylation in empowering organisms to adjust to changing situations [28].

Phosphorylation is additionally fundamental to controlling enzymatic movement in microbial cells. Numerous metabolic proteins are phosphorylated to tweak their catalytic proficiency, guaranteeing that cellular forms are finely tuned to natural conditions or supplement accessibility. For occurrence, key chemicals in central metabolic pathways, such as glycolysis or the tricarboxylic corrosive (TCA) cycle, are regularly controlled by phosphorylation. This energetic control guarantees that organisms optimize vitality generation and asset utilization, especially in nutrient-limited or unpleasant conditions [29].

In expansion to its part in the digestion system, phosphorylation intervenes in stretch reactions in microbial cells. When exposed to unfriendly conditions such as warm, oxidative push, or osmotic changes, organisms depend on phosphorylation-driven signaling pathways to actuate defensive components. For case, phosphorylation of translation components or stress-responsive proteins empowers organisms to quickly alter their quality expression profiles, upregulating stress-induced proteins such as chaperones, cancer prevention agents, or transporters. These adjustments are imperative for microbial survival in extraordinary or fluctuating situations [29].

Developing investigations highlights the centrality of phosphorylation in microbial pathogenesis. In numerous bacterial pathogens, phosphorylation directs harmfulness variables, such as poisons, adhesins, or discharge frameworks, which are basic for colonization and disease. For occasion, phosphorylation of particular proteins may upgrade the gathering or emission of destructiveness complexes, empowering pathogens to avoid having guards or set up diseases. Understanding these phosphorylation-dependent instruments gives important experiences in microbial pathobiology and can advise the advancement of novel antimicrobial methodologies [30].

Examining phosphorylation in microbial cells is challenging but progressively doable due to headways in proteomics and bioinformatics. Strategies like phosphoproteomics, which combines mass spectrometry with enhancement strategies for phosphorylated peptides, permit analysts to methodically outline phosphorylation locales over microbial proteomes. These things not as it were extend our understanding of microbial science but to uncover potential targets for restorative mediation or mechanical applications. As investigation advances, the part of phosphorylation in microbial cells will proceed to be a central point, shedding light on the complex administrative systems that support microbial life.

11.1.7 *Protein–protein interactions and their importance*

Protein–protein interactions (PPIs) are principal to essentially all natural forms, counting flag transduction, enzymatic action control, and auxiliary astuteness in cellular frameworks. Understanding these intuitive is vital for decoding complex atomic systems that oversee cellular capacities. One of the foremost broadly utilized strategies for considering PPIs is co-immunoprecipitation (Co-IP), which permits analysts to separate and distinguish collaboration protein accomplices beneath near-physiological conditions. Co-IP has been demonstrated to be irreplaceable for examining the utilitarian parts of proteins inside organic frameworks [31].

Co-immunoprecipitation (Co-IP) is based on the guideline of antibody–antigen specificity. In this strategy, a counteracting agent particular to a protein of intrigued (the trap protein) is utilized to drag down the protein from a cellular extricate. Any proteins physically connected with the trapped protein (prey proteins) are co-precipitated and can be distinguished through downstream examinations, such as SDS-PAGE, western blotching, or MS. This focused approach gives coordinate prove of physical intuitive between proteins and is especially valuable for approving candidate PPIs recognized through high-throughput strategies.

The quality of Co-IP lies in its capacity to capture protein intuitively in their local states, protecting the physiological setting of the intelligent. This makes Co-IP an effective device for considering transitory or energetic intelligence that is troublesome to distinguish from other strategies. For example, signaling complexes that frame briefly amid cellular reactions to outside boosts can be captured and analyzed utilizing Co-IP. Such thinks are basic for understanding pathways like those included in microbial push reactions or host-pathogen intuition [32].

One of the major applications of Co-IP is in mapping protein interaction systems, which are fundamental for understanding cellular organization and work. By efficiently applying Co-IP to different proteins inside a cell, analysts can develop interaction maps that uncover how proteins work together in complexes or pathways. These interaction maps give experiences into cellular forms, such as translation control, DNA repair, or metabolic flux, and are profitable for recognizing potential helpful targets in pathogens or cancer cells.

Despite its preferences, Co-IP has restrictions, such as its dependence on high-quality antibodies and the requirement for adequate protein expression levels. Furthermore, non-specific official of proteins to the counteracting agent or globules can lead to wrong positives. To address these challenges, Co-IP is regularly combined with complementary strategies, such as yeast two-hybrid screening or nearness labeling approaches like BioID. These procedures offer assistance to approve intelligence and guarantee the strength of what comes about, improving the unwavering quality of PPI.

Propels in proteomics and bioinformatics have assist improved the utility of Co-IP in considering PPIs. For occurrence, coupling Co-IP with MS permits the distinguishing proof of novel interaction accomplices in a fair-minded way, indeed in complex blends. Also, computational tools for interaction network analysis can integrate with Co-IP information with other test datasets, giving an all-encompassing see of cellular forms. As these innovations proceed to advance, Co-IP will stay a foundation method for unraveling the complex web of protein intuition that drives organic frameworks [33].

11.2   Advances and challenges in microbial proteomics

Microbial proteomics has progressed essentially over the past decades, driven by advancements in innovation and computational science. In any case, the field still faces various challenges related to the complexity and differing qualities of microbial frameworks. The study of microbial proteomes is significant for understanding microbial physiology, adjustment, and intuitive with their situations. Experiences picked up from such inquire have wide applications in pharmaceutical, biotechnology, and natural science. This segment highlights the current headways and diligent challenges in microbial proteomics.

One of the foremost critical progressions in microbial proteomics is the advancement of high-resolution MS innovations. Disobedient such as Orbitrap and time-of-flight (TOF) mass spectrometers presently offer unparalleled affectability, determination, and speed. These advancements empower the comprehensive recognizable proof and measurement of proteins, indeed in complex microbial blends. Methods like couple MS (MS/MS) permit nitty gritty peptide sequencing, giving bits of knowledge into microbial protein structures and post-translational adjustments (PTMs).

Label-free evaluation (LFQ) and isotopic labeling methods, such as SILAC and TMT, have revolutionized quantitative proteomics. These strategies empower analysts to degree protein wealth in different conditions, encouraging comparative ponders of microbial reactions to natural changes, stretch, or medicated medicines. Propels in multiplexing innovations permit the synchronous examination of different tests, expanding throughput and empowering large-scale thinking.

Despite these progressions, microbial proteomics experiences a few challenges, especially the complexity and heterogeneity of microbial frameworks. Microbial communities, such as those found in soil, seas, or the human intestine, regularly comprise of different species with shifting protein expression levels. Extricating and analyzing proteins from such complex blends requires vigorous division procedures, such as multi-dimensional chromatography, coupled with progressed MS workflows. Be that as it may, these approaches can be time-consuming and requesting.

Another critical challenge is the inadequate comment of microbial genomes and proteomes. Even though genome sequencing ventures have made huge advances, numerous microbial proteins stay uncharacterized or need utilitarian explanations. This confinement complicates protein recognizable proof and useful investigation in proteomics. Computational devices for de novo peptide sequencing and homology-based comment have somewhat tended to this issue, but assistance progressions are required to progress the exactness and scope of these strategies.

The think about post-translational alterations (PTMs) in microbial proteins presents extra obstacles. PTMs, such as phosphorylation, acetylation, and glycosylation, play basic parts in directing protein capacities and signaling pathways. In any case, their location and investigation require specialized enhancement strategies, such as phosphopeptide improvement or glycan labeling, which can present predispositions or lose low-abundance adjustments. Progresses in PTM-specific mass spectrometry methods and bioinformatics apparatuses are making a difference in overcoming these challenges, but noteworthy crevices stay in our understanding of microbial PTMs.

Information investigation and administration in microbial proteomics moreover posture challenges. The tremendous sum of information produced by high-throughput proteomics tests requires the utilize of progressed computational devices and calculations for preparing, explanation, and elucidation. Database look motors, such as Mascot and MaxQuant, and bioinformatics stages, like Cytoscape and STRING, play an urgent part in distinguishing proteins and mapping interaction systems. In any case, coordination of proteomics information with other omics datasets, such as genomics or transcriptomics, requires modern multi-omics approaches and standardized designs to guarantee interoperability.

Looking ahead, single-cell proteomics is a developing wilderness with extraordinary potential in microbial investigation. Not at all like conventional bulk proteomics, single-cell approaches can uncover heterogeneity inside microbial populaces, giving bits of knowledge into how personal cells react to natural signals or are associated with one another. In any case, single-cell proteomics is still in its earliest stages, with specialized challenges related to affectability, throughput, and information investigation. Continuous progressions in test arrangement, MS instrumented, and computational strategies are anticipated to make this approach more available and impactful within the coming years.

In conclusion, microbial proteomics may be a quickly advancing field with monstrous potential for progressing our understanding of microbial life and its applications. Whereas critical challenges stay, especially in managing complex tests, uncharacterized proteins, and PTMs, progressing developments in innovation and bioinformatics are clearing the way for more comprehensive and precise proteome investigations. These improvements guarantee to open modern bits of knowledge into microbial science, with far-reaching suggestions for wellbeing, industry, and the environment.

11.3   Conclusion

Proteomics approaches have revolutionized microbial research, offering unparalleled insights into the complex molecular processes that govern microbial life. By enabling the comprehensive identification, quantification, and functional analysis of microbial proteins, proteomics has become an indispensable tool for understanding microbial physiology, adaptation, and interactions with their environments. The integration of advanced techniques such as mass spectrometry, chromatography, gel electrophoresis, and computational tools has further enhanced our ability to study microbial proteomes with unprecedented depth and precision.

The chapter highlights key aspects of microbial proteomics, including the role of post-translational modifications, protein–protein interactions, and the application of cutting-edge techniques like co-immunoprecipitation and high-resolution MS. These methods have been instrumental in unraveling the dynamic processes within microbial cells, such as signaling pathways, stress responses, and metabolic regulation. Additionally, the importance of studying extremophilic organisms and their unique proteomes underscores the broader implications of microbial research in biotechnology, medicine, and environmental science.

Despite the significant progress made in microbial proteomics, challenges such as the complexity of microbial communities, incomplete genome annotations, and the detection of low-abundance proteins persist. Advances in single-cell proteomics, PTM-specific analyses, and multi-omics integration are expected to address these challenges, paving the way for more comprehensive and accurate studies. The development of robust bioinformatics tools and standardization of data formats will further facilitate the analysis and sharing of proteomics data across research disciplines.

The applications of microbial proteomics extend far beyond basic research, with implications for developing novel antibiotics, enhancing industrial bioprocesses, and understanding microbial contributions to global ecosystems. By elucidating the intricate roles of microbial proteins, proteomics provides a foundation for designing targeted interventions and harnessing microbial capabilities for human benefit.

In conclusion, proteomics has emerged as a transformative approach in microbial research, bridging the gap between molecular biology and systems-level understanding. The advancements and applications discussed in this chapter highlight the immense potential of proteomics to drive innovation and discovery in microbiology. As technologies continue to evolve, microbial proteomics will undoubtedly play a central role in addressing global challenges and advancing scientific knowledge.

References

[1] Pandey, A., & Mann, M. (2000). "Proteomics to study genes and genomes." *Nature*, *405*(6788), 837–846.

[2] Graham RLJ, Graham C, & McMullan G. (2007). Microbial proteomics: a mass spectrometry primer for biologists. *Microb Cell Fact*. 6, 26. doi:10.1186/1475-2859-6-26. PMID: 17697372; PMCID: PMC1971468.

[3] Long, S., Yang, Y., Shen, C. *et al.* (2020). Metaproteomics characterizes human gut microbiome function in colorectal cancer. *npj Biofilms Microbiomes* **6**, 14. https://doi.org/10.1038/s41522-020-0123-4

[4] Tsakou F, Jersie-Christensen R, Jenssen H, & Mojsoska B. (2020). The role of proteomics in bacterial response to antibiotics. *Pharmaceuticals (Basel)*. 13(9), 214. doi:10.3390/ph13090214. PMID: 32867221; PMCID: PMC7559545.

[5] Curreem SO, Watt RM, Lau SK, & Woo PC. (2012). Two-dimensional gel electrophoresis in bacterial proteomics. *Protein Cell*. 3(5), 346–363. doi:10.1007/s13238-012-2034-5. Epub 2012 May 18. PMID: 22610887; PMCID: PMC4875470.

[6] Reed CJ, Lewis H, Trejo E, Winston V, & Evilia C. (2013). Protein adaptations in archaeal extremophiles. *Archaea*. 2013, 373275. doi:10.1155/2013/373275. Epub 2013 Sep 16. PMID: 24151449; PMCID: PMC3787623.

[7] Chia, X.K., Hadibarata, T., Jusoh, M.N.H. *et al.* (2024). Role of extremophiles in biodegradation of emerging pollutants. *Top Catal*. https://doi.org/10.1007/s11244-024-01919-7

[8] Mesbah NM. (2022). Industrial biotechnology based on enzymes from extreme environments. *Front Bioeng Biotechnol*. 10,870083. doi:10.3389/fbioe.2022.870083. PMID: 35480975; PMCID: PMC9036996.

[9] Rampelotto, P.H. (2024). Extremophiles and extreme environments: a decade of progress and challenges. *Life*, 14, 382. https://doi.org/10.3390/life14030382

[10] Steven R. Shuken (2023). "An Introduction to Mass Spectrometry-Based Proteomics". *Journal of Proteome Research*, 22(7), 2151–2171. DOI:10.1021/acs.jproteome.2c00838

[11] Darie-Ion L, Whitham D, Jayathirtha M, Rai Y, Neagu AN, Darie CC, *et al.* (2022). Applications of MALDI-MS/MS-based proteomics in biomedical research. *Molecules*. 27(19), 6196. doi:10.3390/molecules27196196. PMID: 36234736; PMCID: PMC9570737.

[12] Zhang Y, Wang B, Jin W, Wen Y, Nan L, Yang M, *et al.* (2019). Sensitive and robust MALDI-TOF-MS glycomics analysis enabled by Girard's reagent T on-target derivatization (GTOD) of reducing glycans. *Anal Chim Acta*.

1048,105–114. doi:10.1016/j.aca.2018.10.015. Epub 2018 Oct 9. PMID: 30598139; PMCID: PMC6317096.

[13]   Zhu W, Smith JW, & Huang CM. (2010). Mass spectrometry-based label-free quantitative proteomics. *J Biomed Biotechnol*. 2010,840518. doi:10.1155/2010/840518. Epub 2009 Nov 10. PMID: 19911078; PMCID: PMC2775274.

[14]   Mitulovic G, & Mechtler K. (2006). HPLC techniques for proteomics analysis—a short overview of latest developments. *Brief Funct Genomic Proteomic*. 5(4), 249–260. doi:10.1093/bfgp/ell034. Epub 2006 Nov 22. PMID: 17124183.

[15]   Carr, D. (n.d.). *A guide to the analysis and purification of proteins and peptides by reversed-phase HPLC*. ACE HPLC Columns

[16]   Mitulović G, & Mechtler K. (2006), HPLC techniques for proteomics analysis—a short overview of latest developments, *Briefings in Functional Genomics*, 5(4), 249–260, https://doi.org/10.1093/bfgp/ell034

[17]   Mant CT, Chen Y, Yan Z, Popa TV, Kovacs JM, Mills JB, *et al.* (2007). HPLC analysis and purification of peptides. *Methods Mol Biol*. 386, 3–55. doi:10.1007/978-1-59745-430-8_1. PMID: 18604941; PMCID: PMC7119934.

[18]   Gallagher SR. (2006). One-dimensional SDS gel electrophoresis of proteins. *Curr Protoc Mol Biol*.; Chapter 10: Unit 10.2A. doi:10.1002/0471142727.mb1002as75. PMID: 18265373.

[19]   Curreem SO, Watt RM, Lau SK, & Woo PC. (2012). Two-dimensional gel electrophoresis in bacterial proteomics. *Protein Cell*. 3(5), 346–363. doi:10.1007/s13238-012-2034-5. Epub 2012 May 18. PMID: 22610887; PMCID: PMC4875470.

[20]   Creative Proteomics. (n.d.). Applications of 2D gel electrophoresis. https://www.creative-proteomics.com/resource/applications-of-2d-gel-electrophoresis.html

[21]   Bunai K, & Yamane K. (2005). Effectiveness and limitation of two-dimensional gel electrophoresis in bacterial membrane protein proteomics and perspectives. *J Chromatogr B Analyt Technol Biomed Life Sci*. 815(1–2), 227–236. doi:10.1016/j.jchromb.2004.08.030. PMID: 15652812.

[22]   Aebersold, R., & Mann, M. (2003). Mass spectrometry-based proteomics. *Nature*, 422(6928), 198–207. https://doi.org/10.1038/nature01511

[23]   Steen, H., & Mann, M. (2004). The ABC's (and XYZ's) of peptide sequencing. *Nature Reviews Molecular Cell Biology*, **5**(9), 699–711. https://doi.org/10.1038/nrm1468

[24]   Ong, S. E., & Mann, M. (2005). Mass spectrometry–based proteomics turns quantitative. *Nature Chemical Biology*, 1(5), 252–262. https://doi.org/10.1038/nchembio736

[25]   Na S, Bandeira N, & Paek E. (2012). Fast multi-blind modification search through tandem mass spectrometry. *Mol Cell Proteomics*. 11(4), M111.010199. doi:10.1074/mcp.M111.010199. Epub 2011 Dec 20. PMID: 22186716; PMCID: PMC3322561

[26] Tyanova, S., Temu, T., & Cox, J. (2016). The MaxQuant computational platform for mass spectrometry-based shotgun proteomics. *Nature Protocols*, *11*(12), 2301–2319. https://doi.org/10.1038/nprot.2016.136

[27] Carabetta VJ, & Hardouin J. (2022). Editorial: bacterial post-translational modifications. *Front Microbiol*. 13,874602. doi:10.3389/fmicb.2022.874602. PMID: 35391732; PMCID: PMC8983106.

[28] Ma Q, Zhang Q, Chen Y, Yu S, Huang J, Liu Y, *et al*. (2021). Post-translational modifications in oral bacteria and their functional impact. *Front Microbiol*. 12,784923. doi:10.3389/fmicb.2021.784923. PMID: 34925293; PMCID: PMC8674579.

[29] V. Pancholi, (2022). Editorial: Post-translational modification in response to stresses in bacteria. Frontiers in Microbiology, 13, 844854. https://doi.org/10.3389/fmicb.2022.844854

[30] Malakar B, Chauhan K, Sanyal P, Naz S, Kalam H, Vivek-Ananth RP, *et al*. (2023). Phosphorylation of CFP10 modulates *Mycobacterium tuberculosis* virulence. *mBio*. 14(5), e0123223. doi:10.1128/mbio.01232-23. Epub 2023 Oct 4. PMID: 37791794; PMCID: PMC10653824.

[31] Phizicky, E. M., & Fields, S. (1995). Protein–protein interactions: methods for detection and analysis. *Microbiological Reviews*, 59(1), 94–123. DOI:10.1128/mr.59.1.94-123.1995

[32] Iqbal H, Akins DR, & Kenedy MR. (2018). Co-immunoprecipitation for identifying protein–protein interactions in Borrelia burgdorferi. *Methods Mol Biol*. 1690,47–55. doi:10.1007/978-1-4939-7383-5_4. PMID: 29032535; PMCID: PMC5798869.

[33] Creative Proteomics. (n.d.). Co-immunoprecipitation mass spectrometry: Unraveling protein interactions. https://www.creative-proteomics.com/co-immunoprecipitation-mass-spectrometry-protein-interactions.html

Chapter 12

Enhancing obstructive sleep apnea diagnosis with machine learning: innovations and outcomes

Danisha Verma[1], Vandana Guleria[1] and Avinash Sharma[2]

Abstract

Obstructive sleep apnea (OSA)—a chronic disorder marked by repeated airway collapse during sleep—remains a silent epidemic. Affecting over 25 million adults in the United States alone, OSA fragments sleep, starves the brain of oxygen, and amplifies risks for heart disease, stroke, and cognitive decline. Yet, 80–90% of cases go undiagnosed, often dismissed as "just snoring" or fatigue. The culprit? Traditional diagnostic tools like polysomnography (PSG), while accurate, are costly, inaccessible, and impractical for population-scale screening.

So here, we explore how machine learning (ML) is dismantling these barriers. By merging sleep science with computational innovation, ML offers faster, cheaper, and patient-centric diagnostics, promising to bridge the gap between OSA's prevalence and its alarming under detection.

Keywords: Obstructive Sleep Apnea (OSA); Machine Learning (ML); Polysomnography (PSG); Wearable Devices; Automated Diagnosis; Deep Learning in Healthcare; Predictive Analytics; Telemedicine Integration; Health Informatics

12.1 Introduction

Obstructive sleep apnea (OSA) is a pervasive and serious sleep disorder that affects millions worldwide. It is characterized by repeated episodes of partial or complete obstruction of the upper airway during sleep, resulting in significant interruptions in breathing. These interruptions, referred to as apneas (complete blockage) and hypopneas (partial blockage), disrupt the normal sleep cycle and lead to a reduction

[1]School of Bioengineering and Food Technology, Shoolini University, Solan, Himachal Pradesh, India
[2]Department of Biotechnology, Graphic Era (Deemed to Be University), Dehradun, Uttarakhand, India

in blood oxygen levels, or oxygen desaturation. The brain briefly wakes the individual to restore normal breathing, though often the person is unaware of these awakenings. This cycle can occur hundreds of times throughout the night, leading to severely fragmented sleep [1].

12.1.1   Prevalence of OSA

OSA is highly prevalent and affects people of all age groups, though its occurrence is particularly high among certain demographics. Global estimates indicate that approximately 9–38% of adults experience some form of OSA. The condition is more commonly found in men than in women, with middle-aged men showing a higher prevalence (estimated at 13–33%). The prevalence of OSA increases with age and is strongly correlated with obesity, a major risk factor. In the United States, it is estimated that around 25 million adults suffer from moderate to severe OSA, with similar rates observed in other developed countries. In terms of gender, men are twice as likely to have OSA compared to women, though the risk among women increases significantly after menopause.

Moreover, there is growing evidence that OSA is significantly underdiagnosed. According to some studies, as many as 80–90% of individuals with moderate to severe OSA remain undiagnosed, meaning that millions of people may be living with a potentially dangerous condition without knowing it. This underdiagnosis is often due to a lack of awareness of the symptoms and the misconception that snoring or feeling tired during the day is simply a normal part of aging or lifestyle rather than an indication of a medical condition requiring intervention.

12.1.2   Symptoms of OSA

The symptoms of OSA can vary widely in their presentation and severity, but common symptoms include:

- **Loud-chronic snoring**: Often the most recognizable symptom, snoring occurs due to the partial obstruction of the airway, which causes vibrations in the throat tissues. Partners or family members are usually the first to notice this sign.
- **Episodes of breathing cessation**: This is a hallmark sign of OSA, where the individual's airway becomes fully blocked, causing them to stop breathing for several seconds, followed by a choking or gasping sound as they wake briefly to resume breathing.
- **Excessive daytime sleepiness (EDS)**: Due to the constant interruptions in sleep, people with OSA often experience overwhelming fatigue during the day, which can interfere with daily activities, work performance, and concentration. This excessive sleepiness can also lead to drowsy driving, increasing the risk of accidents.
- **Morning headaches**: Reduced oxygen levels during sleep can lead to morning headaches, which are a common complaint among individuals with OSA.
- **Mood changes and cognitive impairment**: Lack of quality sleep over time can lead to irritability, depression, anxiety, difficulty concentrating, and memory problems.

These symptoms not only degrade the overall quality of life for individuals with OSA but also contribute to the risk of accidents, particularly those involving motor vehicles. Studies have shown that people with untreated OSA are two to three times more likely to be involved in a motor vehicle accident due to drowsy driving.

12.2 Challenges in diagnosing OSA

Despite its prevalence and the severity of its symptoms, OSA remains a highly under-diagnosed condition, largely due to the complexity and limitations of current diagnostic methods. The gold standard for diagnosing OSA is polysomnography (PSG), which is an overnight sleep study conducted in a specialized sleep laboratory with sensors tracking brain waves (EEG), muscle activity (EMG), and blood oxygen levels.

While PSG is highly accurate in diagnosing OSA, it presents several significant challenges that contribute to the condition being underdiagnosed:

- **Accessibility and high cost**: PSG requires overnight monitoring in a sleep lab, with trained technicians and specialized equipment. This makes the test both expensive and inaccessible for many individuals, especially in regions where sleep clinics are scarce. The cost of a single PSG study can range from $600 to $5,000, depending on the healthcare system and location. This cost barrier, coupled with the limited number of sleep centres, prevents many patients from undergoing testing.
- **Long waiting times**: Due to the increasing demand for sleep studies and the limited availability of sleep labs, there are often long waiting lists for patients to undergo PSG. Some patients may wait for weeks or even months for a sleep study appointment, delaying diagnosis and treatment.
- **Inconvenience for patients**: Many individuals find sleeping in a lab environment uncomfortable, which can affect their normal sleep patterns. The presence of numerous sensors, wires, and electrodes attached to the body may also interfere with sleep, leading to results that may not fully reflect the patient's typical sleep experience. This discomfort, combined with the disruption of daily life caused by an overnight stay, deters many people from seeking a diagnosis.
- **Complexity of symptom presentation**: OSA symptoms can vary widely between individuals, and many people, especially those with mild cases, may not realize they have a problem. Symptoms like snoring and daytime fatigue are often dismissed as minor annoyances rather than signs of a serious medical condition. Moreover, many individuals may not have a bed partner to observe the signs of apnea, such as gasping or choking during sleep. As a result, the condition goes unrecognized for years, increasing the risk of complications.

12.2.1 Consequences of underdiagnosed and untreated OSA

OSA is far more than just a sleep disorder; it is a significant health issue that can lead to a wide range of serious medical complications if left untreated.

1. **Cardiovascular disease**: One of the most significant complications of untreated OSA is its impact on cardiovascular health. The repeated episodes of low

oxygen levels during sleep put stress on the cardiovascular system, leading to increased blood pressure and heart rate. This can contribute to the development of hypertension, atrial fibrillation, congestive heart failure, and coronary artery disease. Studies have shown that individuals with severe OSA are 2.5 times more likely to experience a cardiovascular event, such as a heart attack or stroke, compared to those without OSA. In fact, OSA is now recognized as an independent risk factor for hypertension, and the condition is present in up to 50% of patients with heart failure.

2. **Diabetes**: OSA is closely linked to metabolic dysfunction and has been shown to increase the risk of developing type 2 diabetes. The intermittent hypoxia (low oxygen levels) caused by OSA contributes to insulin resistance, a key factor in the development of diabetes. Research indicates that up to 70% of individuals with type 2 diabetes also have OSA, making it a critical condition to address in this population. Moreover, OSA exacerbates blood sugar dysregulation, making diabetes management more challenging.

3. **Stroke**: OSA is a known risk factor for stroke, with studies showing that individuals with moderate to severe OSA are at a significantly higher risk of ischemic stroke, one of the leading causes of long-term disability and death. The repeated drops in oxygen levels during sleep can trigger inflammatory responses, increase blood pressure, and promote blood clot formation, all of which increase stroke risk.

4. **Cognitive decline and dementia**: Chronic sleep disruption caused by untreated OSA has been linked to cognitive impairment, including memory loss, difficulty concentrating, and executive dysfunction. Over time, these cognitive issues may progress to more serious conditions such as mild cognitive impairment (MCI) or dementia, including Alzheimer's disease. Studies suggest that the cumulative effects of poor sleep and reduced oxygen supply to the brain may accelerate cognitive decline.

5. **Mood disorders**: Individuals with untreated OSA are at a higher risk of developing mood disorders such as depression and anxiety. The constant fatigue and poor quality of sleep can lead to significant changes in mood, irritability, and a reduced ability to handle stress. Moreover, the social isolation that can result from OSA-related symptoms, such as loud snoring, can contribute to feelings of frustration and loneliness.

12.2.2   The need for enhanced diagnostic tools

Given the profound health risks associated with OSA and the challenges posed by current diagnostic methods, there is a pressing need for more accessible, accurate, and patient-friendly diagnostic solutions. Machine learning (ML) and technological innovations hold great potential in addressing these challenges, offering the ability to analyze vast amounts of physiological data and predict OSA risk with higher precision. ML-driven tools can help streamline the diagnostic process, making it more scalable, affordable, and effective at detecting OSA in its early stages before more serious complications arise.

By contextualizing OSA in this way, the stage is set to explore how innovations in ML are poised to transform the landscape of OSA diagnosis, leading to better outcomes for patients and healthcare systems alike [2,3].

OSA is a widespread and potentially life-threatening condition, yet traditional diagnostic methods—particularly PSG—face significant limitations in terms of accessibility, cost, time, and patient comfort. Despite being the gold standard for diagnosing OSA, PSG is labor-intensive, expensive, and requires specialized settings. These barriers often result in long wait times, underdiagnosis, and the delayed treatment of many patients who suffer from OSA. ML, with its capability to analyze complex datasets and make accurate predictions, offers the potential to revolutionize the field of sleep apnea diagnosis. ML can provide automated, faster, more cost-effective, and scalable alternatives to traditional diagnostic techniques, thus addressing many of their inherent limitations.

12.2.3 Limitations of traditional diagnostic techniques

- **Resource-intensive nature of PSG**: PSG is a comprehensive sleep study that monitors multiple physiological parameters, including brain activity (via EEG), muscle movements (via EMG), heart rhythm (via ECG), breathing patterns, and oxygen saturation levels. This multi-channel setup requires an overnight stay in a sleep clinic under the supervision of trained technicians. The complexity of the procedure, coupled with the high costs of specialized equipment, makes PSG prohibitively expensive for many individuals. In the United States, the cost of a single PSG study can range from $1,000 to $5,000 depending on the healthcare system, and this does not include follow-up appointments or repeat tests that may be required for borderline or unclear cases. For many patients, this creates a significant financial burden, especially in regions with limited insurance coverage for sleep studies. Additionally, PSG requires substantial human resources—trained technicians and specialists to set up the equipment, monitor the patient, and interpret the results.
- **Limited accessibility and geographic disparities**: Access to PSG is largely restricted to specialized sleep centers, which are not evenly distributed geographically. In many rural or underserved regions, sleep clinics are scarce, making it difficult for individuals to access diagnostic services. Patients often have to travel long distances to reach a sleep center, and the availability of sleep labs is insufficient to meet the rising demand for sleep studies. This has led to long waiting lists, with patients sometimes waiting months for a diagnosis. Delays in diagnosis translate to delays in treatment, which can exacerbate health complications linked to untreated OSA, such as cardiovascular disease, diabetes, and cognitive decline.
- **Patient discomfort and unnatural sleep conditions**: Many patients find it difficult to sleep in a laboratory setting, surrounded by wires and sensors. The unfamiliar environment, combined with the discomfort of multiple electrodes attached to the body, often leads to sleep patterns that do not accurately represent the patient's normal sleep behavior. This may result in less reliable

data, as the quality of sleep observed during the PSG may differ from the patient's sleep at home. Furthermore, the discomfort and inconvenience associated with overnight stays in a sleep lab deter some patients from seeking a diagnosis in the first place.

- **Lengthy data processing and interpretation**: PSG generates a vast amount of data across multiple channels, which requires manual interpretation by sleep specialists. The process of reviewing and interpreting this data is time-consuming, adding to the overall length of the diagnostic process. Manual scoring of sleep stages, respiratory events (apneas and hypopneas), and other sleep disturbances is subject to human error and variability. Different specialists may score the same PSG data differently, leading to inconsistencies in diagnosis. This reliance on expert human interpretation limits the scalability of PSG, as more patients mean more manual reviews, further straining the already limited resources of sleep clinics.

- **Underdiagnosis and limited awareness**: Many individuals with mild to moderate OSA may not be aware that they have a sleep disorder, as they do not experience dramatic symptoms like severe snoring or frequent gasping. OSA is often misinterpreted as simple tiredness, aging, or lifestyle-related fatigue. As a result, many individuals may not seek medical help until the condition has progressed to a more severe stage. Traditional diagnostic techniques like PSG are not suited for large-scale screening due to their cost and complexity, leading to a large proportion of cases going undiagnosed.

12.3   Why machine learning could address these issues

ML, with its ability to process large amounts of complex data, automate decision making, and continuously improve accuracy over time, holds immense potential to address the limitations of traditional diagnostic techniques for OSA. By leveraging algorithms that can detect patterns in vast datasets, ML can make OSA diagnosis faster, more accurate, and more accessible.

Below are the keyways in which ML can overcome the challenges of traditional techniques:

- **Automation of diagnostic processes**: One of the primary advantages of ML is its capacity to automate the diagnostic process. ML algorithms can be trained on large datasets of sleep data—such as airflow signals, oxygen saturation, heart rate, and other physiological parameters—to identify apneas, hypopneas, and other sleep disturbances without the need for manual interpretation by a human expert. This automation dramatically reduces the time and cost involved in diagnosing OSA. For instance, instead of requiring an overnight stay in a sleep lab, ML-powered diagnostic tools can analyze data collected from simpler, home-based sleep monitoring devices, such as wearable sensors or smartphone apps, and deliver a diagnosis within a short time frame. Automated scoring of sleep events eliminates the need for sleep technicians to manually review hours of data, thereby increasing diagnostic throughput and scalability.

- **Faster and scalable diagnosis**: ML-based systems can process sleep data far more quickly than traditional methods. By automating the detection of respiratory events and sleep stages, ML algorithms can significantly shorten the time required to arrive at a diagnosis. This is particularly important in addressing the backlog of patients awaiting sleep studies and reducing long waiting times in sleep clinics. Additionally, the scalability of ML systems means that they can be deployed widely, reaching more patients in both urban and rural settings. Cloud-based ML platforms can analyze data from thousands of patients simultaneously, making OSA diagnosis feasible for large populations at a lower cost and in less time. This scalability is essential for addressing the underdiagnosis problem and ensuring that more patients receive timely care [4].

- **Use of less invasive, home-based monitoring devices**: ML allows for the use of less invasive diagnostic tools, such as wearable devices or smartphone-based applications, which can collect relevant sleep data without the need for a clinical sleep lab setting. For instance, wearables equipped with sensors can track vital signs, such as heart rate, blood oxygen levels, and movement, while audio-based systems can detect snoring patterns and respiratory pauses. These devices can be used by patients in the comfort of their own homes, providing a more natural sleep environment and potentially capturing more representative data than PSG performed in a lab. ML algorithms can then analyze this data remotely to predict the likelihood of OSA with a high degree of accuracy, offering a more patient-friendly and accessible diagnostic approach.

- **Improved accuracy through advanced pattern recognition**: ML algorithms, particularly deep learning models, excel at detecting subtle and complex patterns in data. When applied to OSA diagnosis, ML can enhance the accuracy of identifying apneas and hypopneas by analyzing multiple physiological signals simultaneously, something that may be difficult for a human expert to do consistently. For example, ML can analyze the temporal relationships between airflow, oxygen saturation, and heart rate variability to detect respiratory events that might be missed by traditional scoring methods. Furthermore, ML models can be trained on large and diverse datasets, improving their ability to generalize across different populations and patient profiles. This leads to more consistent and reliable diagnoses, reducing the variability and subjectivity associated with human interpretation of sleep studies.

- **Early detection and continuous monitoring**: ML has the potential to move OSA diagnosis from a reactive to a proactive process. Instead of diagnosing OSA only after the condition has progressed to a severe stage, ML algorithms can be used to screen at-risk populations, even in asymptomatic individuals. By continuously monitoring physiological data over time, ML-powered systems can detect early warning signs of OSA and flag individuals for further testing or early intervention. This capability is particularly valuable for populations at high risk of OSA, such as individuals with obesity, hypertension, or diabetes, as early detection can prevent the development of more severe complications.

- **Reducing costs and expanding access**: By enabling home-based diagnostics, ML-driven tools can significantly reduce the costs associated with traditional sleep studies. Wearable devices and portable monitoring systems are far less expensive than PSG, and when combined with automated ML analysis, they can offer a low-cost alternative that is accessible to a larger proportion of the population. This cost reduction is critical for increasing the reach of OSA diagnostics, particularly in low-resource settings or areas with limited access to sleep clinics. With ML, patients can receive a diagnosis without the need for costly overnight lab studies, reducing the financial burden on both individuals and healthcare systems.

12.3.1 Real-world applications and success stories

The use of ML in OSA diagnosis is not just theoretical—it is already being applied in real-world settings with promising results.

For instance:

- **Wearable technology companies** have developed smartwatches and sleep trackers capable of monitoring sleep patterns, heart rate, and blood oxygen levels. These devices use ML algorithms to predict sleep apnea risk and provide real-time feedback to users. *Apple Watch* and *Fitbit* are examples of wearable devices that incorporate such technology, making OSA screening more accessible to the general public. The *Fitbit OSA Screening Tool*, launched in 2022, uses pulse oximetry and motion sensors to flag apnea risk. In a trial with 5,000 users, it achieved *88% sensitivity* for moderate-severe OSA, democratizing access to underserved communities.
- **ML-based software solutions** have been integrated into portable home sleep test (HST) devices, which are increasingly being used as alternatives to PSG for diagnosing OSA. These devices collect simplified sleep data (such as airflow and oxygen saturation) and analyze it using ML algorithms to detect apneas and hypopneas with high accuracy.
- **Clinical trials and research studies** have demonstrated that ML models can outperform traditional scoring methods in detecting sleep apnea. Several studies have shown that ML can achieve diagnostic accuracy comparable to PSG when analyzing data from home-based monitoring devices, offering a practical solution for widespread OSA screening.

In conclusion, ML offers a powerful set of tools that can transform the diagnosis of OSA. By automating the diagnostic process, improving accuracy, and reducing costs, ML can address many of the limitations of traditional diagnostic techniques like polysomnography. This will enable faster, more accessible, and scalable OSA diagnostics, ultimately leading to better health outcomes for millions of patients worldwide [5].

12.3.2 Innovations in machine learning applied to OSA diagnosis

Recent innovations in ML are significantly transforming the diagnosis of OSA by introducing more efficient, accurate, and patient-friendly solutions. These innovations encompass a range of advanced ML algorithms, new data sources from wearable and portable devices, and improved computational techniques. These advancements are driving meaningful changes in clinical practices and patient care for OSA, making the diagnostic process faster, more accessible, and more affordable.

1. **Deep learning for multimodal data analysis:** Deep learning algorithms, particularly convolutional neural networks (CNNs) and recurrent neural networks (RNNs), have emerged as powerful tools for analyzing multimodal physiological data collected during sleep. Deep learning models excel at detecting subtle patterns and correlations between various signals, such as airflow, oxygen saturation, heart rate, and respiratory effort. These models can be trained to automatically identify apneas and hypopneas by analyzing this data in real-time, which significantly reduces the need for manual scoring by sleep technicians.

   This innovation allows for:

   - **Higher accuracy** in diagnosing OSA, even in complex or borderline cases, due to the algorithm's ability to capture intricate details in the data.
   - **Faster analysis** of sleep data, enabling quicker diagnosis and earlier intervention.
   - **Reduction in human error**, as deep learning models can deliver more consistent results than manual scoring.

2. **Wearable devices and portable home sleep tests:** The use of wearable devices and portable HSTs has transformed OSA diagnosis by enabling patients to be diagnosed in the comfort of their own homes. Devices such as smartwatches, chest bands, and finger pulse oximeters are equipped with sensors to capture vital physiological data, including heart rate, oxygen saturation, and movement patterns. ML algorithms analyze this data to predict the likelihood of OSA.

   Innovations in wearable technology include:

   - **ML-based predictive models** that process data from multiple nights of sleep, improving the accuracy and reducing false positives or negatives.
   - **Integration with smartphones and cloud-based platforms** that allows real-time data transmission and remote analysis, further expanding the accessibility of OSA diagnosis.
   - **Outcomes**: These wearable devices and HSTs have expanded access to OSA diagnosis, particularly in underserved regions, reduced the burden on sleep labs, and lowered diagnostic costs.

3. **Automated scoring and apnea event detection:** Traditional OSA diagnosis requires manual scoring of apnea and hypopnea events by sleep specialists, a labor-intensive process that can lead to variability and delays. ML-driven automated scoring systems are designed to replace or augment human scoring by detecting these events with high precision. These systems use signal processing and ML techniques to automatically classify breathing disturbances, snoring, and sleep stages.

Key innovations in this area include:

- **Supervised learning models** trained on large datasets of annotated PSG studies, which achieve diagnostic accuracy comparable to human experts.
- **Outcomes**: Automated scoring systems reduce the time required to generate diagnostic reports, eliminate inter-scorer variability, and enhance diagnostic throughput in busy sleep centers.

4. **Feature engineering for enhanced diagnostic precision:** Feature engineering involves extracting key features from physiological signals that are most relevant for detecting OSA. ML models are trained to recognize these features, such as the frequency and duration of apneas, oxygen desaturation events, and heart rate variability.

   Recent innovations in feature engineering include:

   - **Unsupervised learning models** that can automatically discover new diagnostic features or risk factors from large datasets without the need for manual feature selection.
   - **Outcomes**: This approach has led to improved diagnostic precision and the identification of novel indicators of OSA severity, which has been particularly useful for detecting mild or atypical cases of OSA.

5. **Predictive analytics for risk stratification:** Predictive analytics, driven by ML, allows clinicians to assess a patient's risk of developing OSA or progressing to more severe stages of the disorder. By analyzing demographic information, medical history, lifestyle factors, and physiological data, ML models can stratify patients into risk categories, helping healthcare providers prioritize those most in need of immediate intervention.

   Key advancements include:

   - **Risk prediction models** that incorporate both clinical and sleep data to estimate the probability of OSA, even before a full diagnostic study is conducted.
   - **Outcomes**: Predictive analytics has enabled earlier detection of OSA in high-risk populations, leading to more timely treatments and the prevention of complications such as cardiovascular disease and stroke.

6. **Cloud-based ML platforms and telemedicine integration:** Cloud computing and telemedicine are becoming integral to OSA diagnosis, especially when combined with ML algorithms. Cloud-based ML platforms allow data from wearable devices or home sleep tests to be analyzed remotely by ML models without the need for a physical visit to a sleep lab. This innovation is particularly beneficial for patients in rural or underserved areas.

   Recent developments include:

   - **ML-powered cloud platforms** that can analyze vast amounts of sleep data in real time and provide instant diagnostic feedback to healthcare providers or patients.
   - **Integration with telemedicine services**, where clinicians can remotely monitor patients' sleep data and provide diagnosis or treatment adjustments.

- **Outcomes**: These innovations are driving a shift toward more decentralized, patient-centered care for OSA, reducing the need for in-person visits and enabling continuous monitoring of patients over time.

7. **Personalized medicine and treatment optimization:** ML models are also being used to personalize OSA treatment plans. By analyzing patient-specific data, such as response to continuous positive airway pressure (CPAP) therapy or other interventions, ML algorithms can help optimize treatment settings, predict adherence, and adjust therapies in real-time.

Some key innovations include:

- **Adaptive ML models** that can recommend personalized CPAP settings or suggest alternative treatments based on individual patient responses.
- **Outcomes**: These systems have improved patient adherence to treatment, increased the effectiveness of therapies, and reduced the trial-and-error process often involved in managing OSA.

8. **AI-assisted sleep laboratories:** AI-driven systems are being integrated into clinical sleep laboratories to enhance the efficiency of OSA diagnosis and treatment. Innovations in AI-powered sleep labs include automated diagnostic platforms, robotic systems for applying sensors, and virtual assistants that guide patients through the diagnostic process. These systems use ML algorithms to process and analyze sleep data in real-time, reducing the workload for clinicians.

Key innovations include:

- **AI-guided sensor placement** for more accurate data collection during PSG.
- **AI platforms that provide real-time decision support** for sleep specialists by highlighting critical events in the data.
- **Outcomes**: AI-assisted sleep labs have streamlined the diagnostic process, reduced human error, and allowed sleep technicians to focus on more complex cases, ultimately improving patient outcomes.

12.3.3 Clinical outcomes driving change

The application of ML in OSA diagnosis has already led to significant clinical outcomes that are transforming healthcare practices:

- **Increased diagnostic efficiency**: Automated ML systems have drastically reduced the time required to analyze sleep data and generate diagnostic reports. This efficiency has shortened waiting times for patients and reduced the strain on overburdened sleep clinics.
- **Greater accessibility**: The use of home-based diagnostic tools, powered by ML, has expanded access to OSA testing for patients who live far from sleep centers or cannot afford traditional PSG. This accessibility is helping to address the significant underdiagnosis of OSA, particularly in underserved populations.
- **Improved accuracy and consistency**: ML algorithms have demonstrated a high level of accuracy in detecting OSA, often matching or exceeding

the diagnostic precision of human experts. The consistency provided by ML models reduces diagnostic variability, leading to more reliable outcomes.

- **Lower costs**: By enabling HSTs and automating the analysis process, ML-driven solutions have significantly reduced the costs associated with OSA diagnosis. This cost-effectiveness is making OSA diagnosis more affordable for patients and healthcare systems alike.
- **Personalized treatment**: ML-powered systems are helping to personalize OSA treatment by optimizing therapies based on individual patient data. This personalized approach improves patient adherence and the overall effectiveness of treatment [6].

In conclusion, the innovations in ML applied to OSA diagnosis are driving substantial improvements in speed, accuracy, cost reduction, and accessibility, which are reshaping clinical practices. These innovations are poised to make OSA diagnosis more efficient, widely available, and patient-centered, ultimately improving long-term health outcomes for millions of individuals worldwide.

12.3.4   Literature review: traditional diagnostic methods for obstructive sleep apnea

The diagnosis of OSA has traditionally relied on a combination of clinical assessments, patient-reported symptoms, and laboratory-based tests. Among these, PSG remains the gold standard for diagnosing OSA. However, other methods such as home sleep apnea tests (HSAT), questionnaires, and alternative diagnostic devices have also been used. In this review, we will summarize the most prominent traditional diagnostic methods, discussing their strengths and limitations in the context of detecting OSA.

1. Polysomnography – the gold standard

PSG is the most comprehensive and widely accepted method for diagnosing OSA. It is an overnight sleep study that monitors various physiological parameters while a patient sleeps in a controlled environment, usually in a sleep lab or clinic. PSG is often viewed as the benchmark against which other diagnostic methods are evaluated.

Strengths:

- **Comprehensive data collection**: PSG is capable of collecting detailed, multi-channel data that provides a holistic view of a patient's sleep patterns. It records several key parameters, including brain activity (via electroencephalography, EEG), eye movements (via electrooculography, EOG), muscle activity (via electromyography, EMG), heart rate (via electrocardiography, ECG), airflow, respiratory effort, oxygen saturation, and snoring.
- **High diagnostic accuracy**: Because PSG captures such a wide range of physiological signals, it provides a highly accurate and reliable diagnosis of OSA. It can precisely identify apneas (complete cessation of airflow) and hypopneas (partial reductions in airflow), along with the corresponding

drops in oxygen levels. The apnea-hypopnea index (AHI), a key measure derived from PSG, is used to quantify the severity of OSA.

- **Sleep staging**: PSG can identify sleep stages (REM and non-REM sleep) and the transitions between them, which is important because OSA events tend to be more frequent and severe during REM sleep. This ability to track sleep stages makes PSG particularly valuable for understanding the full impact of OSA on sleep architecture and quality.

Shortcomings:

- **High cost**: PSG is expensive, with the cost of a single overnight sleep study ranging from $1,000 to $5,000 in many healthcare settings, depending on the region and healthcare coverage. This makes it inaccessible to many patients, particularly those without comprehensive insurance.
- **Resource-intensive**: PSG requires highly trained technicians to set up, monitor, and interpret the study. This resource-intensive nature limits the number of patients that can be tested, contributing to long waiting lists in sleep centers.
- **Limited availability**: Many rural or underserved areas lack the specialized sleep clinics or equipment required to perform PSG. This geographic dis-parity creates significant barriers to care for many patients, particularly in regions where sleep centers are few and far between.
- **Patient discomfort**: PSG is performed in a sleep lab, which can be an unnatural environment for many patients. The presence of numerous sen-sors attached to the body, combined with the need to sleep in an unfamiliar setting, can affect sleep quality during the study. Some patients may find it difficult to sleep normally under these conditions, leading to data that may not reflect their typical sleep behavior.
- **Time-consuming**: PSG studies often require an overnight stay in the lab, and the data must then be manually scored and interpreted by sleep spe-cialists. This process can take days to weeks, contributing to delays in diagnosis and treatment.

2. **Home sleep apnea testing**
   HSAT, also known as portable monitoring, is an alternative diagnostic method that allows patients to be tested for OSA in their own homes. HSAT typically involves the use of portable devices that measure airflow, respiratory effort, and oxygen saturation during sleep. HSAT is often recommended for patients who are at high risk for moderate to severe OSA [7].
   Strengths:

- **Convenience and comfort**: Unlike PSG, HSAT can be performed in the comfort of the patient's own home, leading to a more natural sleep environment and potentially more representative data. Patients are more likely to sleep normally at home than in a lab setting, improving the reliability of the results.
- **Cost-effective**: HSAT is significantly less expensive than PSG, with costs typically ranging from $200 to $600. This makes it a more

accessible option for patients who cannot afford the high cost of a lab-based sleep study.

- **Easier access**: HSAT does not require the patient to visit a sleep center, reducing geographic barriers to testing. This expands the reach of OSA diagnosis, particularly in rural or underserved regions where sleep labs are scarce.
- **Faster results**: HSAT devices are relatively simple to use and generate data that can be quickly analyzed, leading to faster diagnoses compared to the lengthy process of PSG.

Shortcomings:

- **Limited data**: HSAT typically monitors only a few key parameters, such as airflow, oxygen saturation, and respiratory effort, without the additional physiological data collected by PSG (e.g., EEG, EOG, EMG). As a result, HSAT cannot assess sleep stages or the full impact of OSA on sleep architecture. This limits its diagnostic accuracy, particularly for patients with complex or mild OSA.
- **Risk of inconclusive results**: HSAT may produce inconclusive or inaccurate results for patients with mild OSA, complex sleep disorders, or comorbid conditions such as heart failure or chronic obstructive pulmonary disease (COPD). In such cases, a follow-up PSG may be required for a definitive diagnosis.
- **Not suitable for all patients**: HSAT is most effective for patients who are at high risk for moderate to severe OSA. It is not recommended for individuals with mild symptoms, complex sleep disorders, or significant comorbidities, as the limited data collected by HSAT may not be sufficient to diagnose these conditions accurately.

3. **Clinical assessments and questionnaires**
   In addition to PSG and HSAT, clinical assessments and sleep questionnaires are often used as initial screening tools for OSA. Commonly used questionnaires include the **Epworth Sleepiness Scale (ESS)**, the **Berlin Questionnaire**, and the **STOP-Bang Questionnaire**. These tools are designed to evaluate the patient's symptoms, risk factors, and daytime sleepiness [8].
   Strengths:

- **Non-invasive**: Questionnaires are a simple, non-invasive method of screening for OSA. They can be administered in a primary care setting or even completed by the patient at home.
- **Low cost**: These screening tools are inexpensive and easy to administer, making them accessible to a wide range of patients.
- **Early risk identification**: Questionnaires are useful for identifying patients who may be at high risk for OSA and who should undergo further testing with PSG or HSAT. They provide a quick way for healthcare providers to assess OSA risk factors such as obesity, snoring, and daytime sleepiness.

Shortcomings:

- **Subjectivity**: Questionnaires rely on self-reported symptoms, which can be subjective and prone to bias. Patients may underreport or overestimate their symptoms, leading to inaccurate risk assessments.
- **Limited diagnostic accuracy**: While questionnaires can help identify patients at risk for OSA, they are not diagnostic tools. A positive result on a questionnaire does not confirm the presence of OSA, and further testing is required to make a definitive diagnosis. Additionally, questionnaires may not accurately assess patients with mild or atypical symptoms of OSA.

4. Oximetry and other simplified devices

Pulse oximetry is sometimes used as a simplified method for screening or diagnosing OSA. It measures blood oxygen levels and can detect drops in oxygen saturation during apneic events. Some patients are provided with portable oximetry devices to wear overnight, which may help detect severe cases of OSA [9].

Strengths:

- **Simplicity and cost-effectiveness**: Oximetry devices are easy to use and inexpensive, making them a practical option for large-scale screening or for patients who cannot afford more comprehensive testing.
- **Indication of severe OSA**: In patients with severe OSA, significant drops in oxygen saturation can be easily detected by oximetry, making it a useful tool for identifying patients who may need urgent treatment.

Shortcomings:

- **Limited data**: Oximetry only measures oxygen saturation and does not provide any information on airflow, respiratory effort, or sleep stages. This limits its utility for diagnosing mild or moderate OSA, as it may miss apneas that do not result in significant desaturations.
- **False negatives and positives**: Oximetry may yield false-negative results in patients with mild OSA, as their apneas may not cause large enough drops in oxygen levels to be detected. Conversely, conditions such as COPD or heart failure can cause false-positive results by producing oxygen desaturation unrelated to OSA.

Conclusion: strengths and shortcomings of traditional Diagnostic methods

In summary, traditional methods for diagnosing OSA, including PSG, HSAT, and clinical questionnaires, each have distinct strengths and limitations. **PSG** remains the most comprehensive and accurate method but is expensive, resource-intensive, and uncomfortable for patients. **HSAT** provides a more accessible and cost-effective alternative but lacks the detailed data provided by PSG and is not suitable for all patients. **Questionnaires and oximetry** offer inexpensive and non-invasive screening options but cannot definitively diagnose OSA on their own.

The shortcomings of traditional methods—including cost, limited access, patient discomfort, and diagnostic delays—highlight the need for new, innovative approaches to OSA diagnosis. ML and other advanced technologies offer promising solutions to address these challenges by improving diagnostic accuracy, reducing costs, and making OSA testing more accessible and patient friendly.

Table 12.1   PSG involves the use of multiple sensors to monitor a range of physiological parameters, some of which are as mentioned:

Sensors	Descriptions
Airflow	Measures the flow of air through the nose and mouth to detect interruptions in breathing
Oxygen saturation	Measures the levels of oxygen in the blood to assess the impact of apneas and hypopneas
Electroencephalography (EEG)	Records brain wave activity to determine the stages of sleep and detect sleep disruptions
Electromyography (EMG)	Monitors muscle activity, including the muscles involved in breathing
Electrocardiography (ECG)	Tracks heart rate and rhythm, as OSA can affect cardiovascular function during sleep

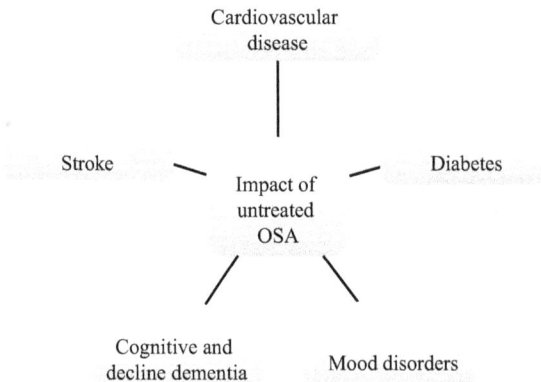

Figure 12.1   Flowchart of several chronic health conditions associated with underdiagnosed or untreated OSA

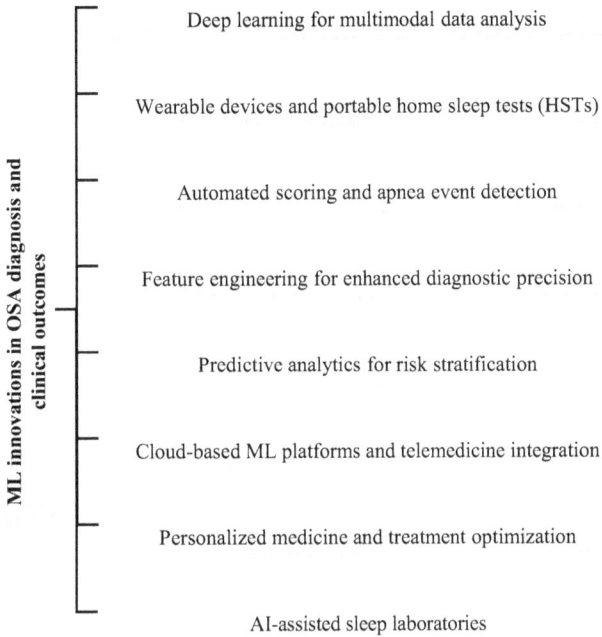

Figure 12.2   Flowchart of detailed overview of the key innovations in ML applied to OSA diagnosis and the outcomes driving these clinical changes

References

[1]   Kang, J. J., and Raghupathi, W. (2023). A comprehensive review on machine learning in healthcare industry: classification, restrictions, opportunities, and challenges. *Sensors*, *23*(9), 4178. https://doi.org/10.3390/s23094178

[2]   Shah, N. H., and Milstein, A. (2017). Machine learning in healthcare: the hope, the hype, and the promise. *Health Affairs*, *37*(10), 1787–1794. https://doi.org/10.1377/hlthaff.2018.05159

[3]   Rajkomar, A., Dean, J., and Kohane, I. S. (2019). Machine learning in medicine. *The New England Journal of Medicine*, *380*(14), 1347–1358. https://doi.org/10.1056/NEJMra1814259

[4]   Esteva, A., Kuprel, B., Novoa, R. A., Ko, J., Swetter, S. M., Blau, H. M., *et al.* (2017). Dermatologist-level classification of skin cancer with deep neural networks. *Nature*, *542*(7639), 115–118. https://doi.org/10.1038/nature21056

[5]   Topol, E. J. (2019). High-performance medicine: the convergence of human and artificial intelligence. *Nature Medicine*, *25*(1), 44–56. https://doi.org/10.1038/s41591-018-0300-7

[6]   Hamet, P., and Tremblay, J. (2017). Artificial intelligence in medicine. *Metabolism*, *69*, S36–S40. https://doi.org/10.1016/j.metabol.2017.01.011

[7]   Levin, S., Toerper, M., Hamrock, E., and Hinson, J. S. (2018). Machine-learning-based electronic triage more accurately predicts patient outcomes than the Emergency Severity Index. *Annals of Emergency Medicine*, *71*(5), 565–574. https://doi.org/10.1016/j.annemergmed.2017.08.005

[8]   Rajkomar, A., Dean, J., and Kohane, I. (2019). Machine learning in medicine. *New England Journal of Medicine*, *380*(14), 1347–1358. https://doi.org/10.1056/NEJMra1814259

[9]   Cambridge University Press. (n.d.). Machine learning and artificial intelligence: Applications in healthcare epidemiology. Retrieved from https://www.cambridge.org

Chapter 13

Ensemble and hybrid machine learning techniques: theoretical foundations, differences, applications and healthcare integration

Gaurav Gupta[1] and Laxminarayana Korada[2]

Abstract

This chapter explores the theoretical foundations, differences, applications, and future directions of ensemble and hybrid machine learning techniques, with a particular emphasis on their application in healthcare. Ensemble learning combines multiple models to enhance accuracy and robustness by leveraging the strengths of individual models, using methods such as bagging, boosting, and stacking. Hybrid machine learning integrates diverse algorithms and paradigms to exploit their complementary advantages, leading to more powerful and adaptable models. We discuss the conceptual and methodological differences between these techniques, highlighting their unique strengths and limitations. Practical applications are examined across various domains, with a specific focus on healthcare applications such as disease diagnosis, medical image analysis, personalized medicine, and treatment optimization. A dedicated chapter on healthcare integration addresses how these advanced techniques can be tailored to meet specific challenges in the healthcare domain, enhancing the accuracy, efficiency, and effectiveness of medical practices. Key challenges such as computational complexity, model interpretability, and scalability are addressed, along with emerging research opportunities and technological advancements, including quantum computing and federated learning. The insights gained provide a comprehensive understanding of these advanced machine learning techniques, guiding their effective application and future development to tackle complex real-world problems, especially in healthcare.

Keywords: Ensemble learning; hybrid machine learning; computational optimization; healthcare applications; predictive analytics; disease diagnosis

[1]Yogananda School of AI, Computers and Data Sciences, Shoolini University, Solan H.P, India
[2]Partner Technology, Microsoft Corporation, Bellevue, WA 98008, USA

13.1   Introduction

Machine learning has become a fundamental aspect of modern decision-making processes that rely on data (Bell, 2022). It has had a significant impact on various fields such as finance, healthcare, and technology (Dahiya *et al.*, 2022; Helm *et al.*, 2020). In healthcare, machine learning has revolutionized how medical professionals diagnose diseases, predict patient outcomes, and personalize treatment plans by allowing systems to learn from data, recognize patterns, and make decisions with minimal human involvement. At its core, ML involves creating algorithms that can learn from data and use that knowledge to make predictions or decisions (Sarker, 2021). Traditional machine learning techniques, including supervised, unsupervised, and reinforcement learning, have proven to be effective in solving a wide range of problems (Wang *et al.*, 2021). Models like linear regression, decision trees, and neural networks have provided robust solutions to many complex problems, including those in healthcare, where they are used for tasks such as medical image analysis, patient monitoring, and predictive analytics (Awan *et al.*, 2020; Riyaz *et al.*, 2022). However, as data becomes more complex and voluminous, single model approaches often face limitations in terms of accuracy, generalization, and robustness.

However, as data becomes more complex and voluminous, single model approaches often face limitations in terms of accuracy, generalization, and robustness. Single-model approaches face challenges in extracting meaningful patterns and insights from complex data due to high dimensionality, heterogeneous attributes, and intricate interrelationships (Pawar *et al.*, 2023). These models struggle to capture nuanced patterns and adapt to diverse contexts, particularly in healthcare, where data can be highly variable and multifaceted. Ensemble and hybrid machine learning techniques offer promising solutions by combining multiple models or integrating diverse learning paradigms (Dasari *et al.*, 2023; Zhou *et al.*, 2023). These approaches offer enhanced predictive performance, improved generalization, and greater robustness in complex and voluminous datasets, overcoming the limitations of single-model approaches (Solomon *et al.*, 2023).

Motivation: Hybrid and ensemble machine learning techniques have become effective approaches to tackle such challenges. By utilizing the advantages of each individual model, ensemble learning integrates several models to increase prediction accuracy and generalization. Conversely, hybrid machine learning combines many machine learning paradigms or algorithms to take use of their complimentary benefits, improving the learning system's overall performance. Developing an understanding of these methods is essential to creating ML solutions that are more scalable, precise, and dependable, particularly in critical domains like healthcare, where accurate and reliable predictions can significantly impact patient outcomes.

Contribution: This study offers a thorough analysis of ensemble and hybrid machine learning approaches, emphasizing their theoretical foundations, methodological differences, and applications, with a specific focus on healthcare. By systematically analyzing and comparing these approaches, we aim to offer valuable insights into their strengths, weaknesses, and suitability for various tasks. In addition, we present new trends and directions for future study and advancement in this dynamic topic, highlighting the potential for these techniques to improve healthcare delivery and outcomes.

Objectives: This study seeks to offer a comprehensive review of ensemble and hybrid machine learning techniques, with a specific emphasis on their theoretical foundations, distinctive characteristics, and application. The specific objectives of this chapter are:

✓ To explain the theoretical foundations of ensemble and hybrid machine learning techniques.
✓ To distinguish between the conceptual and methodological aspects of these techniques.
✓ To evaluate the applications and performance of ensemble and hybrid methods across different domains, with a special focus on healthcare.
✓ To identify current challenges and propose potential areas for future research.

Research questions: The key research questions addressed in this chapter include: (1) What are the fundamental principles underlying ensemble and hybrid machine learning techniques? (2) How do ensemble and hybrid techniques differ in terms of their methodologies and applications? (3) What are the applications of ensemble and hybrid methods, particularly in healthcare? (4) What are the challenges, limitations, emerging trends, and future research opportunities in this area?

This chapter is structured as follows: Section 13.2 provides a detailed overview of the theoretical foundations of machine learning techniques, ensemble learning and hybrid machine learning. Section 13.3 discusses the differences between ensemble and hybrid techniques, highlighting their conceptual and methodological distinctions. Section 13.4 examines the applications of these techniques. Section 13.5 discusses the integration of ensemble and hybrid machine learning techniques into healthcare applications. Section 13.6 addresses the current challenges and future directions in the field. Finally, Section 13.7 concludes with a summary of key points and implications for practice.

13.2 Theoretical foundations

This section provides an overview of the theoretical foundations underlying machine learning techniques, with the focus on ensemble and hybrid machine learning methods. By starting with the fundamental machine either machine learning, learning paradigms or machine learning paradigms, the section discusses ensemble learning and hybrid machine learning. The discussion consists of the basic concepts, types, and mathematical representation of both machine learning approaches, along with their applications in healthcare.

13.2.1 Overview of machine learning techniques

Machine learning comprises a range of approaches that allow systems to acquire knowledge from data and enhance their performance as time progresses. The main paradigms of machine learning reflect supervised learning, unsupervised learning, and reinforcement learning.

Supervised learning (Hastie *et al.*, 2009a): Supervised learning involves training a model using a dataset that has labels, meaning each input is paired with its corresponding output. The goal is to acquire knowledge of the process of associating inputs with outputs in a manner that may be applied to new data that has not been previously seen. Common algorithms include linear regression, logistic regression, support vector machines (SVM), decision trees, and neural networks. In healthcare, supervised learning is extensively used for tasks such as disease diagnosis and patient outcome prediction.

Unsupervised learning (Gentleman and Carey, 2008; Hastie *et al.*, 2009b): Unsupervised learning is the process of training models using data that does not have any labels. The objective is to uncover hidden patterns or structures within the data. Methods such as clustering (e.g., k-means, hierarchical clustering) and dimensionality reduction (e.g., principal component analysis, t-SNE) are commonly employed to detect patterns or reduce the complexity of data. In healthcare, unsupervised learning helps in identifying patient subgroups, discovering new diseases, and analyzing genetic data.

Reinforcement learning (François-Lavet *et al.*, 2018; Sewak, 2019): Reinforcement learning is a field that concentrates on teaching agents to make a series of decisions by engaging with their surroundings. The agent acquires the skill of maximizing the total rewards by engaging in both exploration and exploitation of actions, guided by input received from the environment. Some important methods are Q-learning, deep Q-networks (DQN), and policy gradient algorithms. In healthcare, reinforcement learning is applied to personalized treatment strategies and optimizing clinical pathways.

13.2.2   Ensemble learning

Ensemble learning is the process of merging numerous models to improve forecast accuracy and robustness (Dietterich, 2000). Ensemble approaches strive to generate improved outcomes by combining the strengths of individual models, surpassing the performance of any one model. Ensemble learning utilizes the variety of several models to enhance accuracy, decrease variance, and improve generalization (Sagi and Rokach, 2018; Zhou, 2021). The foundational idea is that a collection of learners with limited individual abilities can together create a powerful learner by effectively merging their predictions (Zhou, 2012). In healthcare, ensemble learning is valuable for improving diagnostic accuracy and creating robust predictive models.

(A)   **Types of ensemble methods**

There are multiple methodologies for implementing ensemble learning, with bagging, boosting, and stacking being the most prevalent and dominant techniques.

Bagging (bootstrap aggregating) (Agarwal and Chowdary, 2020; Wen and Hughes, 2020): Bagging creates several iterations of a training dataset by employing bootstrap sampling and trains an individual model on each iteration. The ultimate prediction is derived by taking the average of the results in the case of regression, or by employing majority vote in the case of classification. Random forests are an often-used illustration of the bagging technique.

In healthcare, bagging is used to improve the accuracy of diagnostic systems by reducing variance and enhancing generalization.

Boosting (Sornsuwit and Jaiyen, 2019; Wen and Hughes, 2020): Sequential boosting involves training a series of models, with each model specifically designed to rectify the errors made by the previous model. Models are aggregated to provide a robust prediction. Some examples of boosting algorithms are AdaBoost, which modifies the weights of incorrectly classified cases, and gradient boosting, which improves a loss function by sequentially adding models. In healthcare, boosting techniques are applied to increase the sensitivity and specificity of disease detection models.

Stacking (Agarwal and Chowdary, 2020; Wen and Hughes, 2020): Stacking is a technique that entails training several base models along with a meta-model. The initial models are trained using the original dataset, whereas the meta-model is learned using the outputs of the initial models. The meta-model is trained to effectively integrate the predictions of the basis models. In healthcare, stacking can integrate diverse diagnostic tools to create a comprehensive system that leverages the strengths of multiple algorithms.

(B) **Mathematical foundations**

Depending on the chosen approach for implementing ensemble machine learning, the mathematical representations will vary accordingly. The foundational mathematical representations for the bagging, boosting, and stacking techniques can be described as follows.

- **Bagging**: Let D be the dataset, and M the number of bootstrap samples. Each sample D_i is used to train a model h_i. The final prediction $H(x)$ is the average or majority vote $h_i(x)$.

$$H(x) = \frac{1}{M} \sum_{i=1}^{M} h_i(x) \tag{13.1}$$

- **Boosting**: Let D be the dataset, and M the number of iterations. Each model h_i is trained on the weighted dataset, with weights adjusted based on errors. The final prediction $H(x)$ is a weighted sum of $h_i(x)$.

$$H(x) = \sum_{i=1}^{M} \alpha_i h_i(x) \tag{13.2}$$

- **Stacking**: Let D be the dataset, and $h_i h_i$ the base models. The meta-model H is trained on the outputs $h_i(x)$. The final prediction is:

$$H(h_1(x), h_2(x), h_3(x), \ldots, h_M(x)) \tag{13.3}$$

13.2.3   Hybrid machine learning

Hybrid machine learning is the integration of various machine learning paradigms or algorithms to leverage their complementing advantages (Solomon *et al.*, 2023; Uthayakumar *et al.*, 2020). Integrating this can improve the performance, scalability, and adaptability of the model. Hybrid machine learning involves the synergistic combination of multiple algorithms or paradigms to create models that benefit from the strengths of each component, thereby improving overall performance (Saravi *et al.*, 2022; von Rueden *et al.*, 2020; Zhou *et al.*, 2023). In healthcare, hybrid machine learning is used to develop robust and adaptable models for complex tasks like disease prediction and personalized treatment.

(A) **Types of hybrid methods**

Combining multiple algorithms: This technique combines various algorithms to use their unique capabilities. Ensemble approaches can enhance the reliability and precision of predictions by using the distinct benefits of different models. Hybrid decision trees can integrate decision trees with other algorithms such as SVMs or neural networks (Lu and Ma, 2020). By leveraging the strengths of each method, this combination can improve generalization and enhance decision boundaries, effectively compensating for their respective deficiencies. In healthcare, combining algorithms can improve the accuracy of medical diagnoses and predictions.

Integration of different machine learning paradigms: This approach involves combining many machine learning paradigms to enhance the performance of the model. Neural networks can be combined with adaptive algorithms, such as genetic algorithms (GA), to enhance the optimization of model parameters and designs (Zhang *et al.*, 2021). Evolutionary algorithms can efficiently navigate the range of possible solutions, resulting in enhanced learning speed and superior solution accuracy for intricate problems. This strategy can improve the overall performance and flexibility of the model by utilizing the advantages of both neural networks and evolutionary algorithms. In healthcare, integrating different paradigms helps in creating models that can handle diverse types of medical data and tasks.

(B) **Mathematical foundations:**

• **Hybrid decision trees**: A hybrid decision tree T might combine a decision tree DT with an SVMs classifier SVM. The final prediction $H(x)$ depends on the outcomes of both DT and SVM. where α and β are weights determined during training.

$$H(x) = \alpha \cdot DT(x) + \beta \cdot SVM(x) \tag{13.4}$$

• **Neural networks with genetic algorithms**: A neural network's parameters θ can be optimized using a GA. The fitness function $F(\theta)$ evaluates the performance, and GA iteratively evolves θ to maximize $F(\theta)$.

Following this introduction to the foundational concepts of ensemble and hybrid machine learning, the subsequent sections will go into depth on their differences, applications, and challenges.

13.3   Differences between ensemble and hybrid techniques

This section delves into the conceptual and methodological differences between ensemble and hybrid machine learning techniques. By comprehending these differences, we will better understand the different advantages and constraints of each method. In addition, we also discuss the strengths and weaknesses of providing a comprehensive comparison, particularly in the context of healthcare.

13.3.1 Conceptual differences

Ensemble and hybrid strategies have different approaches to improve model performance. They use unique procedures to boost predictive accuracy, robustness, and generalization capabilities. Ensemble learning is the fundamental idea of merging numerous models to form a single, stronger model (Zhou, 2021). The primary objective is to combine the results of multiple fundamental models to improve accuracy and generalization (Solomon *et al.*, 2023). Ensemble approaches are based on the concept that combining several models can rectify individual flaws and yield a higher result. In healthcare, ensemble learning is often used to combine various diagnostic tools to increase the accuracy of disease detection and patient outcome predictions. Hybrid learning is the integration of many machine learning paradigms or algorithms into a unified framework (Chakraborty *et al.*, 2020). The objective is to exploit the advantages of many approaches to develop a more robust and flexible model (Singh and Jain, 2023). Hybrid approaches sometimes entail merging models that function based on distinct principles, such as incorporating neural networks with evolutionary algorithms to enhance model performance (Kanchanamala *et al.*, 2023). In healthcare, hybrid models are beneficial for integrating diverse data sources, such as clinical records, medical images, and genetic information, to provide comprehensive predictive analytics.

13.3.2 Methodological differences

The methodologies employed by ensemble and hybrid techniques differ significantly as it represented on Table 13.1. Ensemble methods, for instance, include bagging, boosting, and stacking (Agarwal and Chowdary, 2020; Wen and Hughes, 2020). Bagging involves training multiple models on different bootstrap samples of the dataset and combining their predictions through averaging for regression or majority voting for classification. In healthcare, bagging can improve the stability and reliability of diagnostic systems by reducing variance. Sequentially boosting trains models, with each model focusing on correcting the errors of the previous models, culminating in a final model that is a weighted sum of all individual models, with weights determined by their accuracy. In healthcare, boosting is used to enhance the sensitivity and specificity of predictive models for disease detection. Stacking, on the other hand, trains multiple base models on the original data and a meta-model on the outputs of the base models to make the final prediction. In healthcare, stacking can integrate different diagnostic tools to create a robust system that leverages the strengths of multiple algorithms.

Hybrid methods, in contrast, focus on integrating different algorithms within the same model to exploit their complementary strengths (Ardabili *et al.*, 2020). This can involve combining multiple algorithms, such as hybrid decision trees that integrate decision trees with SVMs or neural networks. Another approach is the integration of different machine learning paradigms, such as combining neural networks with genetic algorithms, to enhance learning efficiency and solution

Table 13.1   Methodological difference between ensemble and hybrid machine learning

Feature	Ensemble techniques	Hybrid techniques
Approach	Combines multiple models	Integrates different algorithms/paradigms
Main goal	Improve accuracy and robustness	Enhance performance and adaptability
Complexity	High due to multiple models	Very high due to integration of diverse methods
Interpretability	Moderate to low	Often low
Computational requirements	High	Very high
Scalability	Scalable with distributed computing	More challenging due to integration complexity
Flexibility	High in combining various models	High in adapting to various problem requirements
Development time	Moderate to high	High due to the need for integrating and tuning different methods
Maintenance	Moderate, needs updating with new models/data	High, requires constant monitoring and updating due to complexity
Data handling	Requires large datasets for training	Can leverage diverse data sources, often needs extensive preprocessing
Model diversity	Typically combines similar types of models (e.g., decision trees)	Combines fundamentally different models (e.g., neural networks and genetic algorithms)
Error reduction	Reduces variance and bias by combining multiple models	Reduces errors by leveraging the strengths of diverse techniques
Adaptability	Good, adapts well to changes in data when retrained	Excellent, highly adaptable to various problem domains and data types
Robustness	High, due to averaging or voting mechanisms	Very high, due to combined strengths and complementary methods
Typical applications	Classification, regression, anomaly detection	Real-time systems, complex problem solving, optimization, personalized recommendations
Learning efficiency	Moderate, due to repetitive training of similar models	High, especially in optimizing model parameters and architectures
Resource utilization	High, requires significant computational and memory resources	Very high, due to the need to support diverse models and algorithms

quality. This method is often used for optimizing model parameters and architectures, leveraging the strengths of diverse techniques to address complex problems more effectively. In healthcare, hybrid models are particularly useful for personalized medicine, where diverse data types and complex relationships need to be modeled to provide tailored treatment recommendations.

13.3.3   Strengths and weaknesses

Both ensemble and hybrid techniques offer unique strengths and have their own limitations, making them suitable for different types of problems and applications.

Strengths of ensemble techniques: Ensemble techniques improve accuracy, reduce overfitting, and enhance robustness by combining multiple models. They achieve higher generalization and accuracy by complementing diverse predictions. They reduce overfitting by averaging biases and variances across models, which is particularly useful in healthcare where data can be noisy. Ensemble methods are more robust to noise and data variations, making them reliable tools for healthcare applications where data inconsistency is common.

Weaknesses of ensemble techniques: Ensemble techniques offer advantages but also face challenges in complexity and interpretability. Combining multiple models increases computational power and memory requirements, which can be a limitation in resource-constrained environments like smaller healthcare facilities. Interpreting ensemble models can be difficult, especially in complex decision-making processes like healthcare where understanding model decisions is crucial for clinical adoption.

Strengths of hybrid techniques: Hybrid methods offer flexibility, enhanced performance, and optimization by combining different algorithms and paradigms, which can adapt to diverse healthcare problems. They can adapt to diverse problems, achieve higher performance in complex tasks, and excel in optimizing model parameters and architectures, resulting in more efficient learning and improved overall performance. Which is essential for tailoring healthcare models to specific patient needs.

Weaknesses of hybrid techniques: Hybrid methods offer advantages but also face challenges in implementation, complexity, and resource intensity. Designing and implementing hybrid models requires expertise and extensive experimentation, which can be challenging in the fast-paced healthcare environment. These models are resource-intensive, requiring substantial computational power and time for training and optimization, which can be a barrier for widespread use in healthcare settings.

13.4   Applications

This section will present the practical applications of ensemble and hybrid machine learning algorithms. This section includes an analysis of how these approaches might be applied to address practical challenges in different fields, through the examination of specific cases and examples. We emphasize the distinctive benefits of each strategy in various circumstances, particularly in healthcare.

13.4.1   Ensemble learning applications

Ensemble learning techniques have proven to be effective in several applications due to their capacity to enhance accuracy, resilience, and generalization. Ensemble approaches can boost overall performance by merging numerous models, allowing them to utilize the strengths of each individual model and mitigate their drawbacks. Ensemble approaches can combine different predictive models, which enables them to perform very well in complicated and diverse tasks across various domains.

- **Classification:** Ensemble methods enhance classification performance by combining predictions from multiple models, reducing errors from single models (Solomon *et al.*, 2023). This is particularly useful in tasks like spam detection, image recognition, and medical diagnosis (Altan and Karasu, 2020). Ensembles combine the strengths of different models, resulting in more accurate predictions and reliable outcomes. In healthcare, ensembles combine the strengths of different models to improve the accuracy of disease diagnosis and patient classification, making them reliable tools in critical medical decision-making processes.

- **Regression:** Ensemble methods improve the accuracy and reliability of regression tasks by combining the outputs of multiple models. This reduces variance and bias, leading to more precise and stable results. Ensemble models are used in financial forecasting, risk assessment, environmental modeling, real estate valuation, demand forecasting, and energy consumption prediction (Pachauri and Ahn, 2022; Von Krannichfeldt *et al.*, 2021). In healthcare, ensemble regression models can predict patient outcomes, treatment responses, and healthcare costs with high precision.

- **Anomaly detection:** Ensemble techniques are crucial for identifying anomalies in data, enhancing security, and preventing losses. They combine the strengths of multiple models, detecting unusual patterns more accurately than individual models. Applications include fraud detection, network security, industrial monitoring, and healthcare (Zhong *et al.*, 2020). Ensemble methods improve detection rates, prevent cyber-attacks, and enable timely interventions in industries relying on accurate identification of irregularities (Jeffrey *et al.*, 2024).

- **Natural language processing:** Ensemble methods improve natural language processing (NLP) tasks by integrating multiple algorithms to process and understand human language more effectively. This approach addresses complexities in sentiment analysis, machine translation, and Named Entity Recognition (Kazmaier and van Vuuren, 2022; Sangamnerkar *et al.*, 2020). In healthcare, combining Long Short-Term Memory (LSTM) networks with traditional classifiers enhances the accuracy of analyzing medical records, patient notes, and research papers, facilitating better clinical decision support (Başarslan and Kayaalp, 2024).

- **Recommender systems:** Ensemble methods combine multiple recommendation algorithms to create robust recommender systems. These systems provide personalized recommendations based on user preferences and behavior. By integrating diverse models, they improve recommendation quality and accuracy. Hybrid recommender systems combine collaborative and content-based filtering, enhancing user satisfaction and engagement (Dong *et al.*, 2020; Forouzandeh *et al.*, 2021). In healthcare, ensemble-based recommender systems can suggest personalized treatment plans, medication adjustments, and lifestyle recommendations to improve patient outcomes.

- **Robotics and autonomous systems:** Ensemble learning enhances the reliability and adaptability of robotics and autonomous systems in dynamic environments (Khaldi *et al.*, 2021). It improves performance and resilience by combining multiple algorithms for path planning and sensor fusion. Ensemble methods also enhance the accuracy and reliability of environmental perception, allowing robots to make informed decisions in real-time (Dutta and Dasgupta, 2017). In healthcare, ensemble methods are applied in robotic surgery and autonomous patient monitoring systems, improving accuracy and decision making in real-time.

13.4.2   Hybrid learning applications

Hybrid machine learning techniques enhance performance and adaptability in complex environments by combining diverse algorithms. These methods leverage the strengths of each algorithm, enabling systems to tackle complex challenges with precision and agility. They are useful in autonomous vehicles, robots, and medical devices, providing versatility and robustness for success in dynamic settings.

- **Real-time systems:** Hybrid methods are ideal for real-time applications requiring swift, precise decision making. They combine diverse algorithms to enhance speed and accuracy. These methods are used in autonomous vehicles, robotic systems, and medical devices, enabling systems to process information rapidly and make informed decisions in real-time, ensuring agility and confidence in complex scenarios (Li *et al.*, 2020). In healthcare, hybrid methods are used in real-time patient monitoring systems and emergency response systems, enabling rapid and informed decision making in critical situations.
- **Complex problem solving:** Hybrid techniques are effective in tackling complex problems that require a multifaceted approach. They integrate diverse algorithms, providing unique insights and perspectives. These techniques are useful in bioinformatics, healthcare, and supply chain optimization (Altan and Karasu, 2020; Solomon *et al.*, 2023; von Rueden *et al.*, 2020). In healthcare, hybrid methods are used in bioinformatics to analyze complex genetic data, in clinical decision support systems to integrate various data sources, and in optimizing healthcare operations.
- **Optimization problems:** Hybrid techniques are effective in tackling optimization problems with multiple objectives or constraints. They integrate diverse algorithms, providing unique insights and capabilities. These techniques are useful in engineering system design, operations research, and financial risk management. In healthcare, these techniques optimize treatment plans, resource allocation in hospitals, and scheduling of medical staff, enhancing overall efficiency and patient care.
- **Multimedia processing**: Hybrid learning methods are essential in multimedia applications, as they handle complex multimodal data sources like images, videos, text, and audio. These methods combine multiple algorithms to analyze and extract insights from diverse data. In healthcare, hybrid models are used for medical image analysis, combining CNNs with RNNs to capture spatial and temporal dependencies, and integrating audio and visual features for better diagnostic accuracy (Ashraf *et al.*, 2023).

- **Recommender systems**: Hybrid recommender systems use a combination of collaborative filtering, content-based filtering, and knowledge-based approaches to provide highly personalized and accurate recommendations (Fararni *et al.*, 2021). Hybrid recommender systems combine various methods to provide robust, accurate recommendations, adapting to diverse user preferences and contexts, enhancing user satisfaction and engagement (Walek and Fojtik, 2020). In healthcare, these systems offer personalized recommendations for treatments, medications, and preventive measures based on patient history, preferences, and clinical guidelines, enhancing patient satisfaction and engagement.

Ensemble and hybrid machine learning techniques are powerful tools for addressing real-world challenges in various fields. They improve performance by leveraging the collective intelligence of diverse models, capturing subtle patterns, handling uncertainties, and adapting to changing environments. These techniques excel in tasks like classification, regression, anomaly detection, and optimization, making them indispensable assets across industries. As real-world problems become more complex, their importance in driving progress is undeniable. Table 13.2 showcases the utilization of ensemble and hybrid machine learning techniques across various domains.

Table 13.2   Common domains where ensemble and hybrid machine learning techniques are applied

Domain	Ensemble techniques applications	Hybrid techniques applications
Finance	Stock price prediction, risk assessment	Portfolio optimization, algorithmic trading
Healthcare	Disease diagnosis, medical image analysis	Personalized medicine, treatment optimization
Manufacturing	Quality control, predictive maintenance	Process optimization, real-time monitoring
Autonomous systems	Object detection, path planning	Autonomous navigation, robotic control
Environmental science	Weather prediction, climate modeling	Ecosystem modeling, resource management
E-commerce	Customer segmentation, fraud detection	Personalized recommendations, dynamic pricing
Marketing	Market trend analysis, customer behavior prediction	Targeted advertising, campaign optimization
Transportation	Route optimization, traffic prediction	Autonomous vehicles, congestion management
Energy	Demand forecasting, energy efficiency analysis	Smart grid optimization, renewable energy integration
Education	Student performance prediction, adaptive learning	Personalized tutoring, educational content recommendation
Aerospace	Flight safety analysis, aircraft maintenance	Flight trajectory optimization, autonomous drones

13.5   Healthcare integration

This chapter focuses on the integration of ensemble and hybrid machine learning techniques into healthcare environment. The discussion mainly explores how both approaches can be used to address specific challenges in the healthcare domain.

- **Disease diagnosis and prognosis:** By combining multiple models, these techniques can improve the performance of the machine learning model, identify better patterns more than single models, and provide more predictions. While ensemble methods such as bagging, boosting, and stacking are used to enhance diagnostic accuracy, hybrid models that integrate neural networks with traditional algorithms can provide robust diagnostic tools.
- **Medial image analysis:** One of the critical areas where both techniques have shown significant promise is image-based diagnostic and aid in early disease detection. In medical imaging, ensemble learning can combine outputs from different image analysis models to improve diagnostic accuracy. Hybrid models that merge CNNs with other machine learning algorithms can enhance image analysis.
- **Personalized medicine:** Both techniques also can play crucial role in identifying and treating of individual patients' characteristics. Developing personalized treatment plans and predicting treatment outcomes can be achieved though ensemble methods that aggregate predictions from various models that consider different patient data types or hybrid models that integrate deep learning with probabilistic models can predict individual responses to treatments.
- **Predictive analysis and risk assessment:** Ensemble and hybrid techniques enhance the predictive power of predictive analysis and risk assessments. Ensemble models can improve the accuracy of predictive analytics by combining forecasts from multiple models. Hybrid approaches that integrate machine learning with statistical models can enhance risk assessment.
- **Real-time monitoring and decision support:** Real-time monitoring and decision support systems benefit greatly from the integration of ensemble and hybrid machine learning techniques. These systems can provide timely and accurate information to healthcare providers, improving patient care. In real-time monitoring, ensemble methods can combine data from various sensors and monitoring devices to provide comprehensive patient health assessments. Hybrid models that combine real-time data processing with predictive analytics can enhance decision support systems.

13.6   Challenges and future directions

The concept of machine learning is undergoing tremendous evolution, ensemble and hybrid techniques leading are contributors in this progress. This section examines the present challenges, developing trends, and future research potential in ensemble and hybrid machine learning.

13.6.1   Current challenges

Ensemble and hybrid techniques enhance predictive performance and model robustness by combining algorithms and paradigms. However, their complexity introduces challenges like computational complexity, model interpretability, scalability, data quality, hyperparameter tuning, communication overhead, and maintenance. Addressing these issues can maximize their potential in machine learning applications including in healthcare environment.

- **Computational complexity:** Ensemble and hybrid models typically demand significant computational resources for both training and interpretation, especially when working with extensive datasets or complex models. Minimizing computational burden while maintaining optimal performance is a primary obstacle.
- **Model interpretability:** Ensemble and hybrid models can be challenging to interpret because of their inherent complexity. Improving the transparency and explainability of models is of utmost importance, particularly in situations where it is important to fully understand the reasoning behind the decisions made. Particularly in healthcare, where understanding the reasoning behind decisions is crucial for clinical adoption.
- **Scalability:** A major difficulty involves ensuring the efficient scalability of ensemble and hybrid methods to handle larger volumes of data and increasingly complicated situations. This comprises the process of improving algorithms and utilizing distributed computing platforms. In healthcare, this challenge is amplified by the need to handle diverse and high-volume patient data.
- **Integration of diverse models:** Successfully incorporating many models and paradigms into hybrid systems necessitates careful preparation and implementation. Achieving optimal performance by balancing the contributions of several components is a challenging task, especially in healthcare, where integrating various types of data (e.g., clinical, imaging, genetic) is essential.
- **Data quality and diversity:** The performance of ensemble and hybrid approaches is greatly influenced by the size and variety of the data utilized for training. It is crucial to tackle data preparation, noise, and bias issues to construct robust models. In healthcare, ensuring high-quality and diverse data is critical for accurate predictions and diagnostics.
- **Hyperparameter tuning:** Ensemble and hybrid models typically require meticulous tuning of multiple hyperparameters. Discovering the most efficient hyperparameter configurations can be a difficult and highly resource-intensive task. In healthcare, the stakes are higher, as optimal tuning can significantly impact patient outcomes.
- **Communication overhead:** In distributed or parallel computing settings, the communication overhead between different models and components may limit performance. Effectively handling this additional cost is crucial for maintaining optimal performance, especially in real-time healthcare applications like patient monitoring systems.
- **Maintenance and updating:** Managing and revising ensemble and hybrid models can be more intricate than handling individual models. As new information

becomes accessible, maintaining the ensemble or hybrid system's integrity and optimal performance could require major effort. In healthcare, continuous updates are necessary to incorporate the latest medical research and clinical guidelines.

13.6.2   Research opportunities

Advancements in machine learning offer numerous opportunities for research and innovation in ensemble and hybrid techniques. The integration of advanced algorithms and paradigms can address challenges and push boundaries, paving the way for novel models, optimization, and innovative solutions in complex problems. The ongoing advancements in machine learning present numerous opportunities for research and innovation in ensemble and hybrid techniques.

- **Novel ensemble methods:** Exploring novel ensemble algorithms that may effectively leverage model diversity and enhance performance across different tasks has great potential as a research path. This involves investigating new approaches to integrate models and enhance their interactions, especially for complex healthcare applications like multi-disease diagnosis.
- **Advanced hybrid models:** Exploring novel hybrid methodologies that combine various machine learning concepts and algorithms can result in more robust and flexible models. This involves integrating deep learning with probabilistic models, evolutionary algorithms, and other sophisticated methodologies. In healthcare, such models can provide more accurate and personalized treatment recommendations.
- **Efficient training techniques:** Exploring solutions for reducing the computational demands of training ensemble and hybrid models, such as parallel processing, distributed computing, and optimization algorithms, is essential for enhancing the accessibility and scalability of these approaches. In healthcare, this can lead to more efficient use of computational resources, enabling widespread adoption in clinical settings.
- **Improving interpretability:** It is crucial to develop methods to improve the transparency of ensemble and hybrid models, which can be achieved using visualization tools, explainable AI (XAI) frameworks, and interpretable model designs. This is particularly important in healthcare, where clinicians need to understand model decisions to trust and use them effectively.
- **Automated model selection and tuning:** Utilizing techniques such as Auto-ML (Automated Machine Learning) to automate the selection and optimization of models inside ensemble and hybrid frameworks can enhance the development process and enhance model performance. In healthcare, automated tuning can streamline the deployment of personalized models.

13.6.3   Technological advancements

Emerging technologies and trends, including quantum computing, edge computing, and IoT, are expected to significantly enhance ensemble and hybrid machine learning techniques. These advancements will drive improvements in performance, scalability, and efficiency, paving the way for more robust and versatile applications in various fields.

- **Quantum computing:** Quantum computing has the potential to greatly transform ensemble and hybrid learning by offering unparalleled computational capacity for sophisticated tasks. Investigations into quantum machine learning have the potential to yield novel and very efficient algorithms.
- **Edge computing:** With the advancement of Internet of Things (IoT) and edge computing technologies, it becomes possible to deploy ensemble and hybrid models on edge devices. This allows for real-time processing and decision making to occur closer to the data sources, resulting in improved latency and reduced bandwidth utilization.
- **Federated learning:** Federated learning, a technique that trains models on distributed devices while keeping data locally, can improve the privacy and scalability of ensemble and hybrid methods. Exploring federated approaches for these techniques has the potential to provide wider and more robust applications.
- **Neuromorphic computing:** Neuromorphic computing, which draws inspiration from the structural design of the human brain, presents a novel approach to creating highly efficient machine learning models. Combining ensemble and hybrid approaches with neuromorphic technology has the potential to achieve significant advancements in both performance and energy economy.
- **Ethical and fair AI:** It is becoming increasingly vital to address ethical problems and ensure fairness in ensemble and hybrid models. Future research should focus on creating frameworks and procedures to effectively recognize and reduce biases, increase transparency, and guarantee ethical application of these techniques.

By tackling these difficulties and investigating these possibilities, the discipline of ensemble and hybrid machine learning might persist in progressing, offering resilient and inventive resolutions to intricate issues throughout various domains. The section summarization is represented on Figure 13.1.

13.7   Conclusion

The chapter discusses the theoretical foundations, differences, application, and future perspectives of ensemble and hybrid machine learning methods, with a particular focus on their applications in healthcare. Ensemble learning is a technique that improves the accuracy and reliability of predictions by combining numerous models and utilizing the unique capabilities of each model. Popular ensemble techniques comprise of bagging, boosting, and stacking. Hybrid machine learning combines various paradigms or algorithms to use their supportive strengths, resulting in more robust and flexible models.

The chapter also explains the differences between ensemble and hybrid techniques, highlighting their strengths and weaknesses. Ensemble methods aggregate

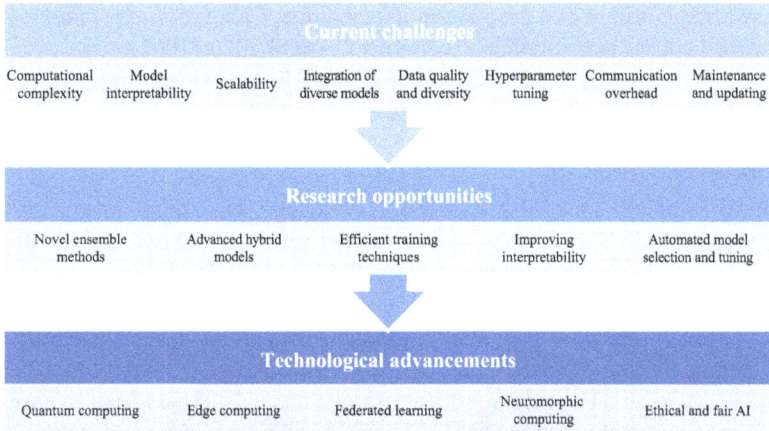

Figure 13.1 Summarization of current challenges, research opportunities and technological advancements

model outputs, while hybrid methods integrate algorithms. They are used in various domains like classification, regression, anomaly detection, real-time systems, and optimization. In healthcare, these techniques are applied to disease diagnosis, medical image analysis, personalized medicine, and treatment optimization, among others.

The findings derived from this study have substantial ramifications for researchers as well as experts in the domain of machine learning. Gaining a comprehensive understanding of the capabilities and constraints of ensemble and hybrid techniques allows for well-informed decision making on their utilization in various situations, particularly in healthcare. Experts can enhance model performance, scalability, and interpretability by choosing the right strategy. The identified issues and future directions serve as a guide for researchers to progress in the subject and create inventive solutions.

Ensemble and hybrid machine learning techniques are advancing models for accuracy and adaptability. As the field evolves, integrating emerging technologies like quantum computing, edge computing, federated learning, and neuromorphic computing holds promise for future development. These techniques will shape machine learning and artificial intelligence.

References

Agarwal, S., and Chowdary, C. R. (2020). A-stacking and A-bagging: adaptive versions of ensemble learning algorithms for spoof fingerprint detection. *Expert Systems with Applications*, *146*, 113160. https://doi.org/10.1016/J. ESWA.2019.113160

Altan, A., and Karasu, S. (2020). Recognition of COVID-19 disease from X-ray images by hybrid model consisting of 2D curvelet transform, chaotic salp swarm algorithm and deep learning technique. *Chaos, Solitons and Fractals, 140*, 110071. https://doi.org/10.1016/J.CHAOS.2020.110071

Ardabili, S., Mosavi, A., and Várkonyi-Kóczy, A. R. (2020). Advances in machine learning modeling reviewing hybrid and ensemble methods. *Lecture Notes in Networks and Systems, 101*, 215–227. https://doi.org/10.1007/978-3-030-36841-8_21/COVER

Ashraf, M., Abid, F., Din, I. U., Rasheed, J., Yesiltepe, M., Yeo, S. F., *et al.* (2023). A hybrid CNN and RNN variant model for music classification. *Applied Sciences, 13*(3), 1476. https://doi.org/10.3390/APP13031476

Awan, F. M., Saleem, Y., Minerva, R., and Crespi, N. (2020). A comparative analysis of machine/deep learning models for parking space availability prediction. *Sensors (Switzerland), 20*(1). https://doi.org/10.3390/S20010322

Başarslan, M. S., and Kayaalp, F. (2024). Sentiment analysis using a deep ensemble learning model. *Multimedia Tools and Applications, 83*(14), 42207–42231. https://doi.org/10.1007/S11042-023-17278-6/METRICS

Bell, J. (2022). What is machine learning? *Machine Learning and the City*, 207–216. https://doi.org/10.1002/9781119815075.CH18

Chakraborty, D., Elhegazy, H., Elzarka, H., and Gutierrez, L. (2020). A novel construction cost prediction model using hybrid natural and light gradient boosting. *Advanced Engineering Informatics, 46*, 101201. https://doi.org/10.1016/J.AEI.2020.101201

Dahiya, N., Gupta, S., and Singh, S. (2022). A review paper on machine learning applications, advantages, and techniques. *ECS Transactions, 107*(1), 6137–6150. https://doi.org/10.1149/10701.6137ECST/XML

Dasari, A. K., Biswas, S. K., Thounaojam, D. M., Devi, D., and Purkayastha, B. (2023). Ensemble learning techniques and their applications: an overview. *Cognitive Science and Technology*, 897–912. https://doi.org/10.1007/978-981-19-8086-2_85/COVER

Dietterich, T. G. (2000). Ensemble methods in machine learning. *Lecture Notes in Computer Science (Including Subseries Lecture Notes in Artificial Intelligence and Lecture Notes in Bioinformatics), 1857 LNCS*, 1–15. https://doi.org/10.1007/3-540-45014-9_1/COVER

Dong, X., Yu, Z., Cao, W., Shi, Y., and Ma, Q. (2020). A survey on ensemble learning. *Frontiers of Computer Science, 14*(2), 241–258. https://doi.org/10.1007/S11704-019-8208-Z/METRICS

Dutta, A., and Dasgupta, P. (2017). Ensemble learning with weak classifiers for fast and reliable unknown terrain classification using mobile Rrbots. *IEEE Transactions on Systems, Man, and Cybernetics: Systems, 47*(11), 2933–2944. https://doi.org/10.1109/TSMC.2016.2531700

Fararni, K. Al, Nafis, F., Aghoutane, B., Yahyaouy, A., Riffi, J., and Sabri, A. (2021). Hybrid recommender system for tourism based on big data and AI: a conceptual framework. *Big Data Mining and Analytics, 4*(1), 47–55. https://doi.org/10.26599/BDMA.2020.9020015

Forouzandeh, S., Berahmand, K., and Rostami, M. (2021). Presentation of a recommender system with ensemble learning and graph embedding: a case on

MovieLens. *Multimedia Tools and Applications*, *80*(5), 7805–7832. https://doi.org/10.1007/S11042-020-09949-5/METRICS

François-Lavet, V., Henderson, P., Islam, R., Bellemare, M. G., and Pineau, J. (2018). An introduction to deep reinforcement learning. *Foundations and Trends in Machine Learning*, *11*(3–4), 219–354. https://doi.org/10.1561/2200000071

Gentleman, R., and Carey, V. J. (2008). Unsupervised machine learning. *Bioconductor Case Studies*, 137–157. https://doi.org/10.1007/978-0-387-77240-0_10

Hastie, T., Tibshirani, R., and Friedman, J. (2009a). *Overview of Supervised Learning*. 9–41. https://doi.org/10.1007/978-0-387-84858-7_2

Hastie, T., Tibshirani, R., and Friedman, J. (2009b). *Unsupervised Learning*. 485–585. https://doi.org/10.1007/978-0-387-84858-7_14

Helm, J. M., Swiergosz, A. M., Haeberle, H. S., Karnuta, J. M., Schaffer, J. L., Krebs, V. E., *et al.* (2020). Machine learning and artificial intelligence: definitions, applications, and future directions. *Current Reviews in Musculoskeletal Medicine*, *13*(1), 69–76. https://doi.org/10.1007/S12178-020-09600-8/METRICS

Jeffrey, N., Tan, Q., and Villar, J. R. (2024). Using ensemble learning for anomaly detection in cyber–physical systems. *Electronics*, 13(7), 1391. https://doi.org/10.3390/ELECTRONICS13071391

Kanchanamala, P., Alphonse, A. S., and Reddy, P. V. B. (2023). Heart disease prediction using hybrid optimization enabled deep learning network with spark architecture. *Biomedical Signal Processing and Control*, *84*, 104707. https://doi.org/10.1016/J.BSPC.2023.104707

Kazmaier, J., and van Vuuren, J. H. (2022). The power of ensemble learning in sentiment analysis. *Expert Systems with Applications*, *187*, 115819. https://doi.org/10.1016/J.ESWA.2021.115819

Khaldi, B., Harrou, F., Benslimane, S. M., and Sun, Y. (2021). A data-driven soft sensor for swarm motion speed prediction using ensemble learning methods. *IEEE Sensors Journal*, *21*(17), 19025–19037. https://doi.org/10.1109/JSEN.2021.3087342

Li, Y., Wang, H., Dang, L. M., Nguyen, T. N., Han, D., Lee, A., *et al.* (2020). A deep learning-based hybrid framework for object detection and recognition in autonomous driving. *IEEE Access*, *8*, 194228–194239. https://doi.org/10.1109/ACCESS.2020.3033289

Lu, H., and Ma, X. (2020). Hybrid decision tree-based machine learning models for short-term water quality prediction. *Chemosphere*, *249*, 126169. https://doi.org/10.1016/J.CHEMOSPHERE.2020.126169

Pachauri, N., and Ahn, C. W. (2022). Regression tree ensemble learning-based prediction of the heating and cooling loads of residential buildings. *Building Simulation*, *15*(11), 2003–2017. https://doi.org/10.1007/S12273-022-0908-X/METRICS

Pawar, A., Manjula Shenoy, K., Prabhu, S., and Guruprasad Rai, D. (2023). Performance analysis of machine learning algorithms: single model vs ensemble model. *Journal of Physics: Conference Series, 2571*(1), 012007. https://doi. org/10.1088/1742-6596/2571/1/012007

Riyaz, L., Butt, M. A., Zaman, M., and Ayob, O. (2021). *Heart Disease Prediction Using Machine Learning Techniques: A Quantitative Review*. In International Conference on Innovative Computing and Communications: Proceedings of ICICC 203, 81–94. https://doi.org/10.1007/978-981-16-3071-2_8

Sagi, O., and Rokach, L. (2018). Ensemble learning: a survey. *Wiley Interdisciplinary Reviews: Data Mining and Knowledge Discovery, 8*(4), e1249. https://doi.org/10.1002/WIDM.1249

Sangamnerkar, S., Srinivasan, R., Christhuraj, M. R., and Sukumaran, R. (2020). An ensemble technique to detect fabricated news article using machine learning and natural language processing techniques. *2020 International Conference for Emerging Technology, INCET 2020.* https://doi.org/10.1109/INCET49848.2020.9154053

Saravi, B., Hassel, F., Ülkümen, S., Zink, A., Shavlokhova, V., Couillard-Despres, S., *et al.* (2022). Artificial intelligence-driven prediction modeling and decision making in spine surgery using hybrid machine learning models. *Journal of Personalized Medicine, 12*(4), 509. https://doi.org/10.3390/JPM12040509

Sarker, I. H. (2021). Machine learning: algorithms, real-world applications and research directions. *SN Computer Science, 2*(3), 160. https://doi.org/10.1007/s42979-021-00592-x

Sewak, M. (2019). Introduction to deep learning: Enter the world of modern machine learning. In *Deep Reinforcement Learning: Frontiers of Artificial Intelligence*, 75–88. https://doi.org/10.1007/978-981-13-8285-7

Singh, V., and Jain, D. (2023). A hybrid parallel classification model for the diagnosis of chronic kidney disease. *International Journal of Interactive Multimedia and Artificial Intelligence, 8*(2), 14–28. https://doi.org/10.9781/IJIMAI.2021.10.008

Solomon, D. D., Khan, S., Garg, S., Gupta, G., Almjally, A., Alabduallah, B. I., *et al.* (2023). Hybrid majority voting: prediction and classification model for obesity. *Diagnostics 2023, 13*(15), 2610. https://doi.org/10.3390/DIAGNOSTICS13152610

Sornsuwit, P., and Jaiyen, S. (2019). A new hybrid machine learning for cybersecurity threat detection based on adaptive boosting. *Applied Artificial Intelligence, 33*(5), 462–482. https://doi.org/10.1080/08839514.2019.1582861

Uthayakumar, J., Metawa, N., Shankar, K., and Lakshmanaprabu, S. K. (2020). Intelligent hybrid model for financial crisis prediction using machine learning techniques. *Information Systems and E-Business Management, 18*(4), 617–645. https://doi.org/10.1007/S10257-018-0388-9/METRICS

Von Krannichfeldt, L., Wang, Y., and Hug, G. (2021). Online ensemble learning for load forecasting. *IEEE Transactions on Power Systems, 36*(1), 545–548. https://doi.org/10.1109/TPWRS.2020.3036230

von Rueden, L., Mayer, S., Sifa, R., Bauckhage, C., and Garcke, J. (2020). Combining machine learning and simulation to a hybrid modelling approach: current and future directions. *Lecture Notes in Computer Science (Including Subseries Lecture Notes in Artificial Intelligence and Lecture Notes in*

Bioinformatics), 12080 LNCS, 548–560. https://doi.org/10.1007/978-3-030-44584-3_43/FIGURES/5

Walek, B., and Fojtik, V. (2020). A hybrid recommender system for recommending relevant movies using an expert system. *Expert Systems with Applications*, *158*, 113452. https://doi.org/10.1016/J.ESWA.2020.113452

Wang, P., Fan, E., and Wang, P. (2021). Comparative analysis of image classification algorithms based on traditional machine learning and deep learning. *Pattern Recognition Letters*, *141*, 61–67. https://doi.org/10.1016/J.PATREC.2020.07.042

Wen, L., and Hughes, M. (2020). Coastal wetland mapping using ensemble learning algorithms: a comparative study of bagging, boosting and stacking techniques. *Remote Sensing*, *12*(10), 1683. https://doi.org/10.3390/RS12101683

Zhang, H., Thompson, J., Gu, M., Jiang, X. D., Cai, H., Liu, P. Y., *et al.* (2021). Efficient on-chip training of optical neural networks using genetic algorithm. *ACS Photonics*, *8*(6), 1662–1672. https://doi.org/10.1021/ACSPHOTONICS.1C00035/SUPPL_FILE/PH1C00035_SI_001.PDF

Zhong, Y., Chen, W., Wang, Z., Chen, Y., Wang, K., Li, Y., *et al.* (2020). HELAD: a novel network anomaly detection model based on heterogeneous ensemble learning. *Computer Networks*, *169*, 107049. https://doi.org/10.1016/J.COMNET.2019.107049

Zhou, F., Fan, H., Liu, Y., Zhang, H., Ji, R., Ma, M., *et al.* (2023). Hybrid model of machine learning method and empirical method for rate of penetration prediction based on data similarity. *Applied Sciences*, *13*(10), 5870. https://doi.org/10.3390/APP13105870

Zhou, Z. H. (2012). *Ensemble Methods: Foundations and Algorithms*. Boca Raton, Fl: Chapman & Hall/CRC.

Zhou, Z.-H. (2021). Ensemble learning. *Machine Learning*, 181–210. https://doi.org/10.1007/978-981-15-1967-3_8

Index

www.ingramcontent.com/pod-product-compliance
Lightning Source LLC
Chambersburg PA
CBHW050510190326
41458CB00005B/1491